TEACHING OF MATHEMATICAL MODELLING AND APPLICATIONS

TEACHING OF MATHEMATICAL MODELLING AND APPLICATIONS

M. NISS Ph.D.
Department of Studies in Mathematics and Physics, Roskilde University, Denmark
W. BLUM M.Sc., Ph.D.
Department of Mathematics, University of Kassel, Germany
I. HUNTLEY B.A., Ph.D.
Head of Open Learning, Sheffield Polytechnic

ELLIS HORWOOD
NEW YORK LONDON TORONTO SYDNEY TOKYO SINGAPORE

First published in 1991 by
ELLIS HORWOOD LIMITED
Market Cross House, Cooper Street,
Chichester, West Sussex, PO19 1EB, England

A division of
Simon & Schuster International Group
A Paramount Communications Company

Printed and bound in Great Britain
by Redwood Press Ltd, Melksham, Wiltshire

British Library Cataloguing in Publication Data

M. Niss, W. Blum and I. Huntley
Teaching of Mathematical Modelling and Applications
CIP catalogue record for this book is available from the British Library
ISBN 0-13-892068-0

Library of Congress Cataloging-in-Publication Data available

TABLE OF CONTENTS

SECTION D. TEACHING OF MODELLING AND APPLICATIONS AT UPPER SECONDARY LEVEL

Preface

The present volume contains 44 articles by 56 authors on the teaching of mathematical modelling and applications. The articles are revised versions of selected contributions to The Fourth International Conference on the Teaching of Mathematical Modelling and Applications, ICTMA 4, held at Roskilde University, Denmark, 3–7 July 1989. Thus the volume forms the Proceedings of that conference. However, in selecting the papers and in giving guidelines for the revision of the papers the editors have strived at creating a volume which is more than the proceedings of a conference. It ought to be of interest and value in its own right to a wide international readership engaged or interested in mathematical modelling and applications in an educational context at any level.

During the last couple of decades, *mathematical applications, models and modelling* have attracted *growing attention* in the theory and practice of mathematics teaching at all educational levels. It is fair to say that this theme has now become an established concern in mathematics education and also in the mathematics community at large. It is no longer necessary to fight for recognition of these aspects of mathematical activity and pursuit amongst mathematics teachers, educators and mathematicians. These aspects are now widely considered to be of crucial importance, whether viewed from an educational, a scientific, a societal or a cultural perspective. One piece of evidence of this may be found in the fact that applications, models, modelling, and applied problem solving have become increasingly central themes during the latest four international congresses on mathematical education, the ICMEs[1].

Against this background, it is not surprising that the majority of the articles presented here report on and discuss *actual experience* with the teaching of mathematical modelling and applications at different levels. Descriptions of curriculum projects, teaching materials or cases which have been, or may easily be, tried out in teaching practice abound in this volume. A number of articles survey or analyse *states and trends* in the field. So, on the one hand, this collection of articles reveals the progress in theory and practice in the field under discussion. On the other hand, articles formulating bold proposals or visionary ideas beyond the scope of current reality are few in this volume. This reflects our belief that for the time being they are less needed than they were a decade ago. However, as new experiences are gained and analysed, new needs and problems may very well emerge that call for the resurrection of utopian creativity.

The present volume is divided into *six sections*. Within each section it has been attempted to order the articles according to increasing specificity, that is the more general articles come first, those having a narrower focus come last. Section A consists of four articles each giving a *survey* of a certain aspect of mathematical modelling and applications in an educational context. They are the papers of the plenary lectures delivered at ICTMA 4, and the flavour and oral style of the lectures have been preserved to some extent. Section B deals with *general* and theoretical aspects of the field, whereas Sections C–E treat the *teaching* of modelling and applications under different perspectives at various educational levels, ranging from the *lower secondary* level (Section C), and the *upper secondary* level (Section D) to the *tertiary* level (Section E). Finally Section F is devoted to the presentation of a number of particular modelling *cases* in different areas of relevance to tertiary level mathematics.

ICTMA 4 was one in a series of biennial international conferences concerned with the teaching of mathematical modelling and applications which was begun on British initiative in 1983. The first two conferences were held in Exeter, UK (in 1983 and 1985), the third in Kassel, FR Germany (1987). The proceedings of the three previous conferences have also been published by Ellis Horwood[2]. Although mathematical modelling and applications have come to form a fairly consolidated component in mathematics teaching, there is still much to be created, explored, developed and analysed. So, there is every reason to hope and to believe that the series of conferences will be continued in the future. The next one, ICTMA 5, will take place in the Netherlands in September 1991.

Mogens Niss Werner Blum Ian Huntley

REFERENCES

[1]See, for instance, the collection of articles from ICME 6 Theme Groups 3 and 6:

W Blum, M Niss, I Huntley (eds). (1989). Modelling, Application and Applied Problem Solving – teaching mathematics in a real context. Ellis Horwood, Chichester.

[2]JS Berry et al (eds). (1984). Teaching and Applying Mathematical Modelling. Ellis Horwood, Chichester.

JS Berry et al (eds). (1986). Mathematical Modelling Methodology, Models and Micros. Ellis Horwood, Chichester.

JS Berry et al (eds). (1987). Mathematical Modelling Courses. Ellis Horwood, Chichester.

W Blum et al (eds). (1989). Applications and Modelling in Learning and Teaching Mathematics. Ellis Horwood, Chichester.

CHAPTER 1

Applied Mathematics as a Social Instrument

PJ Davis
Brown University, Providence, USA

SUMMARY

Each age has preferred modes of prediction. Each mode opens up characteristic possibilities and creates characteristic realities. Mathematical modelling is today's high prestige way of predicting. It is the expression of our age, and it is likely to be around for a long while.

However, we must watch it. We must watch it because mathematical descriptions tend to drive out all others.

1. INTRODUCTION

It is often asserted that applied mathematics is mathematics that has some relationship to the outside world, whether it is the physical world, or to the world of humans and their arrangements. On the other hand, pure mathematics is mathematics in the service only of more pure mathematics. (That is, until the pure mathematicians decide they need government support; then they create another category - applicable mathematics. The pure of heart not infrequently approach the government with the statement that all mathematics is potentially applicable.)

I frequently assert that there is no such thing as pure mathematics, because all mathematics derives from the working of the brain in collaboration with the senses and the outside world. It must have some connection with the outside world, or it is nothing. Put another way, mathematics must have a semantic component as well as a syntactic component. I am not happy with the conception of 'pure brain' operating on 'pure ideas'.

Two centuries ago, applied mathematics was often called mixed mathematics. Applications were the ideal. They were the ultimate goal of all investigations. The individual mathematical subjects each had an outside referrent. The prestige lay in applications. Today (at least within the community of mathematical researchers) it is the reverse – the prestige lies in solving abstruse problems that seem remote from the bottom line of social interest.

There is yet another way of distinguishing between pure and applied mathematics. Applied mathematics often pulls back, relinquishes its claims. An interesting example of this can be found in the writings of Nicholas Bernoulli, who was a lawyer and a mathematician. On his pages probabilistic mathematics is in the service of evidence and judicial decision. This tack was abandoned – but has returned in the last generation.

Pure mathematics seems rarely to pull back on its claims. Its statements seem to be atemporal. In fact, this atemporality is a misconception which results from looking with only one eye at the syntactic element, and not with both eyes at the syntactics and the semantics.

Despite the present emphasis on the superiority of pure mathematics (whatever the word superiority means), our world is becoming increasingly mathematised. It can be characterised as being scientific, rational, and mathematised. By rational, I mean that, by the formalised versions of reason found in mathematics, one attempts to understand and control the world. By mathematised, I mean the employment of mathematical ideas and constructs, either in their theoretical form or in their computer manifestations, to describe, organise, regulate, and to foster our human activities. I use the suffix 'ised' to emphasise that it is we human beings who are putting the mathematisations in place, consciously or unconsciously, and who are affected by them.

2. GOALS OF APPLIED MATHEMATICS

Now what are the goals of applied mathematics? One can distinguish at least three. They are *description, prediction* and *prescription*; that is to say, tell me what is, tell me what will be, tell me what to do about it. Tell me how to change things along certain lines that are perceived to be desirable. These goals are inextricably interrelated, and no preferred order or priority should be implied by my listing.

For further clarification, here are some simple examples. The planets revolve about the earth in Ptolemaic circles and epicycles. There are precisely 17 one-colour two-dimensional patterns in the group-theoretical sense. They have all been observed by anthropologists. There are 46 two-colour two-dimensional patterns.

These are instances of descriptions. Descriptions, not necessarily truths. It is doubtful whether Plato, who was interested in the ultimate truth of things, would have been satisfied by Ptolemy's mess. However, describe it does, and does it very well.

On January 1st, 1995, the population of the country of Erewhon will be 20 121 348 ± 5 981. This is a prediction. Predictions are so common that we hardly need illustrate them.

On the basis of the response to the invitations sent out, you had better order six barrels of beer. On the basis of statistical evidence, the Surgeon General has determined it would be in your best interests to stop smoking. To fly from Copenhagen to Singapore, non-stop, the petrol capacity of a newly designed airplane must be such and such. These are prescriptions. They call for certain actions.

I once asked the world famous British philosopher Sir Alfred Ayer, who died just a few days ago, what his principal dream for contemporary philosophy was. He answered me (without a moment's hesitation) "to describe what *is* and to give reasons for thinking so". Perhaps description is the most important element. If we can decide what *is*, then we are in a better position to tell what to do about what will be.

I have said that these three goals are interrelated. They are also independent, to some degree. We can certainly install policies without finding out what *is*. We can certainly find out what *is*, even project forward to what will be, and still have no idea as to what to do about it. Many social problems are of this nature.

Every mathematisation, whether its goal is description, prediction, prescription, whether it is installed consciously, or whether it evolves organically or culturally, is subject to popular or governmental back pressures. Such a mathematisation may even come under judicial review. This is as it should be, because the ability of mathematics to provide frameworks of reality and of action, and its ability to change what is, is very great. The so-called objectivity of mathematics lends tremendous credence and status to its pronouncements. The arcane quality of mathematics, that is to say, the fact that it is a stuff whose inner workings are hidden or mysterious to the average person, results in the abdication of the processes of social mathematisations to a corps of experts. While government by expertocracies of various sorts is probably an inevitable consequence of our high-tech lives, the process should not be insulated from the checks and balances that democratic systems require.

The tremendous importance of description formation has been summed up brilliantly by David Berlinski, a mathematician, philosopher and polemicist, in his article Prediction and Infection.

"Mathematical descriptions, when applied, tend to drive out all the others. Mathematics is often a matter of bondage, with things in thrall to theories. Strong theories make for weak objects".

I would supplement this with my own observation: once a mathematical description is in place, it is harder to change than moving a graveyard.

3. ASPECTS OF DESCRIPTION, PREDICTION AND PRESCRIPTION

Attempts to determine what *is*, may result in social disapprobation. In the USA, there are now suits claiming that a certain intelligence test descriminates against women. A test, with its associated quantifications and statistical apparatus, has been installed and monitored by applied mathematicians presumably to determine what *is*. The subjective element – or, to be more objective, the socio-political element – in testing is now perceived to be great, and we may have a situation where people are used to determine the nature of the test, rather than the test to determine the nature of people.

Another example. In the research relating to AIDS therapy, placebo testing has been installed. This gave rise to a considerable street demonstration in Boston recently. Placards were carried saying "Give me the real stuff, not placebos". As though we knew what the real stuff was.

Another example of the same sort. A statistically significant study of whether an excess of X chromosomes leads to violent individuals would involve a controlled experiment that would not be socially acceptable.

You may say that the mathematics itself is clean, it is only the questions which are dirty. I say that the very existence of the mathematics, its ease, its unambiguity, leads to a social process which requires constant monitoring. (As Berlinski put it: strong theories, weak objects).

In every country with which I am familiar, there are certain aspects of life (they differ from country to country) where it would be taboo to install a mathematised inquiry to determine what *is*. Ayer possibly is right. What *is*, takes primacy.

The regulation, control, or prohibition of mathematisations has a long history. A book ought to be written about it.

In the days of biblical antiquity, it was forbidden to take a census. (Invasion of privacy. Usurpation of Rights of the Divinity.)

However, the framers of the US Constitution, children of the European Age of Enlightenment and Reason, mandated a census.

A census is part of what *is*. It is critical, in many areas, as part of database formation. Examine this recent news item:

> "US Department of Commerce rejects demands to adjust 1990 census".

A congressman from New York put it this way "I can't believe they would perpetuate a procedure that disenfranchises hundreds of thousands of people and deprives local governments of millions of dollars in aid."

This dispute has resulted in a court suit. The suggestion by minorities and large cities urges the inclusion of people in the national headcount that were not counted.

What *is*? Counting won't or may not be allowed to tell you.

Thus the process of counting, the most elementary, the most fundamental operation in applied mathematics, appears to be a social act chasing after a mathematical ideal.

In the Roman Empire of Diocletian, the profession of Mathematicus was proscribed. Mathematics was identified with Magic. The aura of magic still persists. How many times have I heard "All this stuff of yours, it's all hocus pocus anyway".

Gambling has been illegal, legal, illegal, ... Why? For one reason, gambling usurps the divine prerogative. (Lots used to be cast to determine God's will). Perhaps worse than this, gambling can upset the social order: if a poor man suddenly becomes as rich as a lord, where would this leave an aristocracy defined along blood lines? A more contemporary reason: gambling is a tax on the poor.

Until the 19th century, life insurance was illegal in many countries of Europe (gambling with lives). In the early 19th century, if you wanted to take out a policy, you went to an office where you could also make bets on your favorite ponies.

Interest on loans, which is an interesting mathematisation, was forbidden to believing Christians and to devout Communists. Until a few hundred years ago, the practise of usury (which is the pejorative term for those percentage points we all love and we all hate, depending on whether we're receiving or paying) was relegated to the non–Christian subsets of the community.

Income tax was proscribed in the US until 1912 when a constitutional amendment was found necessary to legalise it.

Men may be sent to jail for practising applied commercial mathematics which employs the wrong kind of axiom system. In some countries, at some period of history, speculators have been shot.

We may yet see laws regulating the prediction of elections.

Individual qualifications for certain national benefits, privileges, based upon personal parameters, are constantly being challenged.

The use of sophisticated IDs backed up by mathematico–computer systems, may be developed and challenged in the very near future.

4. PREDICTION

Description, prediction, prescription: what *is*, what will *be*, what therefore to *do*. Let me now turn more narrowly to the prediction process.

Why do we want to predict? Well, we can't live without prediction. We can't cross the street in the traffic without some kind of prediction. Call it cybernetics, if you want to, and say that prediction and control are intertwined. We need to plan.

A medical diagnosis is a prediction – what is the effect of a drug? How can I make money? Given my talents, what kind of job should I take (or should the tracking system force me into)? Whom should I marry to be happy? Will it rain for Sunday's picnic? Who will win the election? What national policy will alleviate a national problem? We desperately want and need to know the future.

Prediction is everywhere. It makes an interesting exercise to look at the first few pages of your favourite daily paper and count the number of mathematically embedded predictions in them.

Cicero, whose orations I read as a school boy, remarked that two things underlay the Roman government. The first was ritual and the second was divination. It would seem that the Roman politicians, generals, administrators and emperors could hardly make a move without first ascertaining the will of the gods in the matter. The gods spoke through a variety of oracles, augurs and soothsayers, who received their messages variously from divine frenzy, through observation of the flights of birds, the behaviours of the sacred Roman Chickens and an inspection of the entrails of specially slaughtered animals.

The Roman world - if not the whole world of antiquity - was divination - besotted. Consult Cicero's *De Divinatione* on this point.

After the passage of two thousand years we are more or less in the same position as the Romans, the difference being that the art of prediction has undergone a paradigm change. Generally speaking, prediction today, in order to be socially and scientifically respectable must have a mathematical basis. However, this is neither necessary nor sufficient.

Mixtures of ancient and contemporary modes of prediction exist side by side. Sophisticated quantum theorists in certain parts of the world will marry off their children first having had their horoscopes cast. Of course you can claim a strong mathematical basis for classical astrology, and you would be right. However, this is not the reason the practice persists.

There are not many books on the abstract nature of prediction. However I should like to emphasise this: *every time an applied mathematician writes an equation with t for time in it, that mathematician is asserting a claim to oracularity.* Yet, if you look in the index of most applied mathematics books, you will not find the term prediction listed in it. You will not find the term in many good encyclopaedias.

Prediction, at the bottom line, begins with a client. Somebody wants to know the future and is willing to pay for the knowledge. If I were writing a book about prediction, I would certainly discuss the following features of the process: the client, the predictor, the method, the materials required, the medium, the imagery of the predictive statement, the interpretation, the validation, the time interval of validity, the computerisation, the scope, the commercial aspects, the taboos or conflicts raised by the process. The social utility of the act.

This is a very long list. Let me talk about a few of them. I shall do it by telling you about a book on prediction that I read and reviewed recently, *Datawars, the Politics of Modelling in Federal Policy making*, written by a team of economists from the University of California Irvine, and the Institut für Planungs - und Entscheidungssysteme, Bonn, FRG. The book was an eyeopener for me.

Is modelling important in the federal decisionmaking process? You'd better believe it. In 1979, half a billion dollars was spent on the development, use, and maintenance of models.

Now the intellectual subworld in which all this takes place presents a strange mixture of rational economic theory, model formation, model packaging, model promotion, theory pushing towards policy, packages pushing towards sales, policy-need pulling towards model-formation perception of model-bias (ie, liberal, conservative), model-massaging, counter-modelling. Backpressures of the public.

The people who inhabit this subworld are the modellers, the advisers, policy analysts, policy makers. There is cooperation and competition between them. There is interagency rivalry.

There are hundreds of models currently employed in Washington. They have names like CHASE, CLAREMONT, PECKMANOCKNER and so on.

One of the principal goals of the authors of that book was to determine what constitutes *successful* modelling. Now you might say where's the problem? Just select the model that predicts most accurately. How naive! How do we know what predicts best when we don't know what *is* and we can't agree on the meaning of *best*? In the play of the market, what is best tends simply to become what sells best.

At any rate, the authors formulate their conclusions in twenty five propositions. Here are a few. The titles given here are my own, not the authors'.

Old faithful principle
Models based on theories that are relatively well-established, stable, empirically grounded, and widely accepted by members of the discipline, are implemented more broadly and more easily than new or evolving models.

Political neutrality
Don't try to sell a model that has a strong built-in political bias.

If you've paid for it, it must be worth it
Commercial modellers experience greater success than non-profit modellers in implementing their models in user-agencies.

If you believe it's working, don't fix it
Models that are accompanied by relatively high-quality data, measured specially in terms of the *user's belief* in the data, experience greater implementation success than models with poor-quality databases.

5. CONCLUSION

We have given a slight intimation of the workings of the economic modelling establishment. Do you think that the modelling of the physical world works any differently? In some respects, of course, it does, but the theories of Thomas Kuhn and other recent historians of the science point to complex laws of development that are hardly describable by the single word objectivity. In a world fashioned by media and hype, and with the reality granted to the modelling process riding high, the worlds of physical and social modelling may be drawing closer.

Modelling is the way we do things. It is the expression of the age. It is likely to be around for a long while.

However, we must watch it. We must watch it because *mathematical descriptions tend to drive out all others.*

REFERENCES

Berlinski D. *Prediction and Infection.* To appear.

Daston L. (1988). *Classical Probability in the Enlightenment.* Princeton University Press.

Davis PJ and Hersh R. (1986). *Descartes' Dream.* Harcourt Brace Jovanovich.

Davis PJ. (1988). What is a Successful Predictor?. *SIAM NEWS,* January.

Kramer KL, Dickhoven S, Tierney SF and King JL. (1987). *Datawars. The Politics of Modeling in Federal Policymaking.* Columbia University Press.

CHAPTER 2

Applications and Modelling in Mathematics Teaching – A Review of Arguments and Instructional Aspects

W Blum
University of Kassel, FR Germany

SUMMARY

The aim of this paper is a comprehensive presentation of some important *basic* and *general aspects* of the topic applications and modelling, with emphasis on the secondary school level. Owing to the review character of this paper, some overlap with the survey paper Blum and Niss (1989) for ICME-6 in Budapest is inevitable. The paper will consist of three parts. In *part 1*, I shall try to clarify some basic concepts and remind the reader of a few application and modelling examples suitable for teaching. In *part 2*, I shall formulate some general aims for mathematics instruction and, on that basis, summarise the most important arguments for and against applications and modelling in mathematics teaching. Finally, in *part 3*, I shall discuss some relevant instructional aspects resulting from the considerations in part 2.

1. CONCEPTS AND EXAMPLES

1.1 Basic concepts

I would like to commence by giving pragmatic working definitions for some basic concepts and notions such as *modelling* or *application* which sometimes – even in this volume – are used in different senses. Here I follow Blum and Niss (1989, ch 1).

By an *applied (mathematical) problem* or a *real problem situation* I mean an open situation which belongs to the real world and for the solution of which some mathematics might be helpful. By *real world* I mean the rest of the world outside mathematics, that is school or university subjects or disciplines different from mathematics, or everyday life and the world around us. The *applied problem solving process* consists of the entire process of dealing with an applied problem in attempting to solve it. In the following, I would like briefly to recollect some essential steps of this process (see, among many others, Steiner 1976, Pollak 1979, Blum 1985 or Niss 1987). As an illustration, I shall use the case of *income taxes*.

The *starting point* is a real problem situation. In our example, taxes are to be imposed on income, as 'justly' as possible. There are many possibilities for doing this, dependent on different interests and value judgements. In a first step, the situation has to be *simplified, idealised, structured* and *made more precise* by the problem solver, especially by formulating appropriate conditions and assumptions. In our example, such conditions could be "If person A earns more than person B then A ought to pay at least the same percentage of his income for taxes as B" or – perhaps more basically – "The tax should always be less than the income". This leads to a *real model* (here an economic model) of the original situation. The real model is not – as is sometimes misunderstood – merely a simplified but true image of some part of an objective, pre-existing reality. Rather, the first step of the process also structures and *creates* a piece of reality, dependent on intentions and interests of the problem solver.

In a second step, the real model has to be *mathematised*, that is its data, concepts, relations, conditions and assumptions are to be translated into mathematics. Thus, a *mathematical model* of the original situation results. In our example, the model consists of a real function $s: x \mapsto s(x)$, the income tax function (where $x \geqslant 0$ is the income and $s(x)$ is the corresponding tax), together with certain concepts like the marginal tax rate

$$s'(x) = \lim_{h \to 0} \frac{s(x+h) - s(x)}{h}$$

and certain conditions for the function s such as

$$\frac{s(x_1)}{x_1} \leqslant \frac{s(x_2)}{x_2} \text{ for } x_1 < x_2.$$

According to Blechman, Myškis and Panovko (1984, p148), "the ability to choose an appropriate mathematical model lies on the borderline between science and art". Again, the resulting model depends on the intentions of the modeller, whether it's a question of description, of prediction or of prescription (to follow a distinction made by Davis and Hersh (1986), ch 3; see also Davis in this volume).

A remark concerning terminology. Sometimes the notions of mathematising, modelling and problem solving are used synonymously. In this paper *mathematisation* is the process of going from the real model into mathematics, that is the second step above, whereas the process of *modelling* or *model building* comprises the first as well as the second step, and problem solving means the entire process which is being described here, of which modelling is only a part.

This process continues by a third step, that is *choice* of suitable mathematical *methods* and *work within mathematics*, through which certain *mathematical results* are obtained. In our example, these results are concrete income tax functions. In Germany, the actual (1990/91) income tax function is a stepwise polynomial function with degrees 0, 1 and 2. (The function which was effective until 1989 was even nicer for mathematics teaching because it contained, in addition, a polynomial of degree 4).

In a fourth step, these results have to be retranslated into the real world, that is to be *interpreted* in relation to the original situation or to the real model. In doing so the problem solver also *validates* the mathematical model. Here *discrepancies* of various kinds may occur which may lead to a modification of the model or to its replacement by a new one. In Germany, this happens nearly every two years with our income tax, so that we get a new tax function every two years. Thus, the problem solving process may require "going round the loop" *several* times (not necessarily, of course, in the order "step 1, step 2, ..., step 1, step 2" as described above).

Besides such complex problem-solving processes there are restricted links between mathematics and reality: a *direct application* of already developed standard mathematical models to real situations with a mathematical content, for instance the investigation of the present German income tax function by means of differential calculus, or a *dressing up* of purely mathematical problems in the words of another discipline or of everyday life, sometimes in a *whimsical* way (Pollak 1979), such as the investigation of the income tax function of Transsylvania.

I use the term *application* of mathematics to denote all the above-mentioned ways of connecting the real world with mathematics. In this sense real problem situations can also be called *applications*. Eventually, all mathematical models can be seen as belonging to a field which may be called *applied mathematics*. As regards the question what applied mathematics can mean and what its main goals are, I refer to Blechman,

Myškis and Panovko (1984) and to Davis in this volume.

1.2 Some examples

Up to now I have only referred to the case of income tax, which corresponds to the mathematical topics of arithmetic, functions and calculus. However, for nearly every topic in the mathematics curriculum, there is a wealth of application and modelling examples suitable for teaching. Especially in the last 15 years, a great many examples have been developed all over the world which can be used for many different instructional purposes, also for demonstrating the complex interrelationship between the real world and mathematics (see, for instance, numerous examples in the ICTMA-proceedings, Berry et al 1984, 1986, 1987 and Blum et al 1989; many references are given in Pollak 1979 and Bell 1983; see also the extensive bibliography of Kaiser et al 1982/1990). In the following table I shall recollect a few of these well-known examples, all suitable for the *school level*. With this, I intend to help the reader to relate the general considerations in the second and third part of this paper to concrete examples, and thus to give more meaning to these general statements.

Real-world examples	**Related mathematical topics**
various growth and decay processes: population growth, chemical reactions, spread of epidemics, warming/cooling, absorption, decay of beer froth and so on	functions, calculus, linear algebra
rainbow	geometry, calculus
genetics	arithmetic, probability, (linear) algebra
rates of interest	arithmetic, functions, calculus
price index	arithmetic, algebra, functions
*income tax	arithmetic, functions, calculus
*elections	arithmetic, geometry, stochastic
social classes	finite maths, probability, linear algebra
*traffic flow	finite maths, calculus, stochastics
biking	geometry, functions
football	geometry, arithmetic, stochastics
*shot putting	geometry, calculus

chairoplane	linear algebra, functions, calculus
gambling	stochastics, finite maths
painting	geometry, algebra
musical scales	arithmetic, algebra, functions

I look upon the examples marked with a * as particularly appropriate to showing essential features of the applied problem solving process, such as going round the loop several times or building different models for the same situation. I would like, however, to emphasise that *all* the examples just mentioned are suitable for going through this process with learners. For this is essentially a matter of looking at the examples and of preparing them, and not a question of the nature of the examples themselves.

2. ARGUMENTS

2.1 Aims for mathematics teaching

Now, after dealing with basic concepts and citing some examples in part 1, I shall consider in part 2 the *teaching* of mathematics (as a separate subject) and in particular the *role of applications and modelling* in mathematics instruction. It seems to be quite obvious, but it might be worth emphasising once more, that the role of applications and modelling can only be reflected by making explicit reference to some sort of a theoretical *basis*, to a *philosophy* of mathematics instruction. The most important constituent of such a basis is, in my view, the aspired general *aims* for mathematics teaching. Other relevant constituents are basic didactical principles or empirical findings about the learning and teaching of mathematics. Here I can only deal with the question of aims. There is plenty of literature discussing aims and goals for mathematics instruction, and there are countless catalogues of aims. Nevertheless, I will add another catalogue because I regard mine to be a useful basis which I will refer to several times in the following sections.

The discussion of aims for mathematics instruction must be embedded into a more general context with social, political or pedagogical issues, as done for example in D'Ambrosio (1979), Christiansen, Howson and Otte (1986, ch 1/2) and particularly Niss (1981). Roughly speaking, general aims for mathematics teaching are related to higher order *educational* goals, dependent on a certain '*picture of man*'. Such goals are in particular the following. Students are to be educated to be responsible and intelligent citizens, are to be trained towards professionalism at work, they are to master everyday life and to acquire appropriate intellectual attitudes. For all these educational goals *mathematics* may play an important role. In this respect let me formulate the following general aims for mathematics teaching.

1. *Pragmatic aims* : Mathematics teaching is intended to help students to describe relevant aspects of extra–mathematical areas and situations, to understand these better and to cope with them better. Such situations may originate
   (a) from present or future daily life and the world around us, or from other subjects in general education at school,
   (b) from present or future fields of study or profession in vocational education at school or at university.

2. *Formative aims* : By being concerned with mathematics, students should
   (a) acquire important general qualifications, for instance the ability to argue, to solve problems or to translate between reality and mathematics, or attitudes such as openness towards problem situations and willingness for intellectual efforts,
   (b) obtain pleasure and enjoy their activities.

3. *Cultural aims* : Students should be taught mathematical topics
   (a) as a source for philosophical and epistemological reflection, including individual and human self–reflection,
   (b) to generate a comprehensive and balanced picture of mathematics as a science and as a part of human history and culture, including a critical appreciation of actual uses and misuses of mathematics in society,
   (c) to produce knowledge, skills and abilities concerning special mathematical themes.

   By all this students should also acquire meta–knowledge of mathematics.

These aims correspond to different *aspects of human nature*: man as
   (1) an observing, producing, economic being,
   (2) a reasoning, communicating, social being,
   (3) a reflecting, creating, cultural being.

The various aims must be assessed differently according to the *educational history* and context:

A. Mathematics in *schools* offering *general* education, at the lower or at the upper secondary level.
B. Mathematics in *vocational* education at *schools* and colleges, or as a *service subject* in *university* courses for future scientists, engineers, economists and so on.
C. Mathematics in *university* courses for future *mathematicians* or mathematics teachers.

The result of this assessment is shown in the following matrix. The number of * symbols corresponds to the importance I have assigned to the aims.

educational histories		pragmatic aims a b	formative aims a b	cultural aims a b c
maths in general education	lower secondary	***	**	*
	upper secondary	** *	*** **	***
maths in vocational education or as a service subject		* ***	**	**
maths for mathematicians in university		*	**	*** ***

Altogether it may be said that for all levels *all aims* are relevant, though with different emphases in each case.

On that basis, I shall now try to identify and to discuss the most important arguments for and against incorporating applications and modelling in mathematics instruction. I shall commence with counter-arguments.

2.2 Arguments against applications and modelling

Several investigations and inquiries in many different countries have shown that in everyday mathematics teaching applications and modelling very often do not play as important a role as theoretical analyses have assigned to them (see for example the review by Burkhardt 1983). This is not due to ill-will or incompetence of teachers but to certain severe *obstacles*. Let me briefly summarise some of these obstacles in the form of *counter-arguments* (see Blum and Niss 1989, ch 3).

C1 *Counter-arguments from the point of view of instruction.* There is *not enough time* to deal with applications and modelling in mathematics instruction. More fundamentally, applications and modelling *do not belong* to mathematics instruction at all. So, extra-mathematical problems should be treated in the teaching of other subjects, such as the example of elections in social studies, or shot putting in sports.

C2 *Counter–arguments from the learner's point of view.* Applications and in particular modelling make the mathematics lessons and examinations more complex, *more demanding* and less predictable for learners. Students have to learn, for example, not only mathematical concepts such as the derivative but also its real world interpretations as a growth rate, a tax rate and so on.

C3 *Counter–arguments from the teacher's point of view.* Applications and modelling make teaching *more demanding* because additional non–mathematical knowledge and qualifications are required. Applications and modelling work makes instruction *more open* and necessitates types of classroom interaction unusual for mathematics teachers – for instance, in the case of traffic flow, discussions about environmental problems with unpredictable outcome. All these might *undermine the teacher's expert authority.*

2.3 Arguments for applications and modelling

Referring to the aims identified in section 2.1, I shall now identify potential *roles* applications and modelling may play in the promotion of these aims. This will result in various *pro–arguments.* I shall illustrate the arguments by schematic diagrams in which *S*tudents, *R*eal World and *M*athematics are related to each other in a characteristic manner; an arrow → has the rough meaning of 'helps'. Of course, I do not want to quote these pro–arguments in order to convince the reader how important applications are. Fortunately, this is today, in the beginning of the nineties, no longer necessary, in contrast with the situation 15 or 20 years ago. Rather, I would like to systematise the various arguments for applications and modelling which have been invoked time and again during the history of mathematics education, and thus to lay a solid foundation for curricular and methodical conclusions.

P1 *Pragmatic arguments* (Mathematics as an aid for specific real situations) : The ability of students to understand and to master extra–mathematical situations does not result automatically from learning pure mathematics but can only be achieved by dealing, in mathematics instruction, with such real situations. Students can, for instance, be helped to understand better public discussions about the limits to growth by treating ecological growth processes in the context of percentages, geometric sequences or exponential functions. In my opinion, such 'enlightenment' belongs genuinely to mathematics teaching, and not only to the teaching of other subjects.

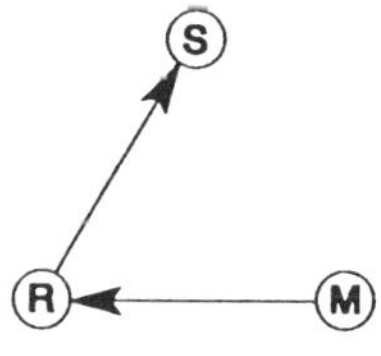

P2 *Formative arguments* (Real-world applications of mathematics as an aid for general qualifications). General competences and attitudes can also be developed by suitable applicational examples. In particular, the ability to translate between the real world and mathematics can only be advanced by means of examples where the entire problem solving process is covered, as was shown in the example of income taxes.

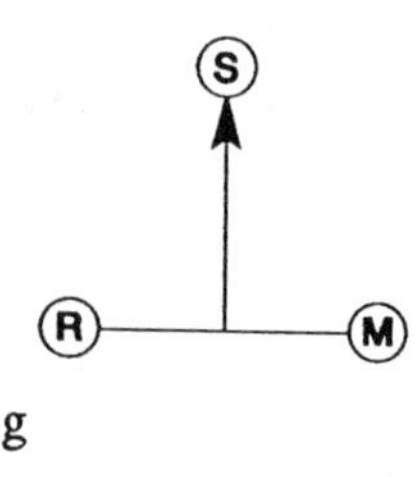

P3 *Cultural arguments* (Real-world applications as a source for reflection and as a component of an appropriate overall picture of mathematics). Reflections on a meta-level can also originate from linking mathematics with the real world. Dealing with the mathematics of perspective drawing, for example, can provoke discussion of the way that man perceives the world. More generally, students can see that – in a certain sense – "modelling is the essential feature of human intellectual behaviour" (D'Ambrosio 1989, p23). Furthermore, applications as well as modelling constitute an essential component of an adequate picture of mathematics. When treating, for instance, the topic of elections, we can illustrate where and how mathematics and mathematicians are needed, and also where they are not needed, by which means blind faith and trust in science can be reduced, and meta-knowledge of mathematics and its relations to applications generated.

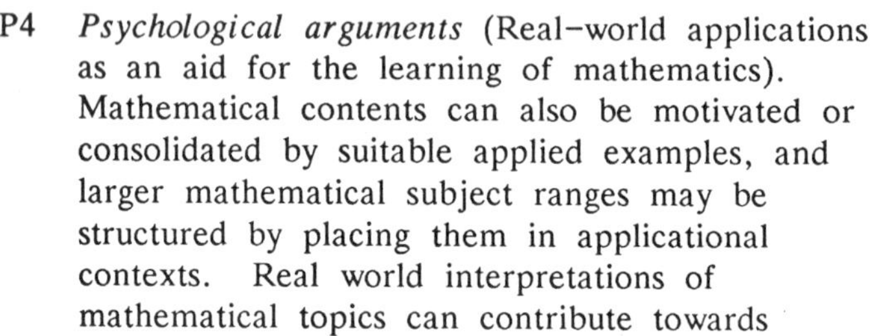

P4 *Psychological arguments* (Real-world applications as an aid for the learning of mathematics). Mathematical contents can also be motivated or consolidated by suitable applied examples, and larger mathematical subject ranges may be structured by placing them in applicational contexts. Real world interpretations of mathematical topics can contribute towards better and deeper understanding and longer retention of these topics, and relations to reality may improve the students' attitude towards mathematics.

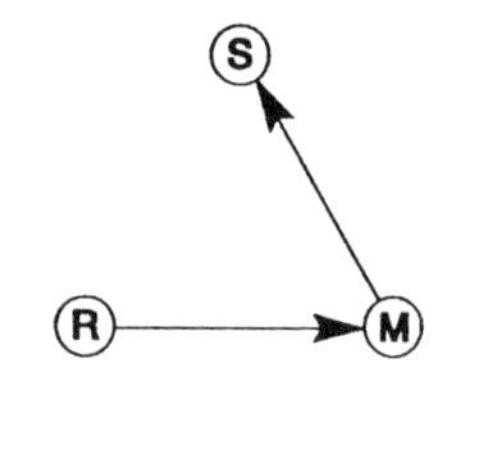

In addition, the inclusion of applications and modelling contributes towards giving more *meaning* to the learning and teaching of mathematics. This is perhaps the most important kind of argument, for (using a statement made by Thom in a slightly different context) "the real problem which confronts mathematics teaching is ... the problem of the development of meaning" (Thom 1973, p202).

I cannot consider the pros and cons in depth here. In any case, in the light of the arguments just mentioned, counter-argument C1 is simply *wrong*. In particular, time spent on applications and modelling is not only justified and necessary but also mathematically profitable according to the arguments just given. So there *has* to be time for applications and modelling, as a 'didactical axiom', if need be by reducing the mathematics programme.

In contrast, counter-arguments C2 and C3 remain largely *correct* - unfortunately! In particular, mathematics instruction does not become easier for students by the inclusion of applications and modelling - on the contrary. This is why applications sometimes fail totally to motivate students to do mathematics.

The pro-arguments have to be weighed according to the importance of the accompanying aims. The result is shown in the following matrix. Again, the number of * refers to the relevance of the arguments.

educational histories	pragmatic arguments	formative arguments	cultural arguments	psychological arguments
lower secondary school	***	**	*	**
upper secondary school	**	***	***	**
vocational education or service subject	***	**	*	**
mathematicians	*	**	***	**

Altogether it may be said that for all levels *all arguments* are of significance, though with different weights.

It is possible now to identify different *schools of thought* within the mathematics education community, corresponding to the aims placed in the foreground in each case and to the arguments given for the inclusion of applications and modelling in mathematics teaching; for details see Kaiser-Messmer (1986b vol 1 and in this volume).

All things considered, the diverse arguments demand *applications and modelling to become and remain essential parts of mathematics teaching* at all levels (whether students or teachers like that or not). Students should (see Niss 1989)

- *know* concrete standard models and applications of mathematics as well as essential steps of corresponding model building, interpreting and applying processes,
- be able to *perform* modelling themselves and critically *assess* models and modelling processes.

Or, in the words of Skovsmose (1989), students should, besides mathematical knowledge, acquire 'technological' knowledge (about models) as well as 'reflective' knowledge (meta-knowledge).

In order to achieve that, every effort should be made to *overcome the obstacles* to applications and modelling mentioned in the counter-arguments C2 and C3. To deal with that question seriously would necessitate several separate papers. Let me make one general remark only. Most important is an adequate pre-service and in-service *teacher education* in order to supply teachers with knowledge, abilities, experiences and, in particular, with attitudes to cope with the demands of teaching modelling and applications. Teachers must have achieved for *themselves* all that they (ought to) demand from their students.

So there is a consensus of opinion in the mathematics education community that applications and modelling should be an integral part of mathematics teaching. Now, what are consequences of such a decision? This question will be addressed in part 3 of this paper.

3. INSTRUCTIONAL ASPECTS

3.1 Approaches to organisation

The following demand is an obvious consequence of part 2.

Demand 1: Applications and modelling should be incorporated in mathematics instruction at all levels.

What is *actually* going on in mathematics education and in the classroom? In Blum and Niss (1989, ch 2) we stated the following.

Trend 1 (Increasing inclusion of applications and modelling). In the eighties, all over the world a *growing* number of mathematics programmes at different levels are *including* applications and modelling. There are, however, some severe *obstacles* to the inclusion of applications and modelling in everyday teaching practice. So, there is a large *distance* between the forefront of research, development and practice in applications and modelling, on the one hand, and applications and modelling in the *mainstream* of mathematics instruction on the other hand. Yet, this distance is in the process of being considerably *reduced*.

Now, in part 3, I shall discuss some curricular and methodical *consequences* of demand 1. That is, suppose we decide to include applications and modelling in mathematics teaching, how should this be done, what kinds of examples should be chosen, and how should they be taught? In the whole of part 3, I shall concentrate on mathematics at *schools*. Again, I have to restrict myself to making only rather *general* remarks. In this first section, I shall address briefly the question of how mathematics teaching should be *organised* if applications and modelling are to be incorporated; for a far more detailed discussion of important aspects and viewpoints for building application-oriented mathematics curricula see Usiskin (1989 and in this volume).

Niss has distinguished between different types of basic *approaches* to the organisation of mathematical and applicational curriculum components (for details see Blum and Niss 1989, p16): the *separation* approach, the *two-compartment* approach, the *islands* approach, the *mixing* approach, and *integrated* approaches. Which of these different approaches is to be chosen depends on the background philosophy, particularly on the aims for mathematics teaching (see section 2.1). The islands approach, for example, may impede the formation of modelling abilities or of a balanced picture of mathematics, and an integrated approach may render the promotion of cultural aims more difficult. For mathematics in *school*, I advocate a certain kind of a *mixing* approach which allows all the aims to advance and takes into account all the arguments for applications and modelling. Here, the mathematics programme should be organised in a consistent way, and many application and modelling examples should be incorporated. What kinds of examples are to be included I shall discuss next.

3.2 Kinds of examples

In my opinion, many local and some global application and modelling examples should be incorporated. By *local* examples I mean comparatively small-scale examples which correspond primarily to the pragmatic and psychological arguments, so that they are more suited to a description as well as a better understanding and mastering of special real situations, and to a motivation or an illustration of certain mathematical topics. Such examples are mostly given as *problems* or *exercises* as they can be found in almost every textbook, for instance parabolic aerials, orbits of comets or planets, or whispering galleries, as applications of conic sections. By *global* examples I mean larger examples which correspond to all the above arguments, in particular to the formative and cultural arguments, so that they are more suited to advancing general abilities (especially modelling) and adequate attitudes, and to developing a balanced picture of mathematics. Such examples are mostly presented as *teaching units* over several hours. As an example, I refer to a unit on income taxes developed in Kassel and taught several times in general and in vocational schools (see Blum 1988).

- Presentation of problem situation by teacher

- Discussion, and formulation of questions by students
- Structuring, simplifying, stating of fiscal policy demands
- Joint mathematising of demands and concepts
- Individual construction of possible income tax functions
- Presentation of actual income tax law by teacher (or providing it by students)
- Individual calculation of income taxes, drawing of income tax function (also with the aid of computers)
- Analytical investigation of income tax function
- Interpretations, comparison with demands
- Discussion, variation of parameters, interpretation of changes
- Methodological reflections

Not only *really real* but also *artificial*, whimsical problems as well as traditional word problems may be used. If their character is made explicitly clear, such unrealistic examples are sometimes better suited to educational purposes than are genuine real world applications, especially for furthering formative aims.

So, the following demand is sensible.

Demand 2: In mathematics teaching, different kinds of real world examples should be incorporated, both local and global ones, also and particularly modelling examples.

What is the actual situation in research and practice? In Blum and Niss (1989, ch 2) we identified the following.

Trend 2 (Broadening of views on applications and modelling). In the eighties, the *relations* between applications on the one hand and modelling on the other hand have become *stronger*, both in research and in practice. That means that different persons and groups doing application-oriented *research* and development in mathematics education have become increasingly aware of and interested in *contacts* among one another. In the *practice* of mathematics teaching or of constructing concrete mathematics curricula, different kinds of examples – not only classical physical or standard everyday applications – are becoming increasingly included, especially *modelling* examples, and the range of application areas has been significantly *broadened*.

There are several reasons for this *opening* of mathematics instruction to new application areas different from the classical one of physics. For instance, some aspects important for application practice and for instruction (such as going round the loop in the applied problem solving process several times or building different models for the same situation) can often be demonstrated much better with non-physical examples. However, I consider it is still important that a *close contact between mathematics and physics* be maintained in our educational system, because relations between mathematics and physics have always been especially intimate and therefore played an outstanding role in cultural history. In some instances, the links in content between a physical situation and its mathematical description are inseparable, for example in the case of natural laws, and the viewpoint of models and modelling tends to create unnatural distances. So students also have to know of physical applications, and to reflect upon this area's peculiar epistemological state (see, for example, Bkouche 1989).

3.3 Some consequences for methods

Another obvious consequence of part 2 is the following.

Demand 3: When incorporating applications and modelling in mathematics teaching, the whole range of arguments for this inclusion - pragmatic, formative, cultural and psychological arguments - should be taken into consideration.

During the history of mathematics instruction and mathematics education this was not always so. Mostly, only *single* arguments were predominant, such as parts of the psychological and of the formative arguments for the traditional, academically-oriented general education in upper secondary schools. Cultural arguments - emphasising metalevels connected with scientific theory and epistemology - were not stressed until the last decade. In Blum and Niss (1989, ch 2) we identified the following actual trend, which is closely connected with the two trends mentioned earlier.

Trend 3 (Widening of the spectrum of arguments). In the eighties, the *complete range* of arguments for including applications and modelling in mathematics teaching is put forward in mathematics education, with different emphases for different educational histories.

This widening of the spectrum of arguments results, on the one hand, in a broadening of examples as described in 3.2. On the other hand, this widening has immediate consequences for the *methods* in mathematics teaching, three of which will be mentioned now.

1. In order to create meta-knowledge (the importance of which is stressed in present-day learning psychology) and to enable students to translate between the real world and mathematics, not only do examples have to be treated in which such translations take place, but also the problem-solving process has to be *analysed*, and the students have to *become* explicitly *aware* of it. This can be done best in the context of global examples, such as the above-mentioned case of income taxes; here I refer to the textbook Griesel and Postel (1988, p208/209).

2. To achieve a comprehensive *understanding of mathematical concepts* – which includes relations to the real world as well – these concepts have to be taught to students in an adequate manner. We in Kassel, especially Gabriele Kaiser–Messmer, have carried out extensive empirical investigations into applications in mathematics teaching in the last few years. Among other things, we have tested the comprehension of the concepts of derivative (see Kaiser–Messmer 1986a) and of definite integral with upper secondary and tertiary students. The following problems are taken from a recently developed *questionnaire* for the concept of *definite integral.*

11. Let x denote the annual income (in DM). Let r(x) denote the corresponding marginal tax rate (in %), that is the local rate of change of the annual income taxes s(x) (in DM) with respect to the income. Herr Y now may earn B DM instead of A DM per year. How can you calculate, using r, how much more income tax Herr Y has to pay per year? Give reasons for your answer.

13. $\int_a^b f(x)dx$ means the volume of blood which has flowed through an artery between time a and time b. Explain the real world meaning of x and of f(x).

16. A factory is diffusing pollutants into the air. In time interval $[t_1,t_2]$ (time t in min) the instantaneous pollution rate p(t) (in g per min) is measured. Interpret the real world meaning of

(a) $\int_{t_1}^{t_2} p(t)\,dt$ (b) $\frac{1}{t_2-t_1} \cdot \int_{t_1}^{t_2} p(t)\,dt$

One of the *results* of our investigations is as follows. The ability to interpret and to apply these concepts in unknown real situations can be achieved extensively with many students, along with an adequate introduction and treatment of these concepts, and this distinctly better than in the case of a pure mathematical procedure. Certain ways (both applied and pure ones) of introducing and treating these concepts can even hinder those interpretation and application abilities. *Adequate* introduction and treatment means, for instance, for the concept of derivative: handling of different kinds of average and local rates of change; non–formal introduction of the concepts of rate of change, of difference quotient and derivative (with an exactification later on); parallel geometric interpretations; re–interpretation of these concepts; applications to new situations. This may sound very natural and totally self–evident, but in most of the cases we have observed, at school and at university, the method of teaching was quite different. This is why our own model classes did significantly better in these tests.

3. Relations to reality may also serve for *proving mathematical results* on a *preformal* level (but rigorously!). A well-known example is the theorem "If f' = 0 in an interval I, then f is constant in I", which is equivalent to the completeness property of the real numbers and therefore not easy to prove formally. This can be made evident immediately by the kinematic argument "If the speedometer is showing 0 all the time, then the car is always standing still". Another interesting example, which I observed recently in grade 12 (18 year olds) and which I have since used several times to test students at school and university, refers to the question whether non-trivial solutions of the differential equation f' = f may have zeros. Most of the students used, or accepted unhesitatingly, an argument which looks very similar to the one just cited, but which now is wrong. See Blum and Kirsch (1991) for a detailed discussion of that problem and for some didactical conclusions.

3.4 The role of computers

For several years now, *computers* have proliferated into many areas in society, including the educational system, and they are also influencing mathematics instruction at all levels (compare, for example, various contributions in Blum et al 1989, section E). Here I shall only make some remarks on computers as a *tool* for mathematics teaching, as a means for performing numerical and algebraic calculations, or for drawing graphs, and as an aid for creating new teaching methods. By computers I mean in particular *pocket computers*; I think it will only be these that will really change mathematics teaching. Generally speaking, one can notice the following (see Blum and Niss 1989, ch 2).

Trend 4 (Extended use of computers). The use of computers in mathematics teaching is being *extended, quantitatively* as well as *qualitatively*, also with respect to applications and modelling.

By the use of computers, new possibilities have become available for making mathematical contents accessible to learners, for promoting the intended aims, or for relieving mathematics teaching of some tedious activities. This also has many implications for the teaching of mathematical modelling and applications. I have to refrain here from quoting examples: see Blum and Niss (1989, ch 3), and see also Huntley in this volume. Instead, let me emphasise that – quite apart from the fact that, in a great deal of numerical contexts, the simple pocket calculator is perfectly sufficient – the computer may also entail many kinds of *problems* and *risks*, especially in relation to applications and modelling in mathematics instruction, some of which can already be observed in the classroom, for instance the following.

- The devaluation of routine computational and graphical skills and the shift in emphasis towards applications and modelling enabled by the computer, may make mathematics instruction *more demanding* for both students and teachers.
- Necessary *intellectual efforts* of students may be *replaced* by mere button pressing, and new kinds of senseless routines may occur.
- Mathematics teachers may become interested in *computers instead of* in *modelling and applications*, and students may be prevented by the computer from reflecting deeply upon mathematical problems by being engaged in outward activities like constructing technically elegant programmes.

One of the most important general recommendations for dealing with these problems is probably that teachers *and* students should become fully *aware* of the problems. This will also contribute towards one of the aims of mathematics teaching mentioned in part 2, namely the acquisition of meta-knowledge of mathematics, including its relation to the real world and the advantages and risks of its tools.

In this sense I would like to finish part 3 by formulating a general proposition.

Demand 4: Computers should be used appropriately in mathematics instruction to improve the learning and teaching of mathematical modelling and applications. When using computers, teachers and students should be aware of various inherent problems and risks.

4. FINAL REMARKS

There are, of course, many more aspects of the learning and teaching of mathematical modelling and applications which I cannot touch upon in this paper, such as the important problem of assessment. I would like to close by making some optimistic remarks. There is a general consensus of opinion that applications and modelling should be an essential part of mathematics instruction at all levels. The literature shows that examples and materials are available to a sufficient extent. Empirical investigations have produced encouraging results with regard to various positive effects of applications and modelling (see, among others, Kaiser-Messmer 1986b, vol 2). On the one hand, they show that mathematics instruction does not become easier for students and teachers by the inclusion of applications and modelling but that, on the other hand, the world around us may be comprehended better, mathematical concepts may be understood more deeply and more extensively, formative abilities may be advanced, and attitudes towards mathematics may be improved by giving more meaning to it. My final remark resumes the end of part 2 and is equally self-evident and highly relevant: the most important factor for achieving such positive effects is the *teacher*. All conceptions and proposals for mathematics teaching stand and fall with the teachers, with their professional abilities, and with their didactical and pedagogical qualifications. Nor is the teachers' vital role to be deplored in my view. On the contrary, this role is utterly desirable and favourable for our work in mathematics education, the ultimate aim of which is to improve permanently the learning and teaching of mathematics.

REFERENCES

D'Ambrosio U. (1979). Overall Goals and Objectives for Mathematical Education. In *New Trends in Mathematics Teaching IV* (UNESCO, ed). Paris, 180-198.

D'Ambrosio U. (1989). Historical and Epistemological Bases for Modelling and Implications for the Curriculum. In Blum, Niss and Huntley (1989), 22-27.

Bell M. (1983). Materials Available Worldwide for Teaching Applications of Mathematics at the School Level. In *Proceedings of the Fourth International Congress on Mathematical Education* (M Zweng *et al*, eds), Birkhäuser, Boston, 252-267.

Berry J *et al* (eds). (1984). *Teaching and Applying Mathematical Modelling*. Ellis Horwood, Chichester.

Berry J *et al* (eds). (1986). *Mathematical Modelling Methodology, Models and Micros*. Ellis Horwood, Chichester.

Berry J *et al* (eds). (1987). *Mathematical Modelling Courses*. Ellis Horwood, Chichester.

Bkouche R. (1989). Mathematics and Physics: Where is the Difference? In Blum *et al*, (1989), 49-54.

Blechman I, Myškis A and Panovko J. (1984). *Angewandte Mathematik*. Deutscher Verlag der Wissenschaften, Berlin-Ost (original in Russian).

Blum W. (1985). Anwendungsorientierter Mathematikunterricht in der didaktischen Diskussion. In *Mathematische Semesterberichte*, **32**, 195-232.

Blum W. (1988). Analysis in der Fachoberschule – Überlegungen zur Konzeption, zu Anwendungsbezügen und zum Rechnereinsatz. In *Technic didact, Schriftenreihe: Diskussionsfeld Technische Ausbildung* (P Bardy *et al*, eds), Leuchtturm, Alsbach, 229–252.

Blum W *et al* (eds). (1989). *Applications and Modelling in Learning and Teaching Mathematics.* Ellis Horwood, Chichester.

Blum W and Kirsch A. (1991). Why Do Non-Trivial Solutions of f'=f Have no Zeros? Examples of and Reflections on Preformal Proving. To appear in *Educ Studies in Maths.*

Blum W and Niss M. (1989). Mathematical Problem Solving, Modelling, Applications, and Links to Other Subjects – State, Trends and Issues in Mathematics Instruction. In Blum, Niss and Huntley (1989), 1–21.

Blum W, Niss M and Huntley I (eds). (1989). *Modelling, Applications and Applied Problem Solving – Teaching Mathematics in a Real Context.* Ellis Horwood, Chichester.

Burkhardt H (ed). (1983). *An International Review of Applications in School Mathematics.* ERIC, Ohio.

Christiansen B, Howson AG and Otte M (eds). (1986). *Perspectives on Mathematics Education.* Reidel, Dordrecht.

Davis P and Hersh R. (1986). *Descartes' Dream – The World According to Mathematics.* Harcourt, Brace and Jovanovich, New York.

Griesel H and Postel H (eds). (1988). *Mathematik heute, Einführung in die Analysis 1.* Schroedel, Hannover.

Kaiser-Messmer G. (1986a). Modelling in Calculus Instruction – Empirical Research Towards an Appropriate Introduction of Concepts. In Berry *et al,* (1986), 36–47.

Kaiser-Messmer G. (1986b). *Anwendungen im Mathematikunterricht,* Vol 1/2. Franzbecker, Bad Salzdetfurth.

Kaiser G, Blum W and Schober M. (1982/1990). *Dokumentation ausgewählter Literatur zum anwendungsorientierten Mathematikunterricht,* Vol 1/2. Fachinformationszentrum, Karlsruhe.

Niss M. (1981). Goals as a Reflection of the Needs of Society. In *Studies in Mathematics Education* (R Morris, ed), Vol 2. UNESCO, Paris.

Niss M. (1987). Applications and Modelling in the Mathematics Curriculum - State and Trends. In *Int J Math Ed Sci Tech*, **18**, 487–505.

Niss M. (1989). Aims and Scope of Applications and Modelling in Mathematics Curricula. In Blum *et al*, (1989), 22–31.

Pollak H. (1979). The Interaction between Mathematics and Other School Subjects. In *New Trends in Mathematic Teaching IV* (UNESCO, ed). Paris, 232–248.

Skovsmose O. (1989). Towards a Philosophy of an Applied Oriented Mathematical Education. In Blum *et al*, (1989), 110–114.

Steiner HG. (1976). Zur Methodik des mathematisierenden Unterrichts. In *Anwendungsorientierte Mathematik in der Sekundarstufe II* (W Dörfler and R Fischer, eds). Heyn, Klagenfurt, 221–245.

Thom R. (1973). Modern Mathematics: Does it Exist? In *Developments in Mathematical Education* (AG Howson, ed). Cambridge Univ Press, 194–209.

Usiskin Z. (1989). The Sequencing of Applications and Modelling in the University of Chicago School Mathematics Project 7–12 Curriculum. In Blum *et al*, (1989), 176–181.

CHAPTER 3

Building Mathematics Curricula with Applications and Modelling

Z Usiskin
University of Chicago, USA

SUMMARY

My perspective is that of a curriculum developer who has for some time been trying to weave applications and modelling into the mathematics curricula of average students. My remarks would be quite different if I were organising courses wholly devoted to applications and modelling, as many have done. I cannot separate applied mathematics from mathematics itself, either in theory or in practice. To me they are both part of the same magnificent edifice.

At international conferences, I am continually surprised by finding that things I thought were the same everywhere are not, but more often things I thought were different in other countries are nearly the same. Still, my perspective was formed in the United States, and I apologise in advance for any parochial views.

The subject of this paper is the building of mathematics curricula with applications and modelling. My remarks and examples are based on my experiences with students in the United States in Grades 7–12, that is, of ages 12–18. For the most part, the students who have been the targets of my work are sitting in classes which they either are required to attend or feel compelled to take for college entrance or success. Most of these students have not made any selection of particular fields of study beyond the notion that they will or will not go to a college or university after completing secondary school. Nearly two–thirds of those who finish secondary school in the United States go to college, almost half of the total age cohort.

I probably do not need to remind you that, although almost all schools in the United States teach nearly the same mathematics curriculum, there is no national curriculum. Furthermore, starting in 9th grade, that is with 14 year olds, the mathematics curriculum is very crowded; only classes with the best students finish their textbooks. This makes change particularly difficult to achieve in grades 9 through 12. Teachers must not only be convinced of the importance of new content, but they have to give up teaching old things in order to have time for the new.

Consequently, when I think of building curriculum in applications and modelling, neither the students nor the teachers I am thinking of are necessarily eager to work with applications and modelling.

The word building has meanings both as a verb and as a noun, and I wish to speak about both these aspects. As a noun, a building is a structure or edifice. If we think of the magnificent edifice of mathematics as a building, with the topics of mathematics being its rooms, then the rooms are interconnected in many ways and this building is very large, perhaps infinite in size. As educators, we have the opportunity to select the doors to the rooms we want to open and the floors of the building upon which we wish students to spend their time. We may lecture on the most interesting features of this building or we may allow students to explore on their own. Some of the rooms of the great edifice of mathematics have doors and windows with wide vistas to the real world; other rooms have small windows to the world and perhaps only one door known to us.

The building of mathematics is very complex, and interrelated with other buildings devoted to the social sciences, the physical sciences, philosophy, business and commerce, and many other domains of human activity. Perhaps a suitable metaphor for the relationships between these buildings is the body. As Mandelbrot (1976) has pointed out, in our bodies there are disjoint networks of arteries, of veins, and of nerves, yet every cell is very close to all of these networks. Similarly, it seems that almost every room of mathematics is close to many many other rooms.

1. THE AIMS OF MODELLING

As a verb, building means constructing. Curriculum theorists generally feel that one should begin the building of a curriculum by agreeing on the aims, goals or objectives of that curriculum. In a plenary session at ICTMA-3, Niss (1989) gave five aims of teaching applications and modelling. Here they are, in a slightly briefer form that he gave them:

(1) to foster creative and problem-solving attitudes, activities, and competences;

(2) to generate a critical potential towards the use and misuse of mathematics in applied contexts;

(3) to provide the opportunity for students to practise applying mathematics that they would need as individuals, citizens, or professionals;

(4) to contribute to a balanced picture of mathematics;

(5) to assist in acquiring and understanding mathematical concepts.

His paper discusses these aims in some detail. I will assume a general agreement with these goals, and will not discuss them further here.

After goals are established, one must decide how to reach them, that is, what to teach, and what experiences to provide for students. Much of the work in the past 20 years in applications and modelling has been devoted to the first step in this process, the collection and exhibition of examples. For instance, at ICME–4 in 1980 my colleague Max Bell was asked to lecture on materials available worldwide for teaching applications of mathematics, and many of the available materials were simply collections of applications (see Bell, 1983). More recently, the proceedings of ICTMA–3 (Blum *et al* 1989) contain numerous articles whose major goal is to point out some beautiful examples of mathematical modelling and applications. The program for ICTMA–4 is similar.

We are in a stage of phenomenal growth of applied mathematics due to the very recent proliferation of personal computers which provide the power to deal with huge amounts of data and heretofore unavailable graphics capability. It is hard to realise that the first personal computers appeared only 12 years ago, in 1977, and the first one–megabyte personal computer is only six years old. Mathematics possesses an accessibility to the general public as never before; that accessibility will cause more and more applications to be found for mathematics. As with pure mathematics, applied mathematics is certain to grow forever. New areas of application are almost certain to be represented at ICTMA–n, for all n.

However, even as this development is in its infancy, the utilisation of applications and mathematical modelling is also beginning its maturity, and as part of that maturation we gain new responsibility. We can no longer be content with the mere display of beautiful applications or of nice examples of modelling. We must organise and sequence the applications so that students gain the general ideas and the power they need to reach the goals summarised by Mogens Niss. If we expect ideas to build in the minds of our students, we must build curricula to match.

Put another way, I believe the time has come for us to consider a longer time frame than just a set of examples or problems, or a single course. If applications and modelling are as important as most of us think they are, then the experiences with these ideas must begin early in a child's education and continue throughout. The questions of selecting, sequencing, and timing these experiences is what I mean by building a curriculum.

2. THE NEED FOR STRUCTURE IN BUILDING A CURRICULUM

I am speaking to the flock, the converted. Not everyone agrees that applications and modelling are so important. These are obstacles to the implementation of applications and modelling. In many countries, including my own, applications of mathematics beyond arithmetic are not part of the standard curriculum. Algebra, geometry, and analysis are taught with few applications. Statistics does not appear. 'Modelling' does not appear even in the index of the books.

There are teachers who think applications are too hard for most students, and so they will teach them only to their best students. And there are teachers who think applications are not really good mathematics, appropriate only for the poorer students as motivation, because the poorer students can't learn good mathematics. The students in the middle, the average students, the majority of students, get fewer experiences with applications than the best or the worst.

I should say that it is from my perspective, perhaps from our perspective, that these students encounter no applications. Many teachers think they are teaching applications through problems like the following, which itself is an example in a widely used 9th grade textbook in the United States.

> A clerk mistakenly reversed the two digits in the price of a marking pen and overcharged the customer 27¢. If the sum of the digits was 15, what was the correct price of the pen?

This problem has at least two characteristics which distinguish it from applications. The first is called *reverse given-find* (Thorndike, 1923). How do you know the sum of the digits is 15 and how do you know the customer was overcharged by 27¢ unless you knew the price in the first place? Reverse given-find is characteristic of many of the word problems which substitute for applications.

Examination of the solution to this problem gives the second characteristic which distinguishes it from an application. The solution begins by letting t = the tens digits and u = the units digit of the sale price. Then it translates the conditions into the equations

$10t + u - (10u+t) = 27$ and $t + u = 15$. After some manoeuvering, this system is solved to find $t = 9$ and $u = 6$. The solution is given in detail – it takes almost a page of the book.

The book does not consider the possibility of using arithmetic. Yet arithmetic provides a more efficient solution. There are only 4 two-digit numbers whose sum of digits is 15; they are 69, 78, 87 and 96. The two of these whose difference is 27 is easy to spot, and so the original price was 69¢ and the overcharge was 96¢. The example *unnecessarily restricts the allowable mathematics,* just the opposite of what would happen in an applied situation.

This problem appears in a 1981 textbook, and I first spoke about it in 1982. I was certain that the publisher would hear about my remarks. You will be interested to know that the problem was changed for the current edition (Dolciani et al, 1986).

> A catalog clerk mistakenly reversed the two digits in the price of a radio fuse and overcharged the customer 36¢. If the sum of the digits was 14, what was the correct price?

It is no wonder that huge numbers of educated adults see no reason to learn algebra, and that there exist newspaper cartoons, in which algebra is portrayed as being only good for one's CV, in which word problems are said to be good not even for the dead, and a famous one (among mathematics aficionados in the United States) in which Hell's library contains many books but all of the same type – story problems.

Most mathematics teachers worldwide received their education when mathematics departments did not have many courses in applied mathematics. Accordingly, to encourage the teaching of applications, it is customary to attempt to acquaint these teachers with large numbers of applications. The result has been virtually imperceptible. Why? Why does knowledge of good applications have so little effect on these teachers? I believe it is because, at the beginning, when they know only a few applications, they do not consider them important. After they learn about greater numbers of applications, there comes a time when they realise that the total number of applications is huge, and that they are relatively ignorant. This overwhelms the average teacher, and every new application merely brings more frustration. The teacher gropes for some way to make applications manageable.

Suppose that just the opposite situation from the present existed, that all the mathematics taught was applied, and none of the abstract properties of operations or of the real numbers were taught. Suppose that we were to desire that some of these properties be learned by students. Suppose we saw some beautiful examples such as the following.

$(a+b)(c+d) = ac + bd + ad + bc.$

$x^n - 1$ is always divisible by $x - 1$.

$1 + nx$ is a good approximation to $(1+x)^n$ if x is small.

Addition is commutative.

If $0 < a < b$ and $0 < c < d$, then $\frac{b}{c} > \frac{a}{d}$

We would be overwhelmed. We need a structure into which these properties fit. (You may already be looking at the last of these and wondering if it is true. You are probably putting it into the structure you already have for these properties.) This is a point I wish to emphasise: *building curricula in applications and modelling requires that we structure the subject.*

3. ORGANISATIONAL STRUCTURES FOR APPLIED MATHEMATICS

We have structures for pure mathematics, structures we associate with famous names: Euclid, Euler, Galois, Boole, and more recently Bourbaki. The structures are logical ones, from postulates through theorems using the vehicles of definitions and proofs. So, although there is a multitude of properties of numbers and operations, people do not feel overwhelmed. We attach new properties to the old structure.

In contract, applied mathematics is described by some speakers as almost the opposite. We hear that applications have exactly what pure mathematics does not have - applications are not well-defined, they are messy, they have many answers, they involve estimation, and so on. This characterisation does not hold universally. There are well-defined, elegant applications with single exact answers. There is pure mathematics that is messy, has many answers, involves approximation, and so on. Emphasising the messiness of some applied mathematics does not help sell this content to traditional teachers.

Teachers need a structure into which they can fit these applications, a structure which is richer than "Here is an application of systems of linear equations". There are many possible structures for applied mathematics. I will describe a few that I have found useful.

Learning Hierarchies

When I first began my work with applications, I wrote a course in which algebra was developed through applications (Usiskin, 1979). The biggest obstacle for my students in learning the applications of algebra was that they did not know the applications of arithmetic. For instance, when discussing population growth and pointing out that a 2% growth rate on a population P yields a population 1.02P, I found that the students had never been given the problem of finding the result of a 2% growth rate on a population of 10,000 or any other size.

A *learning hierarchy* is a network formed in the following way. You examine all the ideas found in an application in order to insure that each component idea is discussed before it is put with the others in the application. For instance, suppose one wishes to consider the graphical representation of velocity as rate of change of distance. You need to know some things about velocity, about rate of change, about distance. Surprisingly many students memorise formulas for the rate of change without having any idea how they are related either to rates or to changes. (In the US we call the rate of change by the word *slope* which disguises the connections; the British term *gradient* is no better.) Their later teachers wrongly assume that the students have learned ideas in earlier courses, ideas which they have never studied. Here is a possible learning hierarchy for this idea.

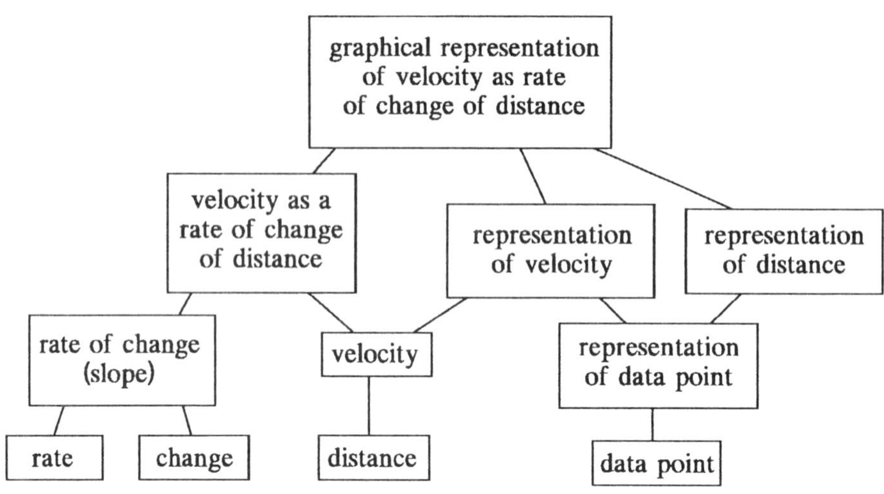

This is a natural hierarchy, but what makes it difficult to implement is that the components occur over many years of a student's schooling. For instance, the idea of change could be taught with subtraction as early as second grade. This may explain why such hierarchies are ignored by teachers. They do not want to go back to first principles, viewing that as a waste of time. Yet, when students cannot do an application, it is often because there is something down the hierarchy which they do not understand or which they have not been taught, and for that student it is pointless to go on.

Use Meanings

The bottom row of the hierarchy contains some ideas which are intimately related to particular mathematical concepts. Rate, as in kilometers per hour, or students per class, or population per square kilometer, is a fundamental use of division. Change is an application of subtraction. Distance can be considered as another application of subtraction; at higher levels distance is an application of absolute value. Data point is a fundamental use of ordered n-tuples. One of the foundations of a curriculum in applications and mathematical modelling has to be the giving attention to the basic uses of the commonly taught or most important mathematical concepts. Active attention to these uses is in the curriculum being developed by the University of Chicago School Mathematics Project (1989).

For example, the idea of point has at least five use meanings, four of which are found in the UCSMP (1990) materials: dot, the meaning held by children and used on computer and television screens; idealised location, the usual meaning in geometry; ordered pair or triple or n-tuple, the usual meaning in numerical applications and in algebra, and the model for data point; node, the meaning found in networks and graphs; and centre of gravity, the meaning needed for work with physical forces. The mathematics of point is different in these various guises, and too often one of these meanings so predominates that the others are considered distasteful. For instance, students may be instructed that dots are not mathematical points. However, for applied mathematics, one must consider pixels as points. The student who thinks points must be locations is thwarted in understanding n-space when $n > 3$. Yet that is a natural occurrence with data points.

There are use meanings of more advanced concepts. Linear functions seem to have two basic use meanings: linear combination and constant increase/decrease. That is, they arise from those kinds of situations in the real world. They also arise from the uses of lines as shortest distances, as outlines or intersections of geometric figures, and as approximations to curves. The trigonometric functions also have at least three basic uses: one has to do with ratios of lengths in triangles; one has to do with circular motion and periodicity; and one has to do with acceleration and differential equations.

Processes also have use meanings. For instance, the process of estimation is used for many reasons: clarity, for ease of understanding; facility, for ease of use; consistency, to agree with other precisions; economy, to save time; safety, as when we estimate the maximum weight an airplane can hold; and finally the many situations in which estimates are forced.

Freudenthal (1983) has stressed the need for multiple characterisations of mathematical concepts if one is to understand their relationships to the real world. I agree completely. When Max Bell and I engaged in an in-depth study of the uses of arithmetic (1983), we developed six meanings for number; count, measure, location, ratio comparison, code, or derived constant. The location category includes addresses, the ordinals, temperatures, and any other numbers on scales. We found that scales like those used for measuring earthquakes or star magnitudes simply did not appear in the curriculum at any level, and when temperature was taught, it was often treated as if it were a measure like volume. We would see problems like the following in books: "If the temperature is 2° and it triples, what will be the new temperature?" Confusions in the use of mathematical objects cannot help if one wishes students to learn applied mathematics.

It is obvious that learning hierarchies and use meanings are somewhat related. The use meanings tell you what to put at the bottom of the learning hierarchy for applications; they tell you what should come first in the curriculum.

Analogies with Pure Mathematics

Building a curriculum in applied mathematics is somewhat of an unsolved problem. Following the advice of George Polya, one way to tackle this problem is to seek analogous problems which we have solved. This suggests that one way to build a curriculum in applied mathematics is to take advantage of its analogies with pure mathematics.

The use meanings discussed earlier are analogous to postulates or definitions in a mathematical system. Combining rate and change to get rate of change is akin in combining postulates and definitions to get a theorem.

I consider the process of modelling as being quite analogous – at least in a pedagogical sense – to proof. Each is at the highest realm of activity. Each is at the highest level of cognitive activity. Do we want students to learn proof, because that is what mathematicians do? If so, then we should want students to learn modelling, because that is what applied mathematicians do. If we do not want students to learn proof, then perhaps we should not want them to learn to model. My own view is that we do want students to learn some aspects of both proof and modelling, to have experiences with both.

We have been teaching (or trying to teach) students proof for quite a bit longer than we have been teaching modelling. What can we learn from that experience? Here are some thoughts.

It is almost impossible to understand proofs with concepts just introduced. For instance, the student who is just introduced to the concept of group finds it very difficult to write proofs about groups immediately. Similarly, modelling in contexts not understood by the students is likely to be a waste of time.

The ability to do proofs in one domain does not necessarily extend to the ability to do proofs in another domain. The same is true for modelling. That suggests that students should learn a variety of kinds of modelling.

Proof learning required the language of if–then statements and justifying conclusions. Modelling requires the same logic – if this model is assumed, then we can deduce ... Thus we should be able to use proof to teach modelling and use modelling to teach proof.

Communication of proofs and modelling both require that students write using both mathematical and non–mathematical terminology. Independent of the mathematical concepts, the writing task itself poses difficulties for many students.

Proof competence comes quite slowly. We should not expect modelling competence to come any more quickly.

Some proofs, like the infinitude of primes or the irrationality of $\sqrt{2}$, are classic. Similarly, some models should be treated with the same reverence, and taught to all. One that would surely qualify is Kepler's modelling of Tycho Brahe's data with ellipses. Another is the simulation of coin–tossing.

Dimensions of Understanding

A fourth structure also relates applied mathematics to other mathematics, through the following question. What is meant by *understanding* a particular concept?

Consider the concept of joint variation, as in the formula $z = kx^2y$. To many people, you understand joint variation if you can calculate the value of k given values of x, y, and z. Understanding is doing; it involves skill and algorithm. A second sort of understanding consists of the mathematical theory: why is z quadrupled when x is doubled and y is kept constant? Understanding is knowing why; it is the mathematical underpinnings dimension. A representational sort of understanding is favored by many psychologists.

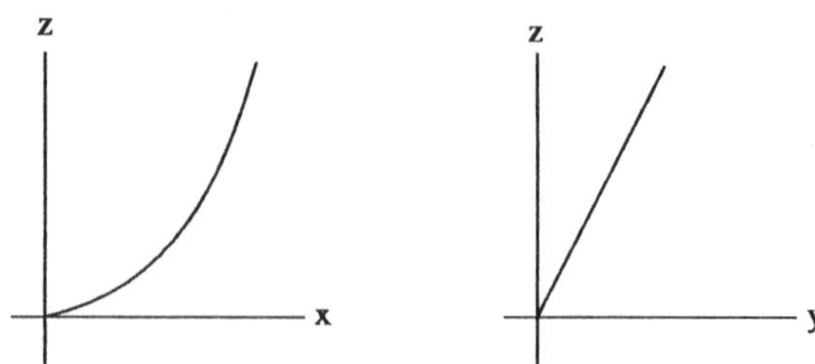

To them, understanding is being able to present or to find metaphors. Many in applied mathematics view a fourth type of understanding as the final goal: Do you know when and how apply the idea? In the case of $z = kx^2y$, one would at least want to consider the formula $V = \pi r^2 h$ for the volume of a cylinder, and discuss all of the other aspects in terms of their real world implications.

These four major dimensions of understanding each have simple aspects, and each have more complicated aspects. The skill-algorithm dimension ranges from memorised basic facts through the carrying out of procedures to the invention of algorithms. The mathematical properties underpinning dimension ranges from being able to name properties to justifications with them and, at the highest level, the discovery of proofs. The use-application dimension ranges from straightforward one-step uses through the applications of mathematical principles to modelling. The representation-metaphor dimension ranges from concrete materials to the use of representations to the creation of new metaphors or representations.

One can view the new math as an attempt to assert the importance of mathematical underpinnings. 'Back to basics' was an attempt to reassert skill. The popularity of Piaget and concrete materials has been a recognition of the importance of representations in understanding. The existence of the ICTMAs and moves towards applications and modelling are the assertion of the importance of the use dimension.

I believe students are best served by the view that all these dimensions contribute to the 'real' understanding of mathematical concepts. Furthermore, viewing understanding as having many dimensions helps to operationalise Niss's last two reasons for teaching applications and modelling, namely it contributes to a balanced picture of mathematics and assists the student in acquiring an understanding of mathematical concepts.

Closeness of Fit of Models

The process of modelling can be described as an attempt to find mathematical concepts which are isomorphic to situations in the real world. The utility of the model is a function of the degree of isomorphism, the closeness of the fit of the model. My experience suggests that one should first consider models which are isomorphic and move gradually to those which are not. For instance, in considering situations modelled by quadratic functions, one might move first from an *exact* model, for example the number of games required for n teams to play each other; to an *almost-exact theory-based* model, for example the path of a thrown, kicked, or batted ball; to what could be called an *impressionistic* model, for example the use of a parabola to model a frequency distribution.

We do not seem to have a language to describe the closeness of fit. We call all of them models, not distinguishing between economic models, which are surely impressionistic given their standard of performance, and the almost-exact models found in the mathematics of celestial navigation. I hope someone can come up with better words than the ones I have used (exact, almost exact theory based, impressionistic). It would help the curriculum if we had a better language for describing the fit of models. (I was told recently that the word *phenomenological* is often used where I have used impressionistic.)

4. BUILDING WITH TECHNOLOGY

No one can consider building a curriculum without considering the role of technology; technology has driven the revolution in applied mathematics. The usual things to say in this regard are: the cost of hardware continues to decrease, the software is getting better and better, there is good stuff out there and bad stuff out there. You will see some wonderful things demonstrated at this meeting. There - I've said them.

I would like to discuss briefly some issues less often discussed: the conflict with traditional ways, logistics, and equity.

The Conflict with Traditional Ways

The conflict with tradition is epitomised in answers given to the following question. What should a student be able to do without the aid of technology? It is a fundamental question, complicated because technology has greatly expanded what students can do. The responses occasionally defy common sense.

There are those who believe that students should be required to have exactly the same paper and pencil arithmetic skills they needed some time ago despite the existence of calculators. There are those who believe that students should still be expected to get integrals by hand even if they have access to computer software which deals with symbols. The justification is that real understanding does not occur without labour.

Yet we know that many students have the skills without any of the other kinds of understandings. One reason for imposing a structure on understanding is to explain the incorrectness of the view that one type of understanding is always necessary for another.

On the other hand, there are those who believe that technology will allow paper-and-pencil skills of students to be maintained even if less time is spent on them. This runs contrary to my own common sense; virtually all educational research supports strong effects for time-on-task.

A compromise position is to require students to do the pure problems by hand but allow technology on the applied problems. This too is illogical. If the technology is there, why not use it all the time?

Curiously, the weaker technological tool – the simple calculator – is seen as more threatening than the computer. But I believe it is because the user-friendly symbol-pushing calculator is not yet here. Once skills from algebra through differential equations become automatic, we will see teachers banning their use. Many teachers are quite willing to have others use technology but unwilling to allow their own students to use it.

Logistics

Logistics, by which I mean the scheduling of student and machine, may contribute to the unwillingness of many teachers to utilise computer technology. In most schools, it is impossible to have all students at computers at all times. The computer technology facilities thus have to be scheduled. Though it is more difficult to schedule sporadic use of a computer than to schedule continual use, the scheduling of continual use may not be warranted by the range of the mathematics content in the course.

The logistics problem is particularly critical on timed examinations. There is a natural tendency to give harder tasks to students when the students have access to calculators or computers. Otherwise, why use this equipment? However, working the technology takes time, and if the technology does any of the work, it takes quite a bit of time for the students to explain what was done. The management of evaluation with technology is in its infancy.

The logistics problem also appears with out-of-class assignments. If a student is living in a place with no access to computers, how can an assignment be given that requires computers?

Equity

Pencils and paper are inexpensive; if a student does not have them, they can usually be obtained. But a computer is much more expensive, and a student in today's mathematical world who has a computer equipped with sophisticated software at home is at an enormous advantage over one who doesn't. Thus the greater accessibility of mathematics has brought with it the troubling thought that, until all in a society have access to technology, the poor will be even more disadvantaged.

Hardware

When both a calculator and computer can do a job, even though the calculator does not do it as well, the wisdom is beginning to be amassed to use the calculator. For instance, Bert Waits and Franklin Demana at Ohio State University now encourage use of a calculator function grapher rather than their own very fine computer function graphing software for secondary school classes, because every student can take home a calculator. This also serves to solve the equity issue.

Some lap-tops are not full personal computers, have limited memory, and are meant to be used in connection with a larger computer. These are the modern-day slates and may become the student calculator of the future.

5. CLOSING REMARKS

I would like to put the computer revolution in perspective. It is truly a revolution, unlike anything in mathematics since the development of the written algorithms for arithmetic in the 15th century. At that time, algorithms for addition and subtraction of whole numbers had been known for some time, but algorithms for multiplication and division like those we now teach in elementary school had just been invented. Before this time, it is reported (Schwetz, 1987) that some arithmetic students were required to learn the multiplication tables from 1 × 1 to 99 × 99 because they had no other option for quick multiplication. Their only other option was to use lots of addition.

It was in Italy that they first taught these new algorithms for multiplication and division, and they were taught to students of secondary school and university age. This was the applied mathematics of the time, driven by the needs of the merchants and traders of the Italian city-states.

If these algorithms were the new mathematics, the new technology was printing. Before printing, universal literacy was non-existent. There weren't enough books to teach the masses how to read. Reading was considered a difficult chore, capable of being mastered only by an aristocracy or by the specially ordained. We all know that stories were passed down orally from generation to generation. I've wondered why they weren't written down - they seem to have been passed down even in cultures that had some writing. I'm convinced that people felt that something would be lost in the writing – the meter, the pauses, the music possibly. Those who could not read and write did not trust this new technology of reading and writing.

But printing changed all that. Now everyone could have books, even in their homes. And so the fear and mistrust of reading and writing waned, and people could aim for full literacy.

I believe the same will happen for mathematics. Calculators already are in almost every home in the industrialised countries, and computers in increasing percentages of them. These are the 20th century equivalents of books, and they are taking the fear out of numbers, out of graphs, and out of symbolism. The public is slowly learning that mathematics is simply a particular kind of language, the language of the logic and structure of phenomena in the universe – and that when mathematics is around, mathematics can be learned by all who learn to read as early as they learn to read.

Thus we can expect greater and greater amounts of applied mathematics to be a part of the experiences of all people whether or not they learn it in school. We all want students to be able to recognise situations that call for mathematical techniques, to encourage the wise use of the mathematics, and to discourage improper use. The fundamental principles must begin early, for a building is no stronger than its foundation. I hope that my remarks have given a few ideas on how that building might be arranged.

REFERENCES

Bell MS. (1983). Materials Available Worldwide for Teaching Applications of Mathematics at the School Level. In Zweng M *et al* (eds), *Processings of ICME-4*. Birkhäuser, Boston.

Blum W *et al* (eds). (1989). *Application and Modelling in Learning and Teaching Mathematics*. Ellis Horwood, Chichester.

Dolciani MP *et al*. (1986). *Algebra 1*. Houghton Mifflin, Boston.

Freudenthal H. (1983). *Didactical Phenomenology of Mathematical Structures*. Reidel, Dordrecht.

Mandelbrot B. (1976). *Fractals: Form, Chance and Dimension*. WH Freeman, San Francisco.

Niss M. (1989). Aims and Scope of Applications and Modelling in Mathematics Curricula. In Blum W et al (eds), *Applications and Modelling in Learning and Teaching Mathematics.* Ellis Horwood, Chichester.

Schwetz F. (1987). *Capitalism and School Arithmetic: The New Math of the Fifteenth Century.* Open Court Publ Co, La Salle, Illinois.

Thorndike EL *et al.* (1923). *The Psychology of Algebra.* Macmillian, New York.

University of Chicago School Mathematics Project. (1989, 1990). *Transition Mathematics. Algebra. Geometry. Advanced Algebra.* Scott, Foresman, Glenview, Illinois.

Usiskin Z. (1979). *Algebra Through Applications.* National Council of Teachers of Mathematics, Reston, Virginia.

Usiskin Z and Bell M. (1983). *Applying Arithmetic: A Handbook of Applications of Arithmetic.* Available from the ERIC Reproduction Service, document nos ED 264 087, ED 264 088, ED 264 089.

CHAPTER 4

Computing Mathematics – A New Direction for Applied Mathematics?

ID Huntley
Sheffield City Polytechnic, UK

SUMMARY

Many applications of mathematics nowadays involve the use of computers as an integral part of the solution. Increasingly, though, these applications have less to do with numbers and far more to do with structures – be they data structures, types, or a way of thinking that is endemic to the computer language chosen.

This paper surveys some of the developments taking place in the UK under the title of computing mathematics, and suggests how they might affect the curriculum over the next few years.

1. INTRODUCTION

Despite many conferences and journal articles in connection with new technologies, there is still an undercurrent of suspicion in many quarters. I have been encouraging the use of micros in teaching for over a decade, and this paper brings together many of these ideas – ideas which present tremendous opportunities and challenges for curriculum designers, at secondary as well as at tertiary level.

Here's a checklist I used some years ago when talking about the use of micros in maths teaching – it lists the various ways in which micros could then be used.

- electronic blackboard
- drill and practice
- simulation
- investigations
- number cruncher
- utilities

I imagine that these terms, and the associated ways of using micros in class teaching, are familiar to most readers.

However, the list is showing its age somewhat now, and ought probably to be extended.

- interactive video
- algebraic manipulators
- special purpose languages
  - functional
  - logic
  - parallel
  - specification
- integer manipulation
  - codes
  - ciphers
  - factoring primes

The paper will take each of these headings in turn, give a few examples of what is meant, and make some comments.

2. INTERACTIVE VIDEO

This is probably the most exciting topic in the whole of this area – linking a micro to a videodisc in order to give the learner complete control over the learning package, linking the visual stimulation of the video images with the computational power of the micro.

Some very interesting work has been carried out at secondary level – the Domesday discs and the DES–funded Interactive Video In Schools project are probably the best–known examples in Britain. However, as yet there has been little interest from tertiary education, and it is to be hoped that the present emphasis on new delivery systems for open learning will encourage future investment into this (currently expensive) area.

3. ALGEBRAIC MANIPULATORS

Programs such as muMath, Maple, Macsyma, Reduce and MicroCalc have been available on mainframe and PC–compatible computers for some years. However they are not very user friendly, requiring users to learn a complete new language before a meaningful application can be carried out, and so have only had a limited success.

Several much more user-friendly packages are now coming onto the market – the best-known one being Derive – which are driven entirely from menus, and these allow even naive users to complete quite sophisticated applications.

The newer packages are also available on a range of micros, which means that they are beginning to appear in schools as well as in polytechnics and universities. Thus we are all now being required to think very seriously about what we teach, why we teach it, and how we will teach it in the future – the teaching of topics such as trig expansions and integration may well be due for a major overhaul very shortly.

4. FUNCTIONAL LANGUAGE

Most of the common computer languages – Basic, Fortran, Pascal, C – are imperative; they consist of procedures containing instructions to the machine concerning which variables to manipulate, when, and how.

Functional languages – such as Hope, ML, Miranda – are not like this; they don't consist of machine instructions, but mathematical definitions of functions. At Sheffield, we find that the use of such languages motivates the study of discrete mathematics, and we anticipate that their use will really take off as this style of programming becomes better known and more computationally efficient compilers become available.

The following example of a function declaration is written in Hope, although it would look fairly similar in the other languages mentioned, and shows the essential features of functional languages.

```
dec time24 : num → num × num × num
--- time24 ≙ (s div 3600,
              s mod 3600 div 60,
              s mod 3600 mod 60);
```

The function clearly takes in a num(ber) – a time interval in seconds – and converts it into a triple (hours, minutes, seconds). So, typing in

```
time24(12345)
```

returns the answer

```
(3,25,45) : (num × num × num)
```

which, as well as providing the required time, also provides a reminder of the types involved. This, then, begins to show the real power of this sort of language – making a mistake such as typing in only time24;

returns the helpful information

time24 : num → num × num × num

and allows users to concentrate on the errors in their mathematics rather than in their computer programming.

A more significant example is the following.

```
dec firstword : list(char) → list(char) × list(char);
--- firstword(nil) ≙ (nil,nil);
--- firstword(c | s) ≙ if c = ' '
                         then (nil,s)
                         else ((c | w,r) where
                         (w,r) == firstword(s));
```

Here the function firstword is defined recursively, and strips the first word off a list of char(acters) provided as input. The infix operator | is called concatenate, and merely indicated that the two operands are placed one after the other in the list – add one list to another, in effect. Thus

firstword("Ian Huntley");

is evaluated via

= I | firstword("an Huntley")
= I | a | firstword("n Huntley")
= I | a | n | firstword(" Huntley")
= I | a | n | (nil,"Huntley")
= ("Ian","Huntley")

More impressively

Firstword(42);

returns

Actual parameter type does not match formal parameter type
Formal parameter type : list(char)
Faulty expression : firstword(42)
Actual parameter type : num

again showing the significant amount of help available to the user.

5. LOGIC LANGUAGES

Prolog is the best-known logic-based language. Logic programming, like functional programming, does not concern itself with machine instructions, but directly with the logic and mathematics behind the problem. A few examples will illustrate how it can be used.

Figure 1 shows a Prolog program written to describe the (real) roots of a (real) quadratic equation. Although it's not entirely trivial to write (how many of *your* students hand in programs which crash because of the lack of simple error trapping?), the main problem with using this as a teaching aid is that students tend to *think* that the problem is trivial and so lack the motivation to treat it seriously.

```
predicates
    roots(real,real,real)

clauses
    roots(A,B,C):-
        A=0,B<>0,X=-C/B,
        write("a is zero, root is", X).

    roots(A,B,_):_
        A=0,B=0,
        write("no solution possible").

    roots(A,B,C):-
        A<>0,R=B*B-4*A*C,R<0,
        write("no solution possible").

    roots(A,B,C):-
        A<>0,R=B*B-4*A*C,R>=0,
        X1=(-B+sqrt(R))/(2*A),
        X2=(-B-sqrt(R))/(2*A),
        write("roots are",X1,"and",X2).
```

Figure 1

The program shown in figure 2 overcomes this motivation problem. There is an enormous amount of serious modelling going on under the guise of the 'trivial' task of specifying a vending machine, and the computer is being used in a subsidiary role – merely to cut down on the manipulations required. We find it relatively easy to excite students with such applications from everyday life.

```
domains
    cash, button, change, cost = integer
    good = symbol

predicates
    vend(cash,button,good,change)
    price(button,cost)
    article(button,good)
    valid(cash)

clauses
    price(1,20).
    price(2,15).
    price(3,25).
    article(1,mars).
    article(2,twix).
    article(3,choc).
    valid(Cash) if Cash>=0,Cash<=50.

    vend(C,B,G,Ch):-
         price(B,Cost),C>=Cost,Ch=C-Cost,
         article(B,G),valid(C).

    vend(C,B,G,Ch):-
         price(B,Cost),C<Cost,Ch=C,
         G=nothing_sorry,valid(C).
```

Figure 2

The problem posed below is a slightly different type of example – again, typical of that used with first-year students on our BSc Computing Maths degree.

> Three friends came first, second and third in a programming competition.
>
> Michael likes basketball, and did better than the American. Simon, the Israeli, did better than the tennis player. The cricket player came first.
>
> Who is the Australian?
> What sport does Richard play?

Figure 3

For a small-scale problem such as this, students can usually come across a solution in only a few minutes, but they are not always sure quite how they achieved the answer and can rarely use the same methodology on larger problems – so the example provides an extremely good object lesson in modelling.

This problem is, of course, identical in design to those provided in most puzzle books, intended to be solved by a process of logically turning data (the story) into information (the facts). Figure 4 shows the sort of approach intended in the puzzle books, applied to the simple problem above, when just the first two pieces of data have been utilised.

		Nationality			Sport			Position		
		Israeli	Australian	American	Cricket	Tennis	Basketball	First	Second	Third
Name	Michael			x	x	x	✓			x
	Simon						x			
	Richard						x			
Position	First			x						
	Second									
	Third									
Sport	Cricket									
	Tennis									
	Basketball			x						

Figure 4

However, although this methodology works extremely well for small problems, it would not be suitable for a realistic-sized problem. Figure 5 shows a Prolog solution to this same problem, where a mechanised solution method is given which would be suitable for problems of any size. The program is intended to be self-explanatory; note the use of recursion in the definition of member - [...] here indicates a list. This is a technique which students find easy to master, although I suspect that not all maths teachers find it quite so easy! However the importance is that the computer is again helping us to solve problems in mathematical modelling.

```
domains
        person=person(symbol,symbol,symbol)
        list=person*

predicates
        puzzle(person,person,person,symbol,symbol)
        clue1(person,person,person)
        clue2(person,person,person)
        clue3(person,person,person)
        query1(person,person,person,symbol)
        query2(person,person,person,symbol)
        member(person,list)
        first(list,person)
        sport(person,symbol)
        nationality(person,symbol)
        name(person,symbol)
        did_better(person,person,list)

clauses
puzzle(A,B,C,X,Y):-
        clue1(A,B,C),
        clue2(A,B,C),
        clue3(A,B,C),
        query1(A,B,C,X),
        query2(A,B,C,Y).
clue1(A,B,C):-
        did_better(Man1Clue1,Man2Clue1,[A,B,C]),
        name(Man1Clue1,michael),
        sport(Man1Clue1,basketball),
        nationality(Man2Clue1,american).
clue2(A,B,C):-
        did_better(Man1Clue2,Man2Clue2,[A,B,C]),
        name(Man1Clue2,simon),
        nationality(Man1Clue2,israeli),
        sport(Man2Clue2,tennis).
clue3(A,B,C):-
        first([A,B,C],ManClue3),
        sport(ManClue3,cricket).
query1(A,B,C,X):-
        member(Q1,[A,B,C]),
        name(Q1,X),
        nationality(Q1,australian).
query2(A,B,C,Y):-
        member(Q2,[A,B,C]),
        name(Q2,richard),
        sport(Q2,Y).

did_better(A,B,[A,B,_]).
did_better(A,C,[A,_,C]).
did_better(B,C,[_,B,C]).
name(person(A,_,_),A).
nationality(person(_,B,_),B).
sport(person(_,_,C),C).
first([X|_],X).
member(P,[P|_]).
member(P,[_|Tail]):-
        member(P,Tail).
```

Figure 5

6. PARALLEL LANGUAGES

This is not the appropriate place to expound the details of parallel languages – ones that utilise the power of several processors running in parallel, such as Parallel Fortran or C, or are designed especially for the transputer, such as Occam.

However, my guess is that most of us are still having problems driving the processor in one micro effectively, and haven't really begun to think about what entirely new applications would be opened up by having access to several processors in parallel. The opportunities for simulation and for new algorithms are clearly very exciting.

7. SPECIFICATION LANGUAGES

Figure 6 shows a very simple application of a specification language – a program written to describe precisely a real-world problem. This example specifies some of the facilities you would require from a computerised birthday book – the ability to add, delete and change names and (birth)dates.

```
┌──Book──────────────────────────────────────
│ birthdays : P(Name × Date)
└────────────────────────────────────────────

┌──Adding to─────────────────────────────────
│ birthdays  : P(Name × Date)
│ birthdays' : P(Name × Date)
│ birthdays? : P(Name × Date)
├────────────────────────────────────────────
│ birthdays? ∉ birthdays
│ birthdays' = birthdays ∪ (birthday?)
└────────────────────────────────────────────

┌──Deleting──────────────────────────────────
│ ΔBook
│ birthday? : Name × Date
├────────────────────────────────────────────
│ birthday?  ∈ birthdays
│ birthdays' = birthdays - {birthday?}
└────────────────────────────────────────────

┌──Changing──────────────────────────────────
│ ΔBook
│ birthdayold?, birthdaynew?
├────────────────────────────────────────────
│ birthdayold? ∈ birthdays
│ birthdays'   = (birthdays - {birthdayold?}) ∪ {birthdaynew}
└────────────────────────────────────────────
```

Figure 6

The example shown in figure 6 is clearly very small, but even this is able to show how a problem can be broken down into smaller units for clarity (called schemas in Z, the language used here) – good modelling or programming practice.

A realistic problem might be to specify how a new cashcard for a bank might work, with all the difficulties associated with passwords, security, and the like, for which the natural language specification would be several volumes thick. The Z specification would not only be substantially smaller than this, but would also be substantially more precise – it would have the sharpness engendered by a mathematical, rather than a natural language, statement.

The Z Toolkit, which is actively under development for use on realistic-scale problems, will then check automatically for consistency and syntax, and will tell the user whether there are any errors in the specification. It will also allow What If questions to be asked. It is hoped eventually to have an executable version of Z, so that not only can the specification be written and tested in Z, but also the resulting program can be executed straight from the Z specification.

8. CODES

Figures 7 and 8 show some of the Spode Group's examples on codes, which are designed to be used in secondary schools as the first stages in making pupils more aware of the issues involved in coding theory.

APPLIED COURSEWORK 8

Morse Code

Sheet 1

In sending a message by Morse Code suppose that we call the time for a dot the 'dot time'. Then the time for a dash is three 'dot times'. The space between two parts of a letter (say the dot and dash in letter A) is one 'dot time'. The space between letters is five 'dot times'.
All this is shown in a time-line below, which shows MORSE in Morse Code.

A ·—	J ·———	S ···
B —···	K —·—	T —
C —·—·	L ·—··	U ··—
D —··	M ——	V ···—
E ·	N —·	W ·——
F ··—·	O ———	X —··—
G ——·	P ·——·	Y —·——
H ····	Q ——·—	Z ——··
I ··	R ·—·	

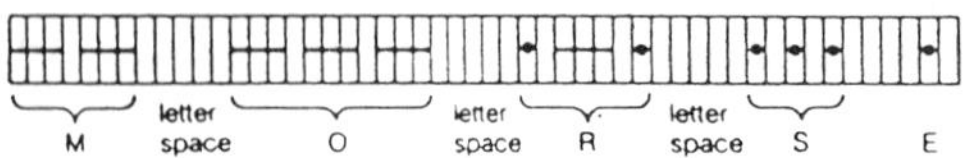

Problem 1

Draw time-lines like the one shown for

(a) SOS (b) HELP ME

Squared paper will be useful.

Problem 2

When a ship's signaller is tapping out Morse the 'dot time' is $\frac{1}{24}$ second. How long would it take him to tap out the messages in Problem 1?

Problem 3

Complete this table, finding 'dot times' for all the letters of the alphabet.

Letter	Number of 'dot times'	Total
A	1 + 1 + 3	5
B	3 + 1 + 1 + 1 + 1 + 1 + 1	9
C	3 + 1 + 1 + 1 + 3 + 1 + 1	11

Problem 4

Find how long it would take the ship's signaller to send this message?

TITANIC SINKING FAST

Figure 7

APPLIED COURSEWORK 8

Morse Code

Sheet 2

Problem 5

Complete this table, showing all possible arrangements of the signs dot and dash, taken 1, 2, 3 and 4 at a time.
The first two columns are complete. Finish columns 3 and 4, working systematically and writing in the letters, as shown.

Number of Signs			
1	2	3	4
. E - T	. . I . _ A _ . N _ _ M	. . . S . . _ U . _ . R _ . . D . _ _ W	. . . . H . . . _ V

Problem 6

Find an article in a newspaper containing between 500 and 1000 words of ordinary English prose. (A page of a novel may be used instead.) Write the letters of the alphabet down the left side of a page and use tally-marks to count the number of times each letter occurs in the article. Your first line will look something like this.

A ~~1111~~ ~~1111~~ 11 (12)

You should find that the letter E is the most frequent. Now list the letters again in order of frequency, starting with E.

Problem 7

Suppose that when designing his code Samuel Morse intended
(i) to use all arrangements of 1, 2, 3 or 4 signs before using 5 signs,
(ii) that the transmission times (numbers of 'dot times') for letters should increase as their frequencies decreased.

(a) Give one reason in each case why these aims are sensible. Are there any reasons why they might not be sensible?
(b) Was intention (i) achieved?
(c) Was intention (ii) achieved?

Problem 8

How could you improve the Morse Code?

Figure 8

These examples concentrate heavily on an investigational approach to learning, and the importance of discussion in the classroom. Such discussion often leads to the suggestion *from the pupils* that we should design a code which is even more sensible than the Morse Code – which merely tries to ensure that frequently–occurring letters have a short representation – and has its message length minimal.

This suggestion is superb – it not only comes from the learners, with all the increased motivation that negotiated learning can bring, but also leads into the theory of Huffman codes and the like, which are up–to–date examples of mathematics applied to computing. For those who worry about such things, the mathematics can also become as difficult as anyone would want!

Figures 9 and 10 are also examples which can be used at both secondary and tertiary level – often a mark of a good problem area. These are intended for students who already know something about ASCII codes – as used in the majority of present micros – and parity bits, but who feel that these topics are nothing to do with everyday life. They all know something about bar codes on groceries and ISBN codes on books, and are intrigued to find out that these familiar ideas are intimately tied up with matrices and mathematics. Motivation again.

To check whether there has been a transmission error, we often add extra characters to a message **u** to give the transmission **v**.

Bar codes use the decoding matrix

$$D = [1\ 3\ 1\ 3\ 1\ 3\ 1\ 3\ 1\ 3\ 1\ 3\ 1]$$

and evaluate D**v** on receipt of the message.

Try this out on some of the bar codes below, and try to find out what feature indicates a successful transmission.

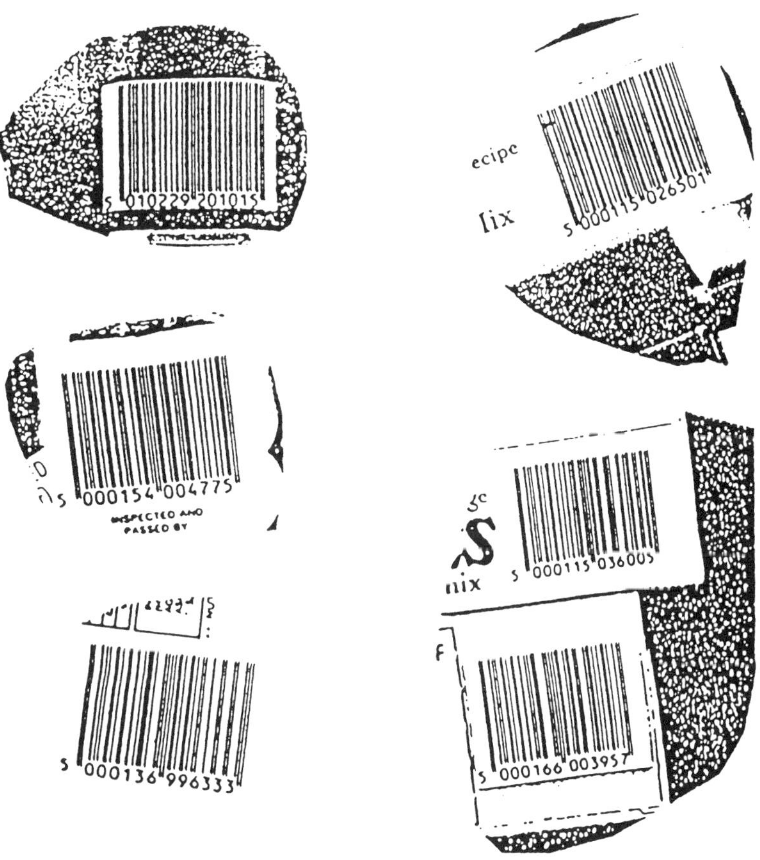

Figure 9

ISBN codes, which appear on most books nowadays, use the decoding matrix

D + [X 9 8 7 6 5 4 3 2 1]

with X indicating 10.

Pick a few books off your bookshelf, and try to find out what feature indicates an error-free code.

Figure 10

9. CIPHERS AND PRIME NUMBERS

There are very few of us – whether children or adults – who can resist the appeal of a secret code, and so this is a very fruitful area for interesting applications.

The following shows a lovely example from Geoffrey Mathews' book Contemporary School Mathematics, written in 1962, which younger children enjoy. In this case the motivation is supplied by the context, and this motivation is easily transferred to the mathematics.

Message is DICE

With A = 1, B = 2, ... this is written as

$$M = \begin{bmatrix} 4 & 9 \\ 3 & 5 \end{bmatrix}$$

Encoder is

$$E = \begin{bmatrix} 2 & 3 \\ 1 & 2 \end{bmatrix} \text{ for instance}$$

Message is sent as

$$ME = \begin{bmatrix} 2 & 3 \\ 1 & 2 \end{bmatrix} \begin{bmatrix} 4 & 9 \\ 3 & 5 \end{bmatrix}$$

$$= \begin{bmatrix} 17 & 33 \\ 10 & 19 \end{bmatrix}$$

Decoder is then

$$D = \begin{bmatrix} 2 & -3 \\ -1 & 2 \end{bmatrix}$$

Then

$$D(EM) = \begin{bmatrix} 2 & -3 \\ -1 & 2 \end{bmatrix} \begin{bmatrix} 17 & 33 \\ 10 & 19 \end{bmatrix}$$

$$= \begin{bmatrix} 4 & 9 \\ 3 & 5 \end{bmatrix} \text{ as required}$$

Now use the system

A	B	C	D	E	F	G	H	I	J	K	L	M
1	2	3	4	5	6	7	8	9	10	11	12	13

N	O	P	Q	R	S	T	U	V	W	X	Y	Z
14	15	16	17	18	19	20	21	22	23	24	25	26

with encoder $E = \begin{bmatrix} 5 & 2 \\ 2 & 1 \end{bmatrix}$.

What happens to the message TERM if the encoder is used in the wrong order?

However, an added advantage of this subject area is that it is presently one of the most active research areas of modern mathematicians – all too often we only manage to teach material which was discovered last century, if not earlier!

The RSA cipher in figures 12a and 12b, for instance, were only developed in the late 1970s; it is a method for encrypting messages using a public key (ie a process which is not secret), but which only the decrypter can understand. The example comes from the original paper by Rivest, Shemir & Adleman (Comm ACM 1978 **21** 120–126). In essence the method works because of the time that even the largest computers take to factorise very large prime numbers – until recently a 200 digit prime still took about a billion years to factorise! However, advances in parallel languages and the associated algorithms – enabling large primes to be factorised orders of magnitude more quickly than this – look as though they may make the method less useful than anticipated.

RSA cipher			RSA example		
Step 0	:	Turn message into numbers	Step 0	:	ITS ALL GREEK TO ME 0920 1900 0112 . . .
Step 1	:	Choose the two primes p and q Let n = pq Find d the coprime to (p-1)(q-1) - the private key Find e from ed = 1 mod (p-1)(q-1)	Step 1	:	p = 47, q = 59, n = 2773 d = 157 (a prime) 157e = 1 mod 2668 Euclid's algorithm gives e = 17
Step 2	:	Announce values of n and e - the public key	Step 2	:	Publish
Step 3	:	Encrypt message M using $C = M^e \bmod n$ and send	Step 3	:	$C_1 = (0920)^{17} \bmod 2773$ = 948 = 0948 and so on
Step 4	:	Decrypt message, using $D = C^d \bmod n$	Step 4	:	$D_1 = (0948)^{157} \bmod 2773$ = 920 mod 2773 and so on

a **b**

Figure 12

10. CONCLUSIONS

Even though they do not resemble the problems that many of us learned as undergraduates, there are huge application areas in the real world which involve the application of mathematics to computers and vice versa. Many of these problem areas are totally different from traditional maths applications, and their value in motivating students can be enormous.

Perhaps the moral is that computing maths can open up (new) areas of interest for (old) applied mathematicians!

[The author would like to thank his colleagues at Sheffield City Polytechnic, who have contributed to many of the examples discussed above.]

CHAPTER 5

A Modelling View of Mathematics

CP Ormell
University of East Anglia, Norwich, UK

SUMMARY

Mathematical modelling used to be a sideline of mathematics. Now, with the help of the digital computer, it has become a major aspect of the subject. We can use it as the basis for a new, naturalistic, de-mythologised view of mathematics.

1. INTRODUCTION

It is clear that mathematics can be seen in a large number of slightly different ways. Each such variant 'way of seeing' mathematics is a different view of it. The idea of looking for a new view of mathematics which takes its modelling capabilities fully into account is that we might establish a perspective which simultaneously increases the motivation of students and clarifies the substantial, long term social purposes of the subject. The two parts of this condition are equally important. Some varieties of New Maths increased the motivation of students, but succeeded only in turning the subject into a kind of art form or one person game. Naturally this produced dismay among those who were aware of the substantial, useful, *applicable* mathematics needed in technology, science and industry. Motivating students to learn a quasi-mathematical subject which does not deliver the good which society expects mathematics to deliver is self evidently absurd (see Sawyer, 1966).

2. WHY DO WE NEED A MODELLING VIEW OF MATHEMATICS?

The main point of creating a Modelling View of Mathematics is, however, that we need to establish our identity as mathematical modellers, and our subject's identity as distinct from science. Having a clear view of mathematics as a modelling instrument provides such an identity: it can secure our own and our students' self-esteem when we are berated by those who would prefer mathematics to be seen merely as the handmaiden of the sciences.

We are certainly not opposed to scientific aims, but we should not, I believe, totally subordinate our aims to those of science. Mathematical modelling can be applied to *four* main sectors of life – those of science, technology, development and personal/corporate organisation. Science is only one of these. Science, however, has been the dominant target for applicable mathematics since the time of Ancient Greece. This is not simply an imposition of scientific values onto mathematics, though it has tended to have that effect. The applications of mathematics to science were perceived by mathematicians themselves to be the serious ones. In contrast the mathematics needed in the bazaar was, to use GH Hardy's phrase (1948), trivial mathematics. So mathematicians naturally valued the kind of modelling which provided the greatest challenge and called for the most impressive response. This meant that mathematicians went along with scientific aims. However, if we subordinate our aims to those of science we inevitably under-represent the equally legitimate non-scientific sectors – and there is a hidden bias in the way historians describe the record. Since the end product of mathematical modelling in science is the creation of accredited scientific theory (reliable, fully researched, tested, unfalsified mathematical models) these scientific models have been preserved, studied, taught and generally envalued. In the other three sectors of mathematical modelling the end-product is not a special, knowledge-enriched *model*, but a physical artefact, gadget, system, arrangement or event. So the apt and successful practical use of mathematical modelling is usually seen, not as a triumph of mathematical modelling, but as part of a triumph of practical activity – rocketry, motor cycle engineering, bridge building, information technology, architecture, environmentalism, personal impressario-type flair. The situation is even worse in *negative modelling*, for example where mathematical modelling has told us that something is *not* possible, or will not work. In this area what we should hail as resounding *successes* of mathematical modelling – because they led people to avoid costly, protracted fiascos – are commonly recorded as *failures* of technology, development or personal nerve (see Ormell, 1990a).

This cultural subjugation of mathematical modelling is what we need a Modelling View of Mathematics to counteract. We need it, in a word, to encourage our students to be proud of their work, to give them a sense of identity. Instead of letting the successful mathematical modelling of the past be thrown into the waste paper baskets of history, we ought to mount a campaign to preserve, record, catalogue and celebrate it. How we see phenomena depends heavily on the concepts we apply to them. The concepts, in turn, depend on an overall view. We need an overall view of mathematics as a modelling instrument – that is, in effect, chiefly as an *exploratory* and *previewing* discipline.

3. COMPUTABILITY: A MASK COVERING APPLICABILITY

One of the chief obstacles to forming a clear overall view of mathematics as a modelling instrument used to be the strangely idiosyncratic nature of the applications of mathematics. This occurred, we now know, because successful applications necessitated that the answers be *computable*, and computability, prior to the 1950s, was almost entirely dependent on clever conceptual simplification. Some equations could be solved by classical methods, others could not. Some integrals could be performed, others not. Computability hinged, in the end, on the adroit use of manipulative tricks. There was no overall rhyme or reason in the pattern of computability, so there was no overall rhyme or reason in the successful application of mathematics. The natural contours of applicability had the contours of computability superimposed on them. The result was an inexplicable pattern of success and failure, illumination and shadow (see Ormell, 1990b).

The arrival of the digital computer has simplified the scene dramatically. Now computability, to the level of accuracy required by an empirical context, need hardly ever present a problem. We can now see, sufficiently clearly, that mathematical modelling is a general instrument for mimicking reality, and thereby telling us the hidden potential (in terms of maxima, existences, shortcuts and pitfalls) of interesting theoretical or practical possibilities.

4. TWELVE LEVELS OF APPLICATION

To form an overall view we need to look closer than this, and here we find a baffling complexity of different *kinds* of applications of mathematics. The sheer scale on which mathematics is applied to the real world, from counting one's change at bedtime to simulating rival cosmological hypotheses, is a formidable obstacle to progress. I have argued elsewhere (1990a) that we need to recognise at least twelve different levels of *purpose* in applying mathematics. We begin with our feet on the ground: the most basic level of purpose seems to be the use of elementary arithmetic to check the results of everyday transactions like spending money, accounting for time, looking at consumption of consumables (for example petrol). From this domestic, personal level we rise by degrees (all of which are concerned with planning to do and make things) to the design of scientific experiments on Level 7. Level 8 uses mathematics to search for new scientific theory, so between Level 7 and Level 8 there is a fundamental change from the practical to the theoretical. Levels 9 and 10 explore, extend and consolidate the mathematics used on Levels 1–8. Here the exercise of scientific curiosity is applied to mathematics itself. Why is $^{1}/_{x}$ the only power of x which, on being integrated, gives a non–polynomial result? Why do the circle, the ellipse, the parabola and the hyperbola have certain properties in common, for example, pole–polar properties? Level 11 attempts a complete systematisation of the language and concepts of mathematics at a working level of logical rigour, while Level 12 attempts, ambitiously, to do the same thing at the level of rigour associated with mathematical logic.

We may show the twelve levels thus (see Ormell, 1983).

		levels	**purpose**
12		Meta-mathematics	philosophical/academic
11		Modern Pure Mathematics	academic/solving puzzles arising from levels 9/10
10		Modern Applicable (discrete) Mathematics	solving problems arising on levels 6/8
9		Traditional Applicable (continuous) Mathematics	solving problems arising on levels 1-8
8	W	Scientific Theorising	explaining a scientific anomaly
7		Scientific Research	obtaining scientific information
6	U	Technological Design	exploring innovative technological ideas
5		Technical Planning and Development	working out the consequences of a particular new development
4		Corporate Planning and Development	working out the consequences of economics, marketing, etc
3		Planning to make things	looking ahead in DIY work
2		Planning to do things	avoiding slip-ups
1		Everyday Check-ups	ensuring fairness in everyday transactions

Levels 1–6 are, in ordinary language, levels on which mathematics is useful practically. Levels 1–8 apply mathematical modelling to the real world. Levels 9–12 apply mathematics back onto itself.

Of course, as we move from Level 1 to Level 12 we move up from the ground to more and more abstruse levels of thought, so that, at the end, we are in the 'stratosphere' where burning questions involve such dilemmas as whether there is a transfinite number between $_{0}$ and $_{1}$. However, the levels are not levels of *mathematical* sophistication as such. 2 + 2 = 4 can be used on all twelve levels.

5. MODELLING AS SIMULATION

If we begin by looking at Levels 1 to 7 we may see mathematics as a way of mimicking, that is, simulating, proposed acts and constructions in the real world. These are the levels on which mathematics makes a direct, physical contribution to the well-being of individuals, groups and society as a whole. The point of simulating these constructions is to explore their implications *before* anyone is committed to building them. Mathematics, on these physical levels is sometimes described as a way of making things happen. Of course the mathematics as such does not actually make anything happen. Human willpower is needed to do that. Mathematics' role is, rather, to give the constructor the *confidence* first to embark on the enterprise, secondly to will the construction process, and thirdly, to sustain the effort through to a successful conclusion. Mathematical modelling, by enabling us vividly and accurately to preview the construction, enables us to communicate intention, steady nerves and cement resolve. One might say that it makes the conditions happen which then, in turn, make things happen.

6. USEFUL MATHEMATICS, QUEEN OF THE POSSIBILITY-SIMULATING DISCIPLINES

Mathematics, pursued for these purposes, belongs to a class of 'possibility-simulating disciplines'. It is indeed the Queen of the possibility simulating disciplines, because it combines maximum flexibility and adaptability with minimum inertia and cost (see Schools Council, 1975).

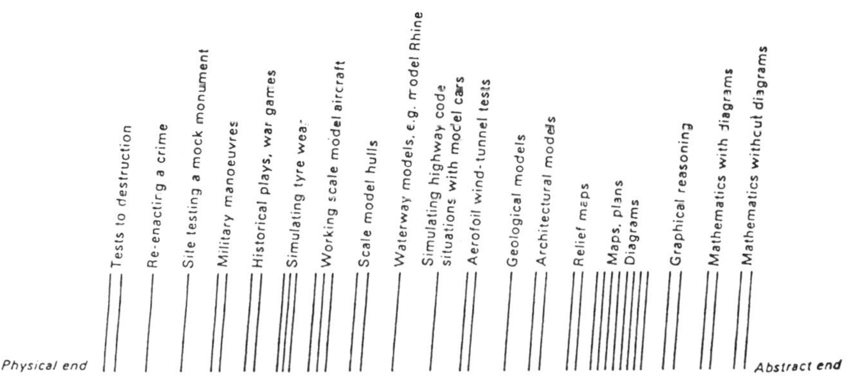

The Spectrum of Possibility Simulation

All possibility-simulating disciplines require a written record of the manipulations which have been performed. So we have, typically, some apparatus (which is manipulated) and a record of the manipulations. In the case of mathematics, though, the apparatus as such disappears. Its rolé is taken over entirely by the symbols (the record). This is a striking characteristic of mathematics – that it is the possibility-simulating discipline which uses *zero* apparatus, that is, literally *nothing* in addition to, the symbolic record (see Ormell, 1972).

7. HIGHER MATHEMATICS, SCIENCE OF THE QUEEN OF THE POSSIBILITY SIMULATING DISCIPLINES

Mathematics on the higher levels (8–12) clearly consists of the *science* of the mathematics of the practical and scientific Levels 1–8, that is, the Science of the Queen of the Possibility-Simulating Disciplines! In earlier times mathematics was often described as The Queen of the Sciences. However, most of it is not a science at all, and that which is a science is not a science telling us truths about the outside world, but a science telling us truths about the system of zero apparatus we have corporately slowly and painfully contrived to mimic the real world and our purposive interventions in it.

Such an account of mathematics succeeds, I believe, in placing it in the spectrum of human activities. It gives it a special exploratory role, different from science and different again from technology. It explains mathematics' epistemological primacy. It can serve as a rhetoric capable of giving mathematics-as-a-modelling-instrument a self-evidently worthwhile and distinctive identity. This is centrally important in presenting the subject to our students; they need to know what they are at.

8. PLATONISM, THE ULTIMATE MODELLING VIEW OF MATHEMATICS?

What is being claimed here may be put into still sharper perspective by being contrasted with the chief alternatives, Platonism and formalism, though the latter may be regarded as an off-shoot of the former. Platonism, contrary to many initial impressions, does not *deny* the capacity of mathematics to represent the real world. Indeed we may characterise Platonism by the belief that mathematics is the highest form of knowledge, and therefore the form of knowledge God must have used when he created the universe. Platonism, then, is like the attitude of an unbelievably rich person, who has lost any detectable drive to acquire more money. Since Platonists have already accepted that reality is mathematical in character, they tend to feel little urgency concerning whether any particular patch of reality can be swiftly, effectively and neatly modelled.

Is Platonism, then, the modelling view of mathematics we need in a teaching context? Unfortunately, Platonism is an idea whose time is past. Its credibility has suffered a series of devastating blows in modern times – from the discovery of the contradictions of set theory, Gödel's incompleteness theorem and quantum theory, to the popular rejection of its historicist and elitist social assumptions.

Two steps need to be taken, I believe, to clinch the argument and hence to bury Platonism as any kind of viable modelling view of mathematics. The first is to create a fully naturalistic, commonsense account of the contradictions of set theory and self-reference. The second is to recognise that the transfinite is a myth.

Both are long overdue. Richard saw plainly enough that a complex lexicographic ordering of defining formulas limits mathematics to, at most a countable infinity of defined objects, and this was in 1905 (see Ormell, 1985)! I offered a naturalistic view of self-reference in 1958. It depends on the view that self-reference enables us to create, for the first time dynamic contradictions, as opposed to the static contradictions of classical logic (see Ormell, 1985). It enables one to deal with the paradoxes in a commonsense way, and hence to lay the foundations of a de-mythologised view of mathematics.

REFERENCES

Hardy GH. (1948). *A Mathematician's Apology*.

Ormell CP. (1990a). The New Applicability of Mathematics. *MAG*.

Ormell CP. (1990b). The End of the Defensive Era in Mathematics. In Johnson and Loomes (eds), *The Mathematical Revolution Inspired by Computing*. Oxford University Press.

Ormell CP. (1985). Looking for Meaning in Mathematics. *MAG*.

Ormell CP. (1983). Maths with Bite. *Education*, 3–13.

Ormell CP. (1972). Mathematics, Science of Possibility. *Int J Math Ed Sci Tech*, 329–341.

Sawyer WW. (1966). *A Path to Modern Mathematics*. Penguin.

Schools Council. (1975). *Mathematics Changes Gear*. HEB.

CHAPTER 6

Against Ill-founded, Irresponsible Modelling

B Booss-Bavnbek
Roskilde University, Denmark

SUMMARY

A set of criteria is developed for the qualitative assessment of mathematical modelling. Arguments are advanced against the pedagogical goal of 'helping the students to love mathematical modelling', against the political goal of 'higher acceptance of mathematical modelling in industry and administration', and against the mathematics-based illusion of controllability where uncontrollable complexity is introduced by ill-founded, irresponsible modelling. An extension of the Hippocratic Oath to mathematical scientists in relation to mathematical models for risky decisions and technologies is suggested.

1. INTRODUCTION

I ought to begin by saying that I am not an educational expert but only a practitioner. My own research is in partial differential equations of mathematical physics, with special emphasis on elliptic boundary value problems. Over the years I have consulted and discussed with, and on behalf of, engineers, chemists, geologists, biologists, and economists on various matters, and with non-experts on mathematical and political aspects of military systems and environmental issues. My teaching in Roskilde is mainly directed towards systematic accounts of the strengths and limitations of mathematical modelling (in our interdisciplinary Natural Sciences Basic Education Programme and, at graduate level, in mathematical modelling and mathematics education).

In the first two of the following sections I shall discuss the deceptive similarity of mathematical models, and I shall try to draw attention to various qualities of mathematical modelling which are of fundamentally different character. In section 4 the great potential of theoretical mathematics for handling complexity is emphasised. In section 5 some warnings are given against ill-founded, irresponsible modelling. In the closing section, I want to draw some conclusions concerning the adoption of criteria for useful and harmful effects of mathematical modelling, a statute for benign systems. I shall propose a moratorium on new mathematical models supporting risky decisions or the further introduction of insecure technology. I shall close with a view of the major problems in following this goal. An elaboration of the concept of this lecture is presented in Booss-Bavnbek *et al* (1989).

A major source of inspiration for the present paper is the work carried out in our department, IMFUFA, over more than a decade. Thus I am grateful to several colleagues in the department for both discussions and examples.

2. WARNING: BEWARE OF DECEPTIVE SIMILARITY

In this section I want to show that, in spite of their superficial similarity, mathematical models often display remarkably different characteristics with respect to their empirical foundation and scientific extramathematical status, the extent of mastery of the mathematical concepts involved, and the pragmatics of the application situation. These differences are not necessarily decisive for the scientific or practical value or for the credibility of the model, but the adequate choice of the way of assessing their value and credibility requires attention to all of these aspects. This I want to explain by the following distinctions and examples.

2.1 Example : the geostationary height of a TV satellite

Finding the geostationary height for a telecommunication satellite is a mathematical exercise in elementary celestial mechanics which could have been done by Newton himself with the required accuracy.

Let us look at the procedure of solving the problem. To make the presentation easier I will discuss the model as prototype only, and thereby neglect many of its concrete features. The determining equations are Newton's gravitational law

$$F = GmM/r^2$$

and Newton's second law

$$F = ma$$

for masses M and m of the earth and the satellite respectively. Mathematical manipulations lead to a literal expression for the required unknown height

$$r = \left[\frac{GM\ (\text{one day})^2}{4\pi^2}\right]^{1/3}$$

The final answer

$$r = 42\ 200\ \text{km}$$

is then obtained by inserting numerical values.

Solving the next problem is a similarly elementary exercise from a mathematical point of view.

2.2 Example: the time of maximal weight output in a trout pond
Here we have the mortality equation

$$\frac{dN}{dt} = -\ MN$$

and the metabolic equation

$$\frac{dW}{dt} = hW^{2/3} - kW$$

for the number N and weight W of the fishes. Neglecting all other features of this model we obtain the optimal delay time

$$t_0 = \frac{3}{k}\ \ell n(M+k/M) = 4.6\ \text{years.}$$

2.3 How can we compare the two models?
Formally the procedures of solving the two problems look very similar. In spite of their formality both models build upon informal issues, the intuition or "modelling insight, which is beyond the reach of a formal approach" (Naur, 1989). The items of the two models are not inherent in the aspects of the world under consideration, but are human constructions. The modelling process can hardly be automatised. Also in this aspect the two models are similar.

However, there *is* a strong difference in credibility. The first model permits the calculation of the altitude of the orbit with the degree of precision required by the members of an investment syndicate for the safety of their investment. The second model, however precisely the calculations are made and however valuable the results are considered by marine ecologists pond owners and others, has only a tentative character, not taken truly seriously by anybody, subject to continuous change and adaption and supporting nothing on a firm basis.

We should be aware that the explanations and distinctions repeatedly given in the assessment of mathematical models do not suffice to understand fully the differences in the reliability of our two prototype models. Firstly, could there be a more radical simplification of a complex large-scale system than the reduction of the earth to a mass point with no extension? Secondly, difficulties in identifying model parameters (calibration) and errors in measurement could be substantial even for the Newtonian model, if one wants to determine the parameters in the law of gravitation by calibration only. Within the limits of error in astronomical measurements at that time, a gravitational force proportional to $r^{-1.999}$ or $r^{-2.001}$ was as probable as r^{-2}. Finally, the impossibility of experimental verification due to complexity and variability of large technical and biological systems prevented the empirical verification of the Newtonian calculations in 300 years.

The decisive difference is that the extra-mathematical scientific status of both calculations is not the same. Our first case belongs to a *theoretical universe* built on the accumulated experience of a great number of generations and certified in a great variety of contexts (free fall, pendulum and planetary motion, ballistics, potential theory and so on). Therefore, the examination can be confined to a theoretical check. The legitimacy of idealisations such as disregarding the gravitational forces from the moon and the sun can be derived from theoretical analysis of the centrifugal forces within the Newtonian theoretical universe. An experimental verification is not needed, even though possible nowadays.

Our second case, the fish pond model, stands or falls with practical examination, though never standing firmly and never falling definitely because it is built on *ad hoc* assumptions.

Models of high mathematical transparency and precision, but based on *ad hoc* assumptions, lacking an embedding in a theory universe, are quite common. They are not only designed for determining optimal rules for fish feeding and harvesting in a pond, but also for fiscal policy to control unemployment, budgetary deficits and monetary problems, for optimal use of various chemicals in agriculture, horticulture and forestry, for estimating reliability and life times in material sciences, and for modelling emissions and consequences in environmental studies.

2.4 Supplementary theses

Well founded applied mathematics generates prestige which is inappropriately generalised to support these quite different applications. The clarity and precision of the mathematical derivations here are in sharp contrast to the uncertainty of the underlying relations assumed. In fact, similarity of the mathematical formalism involved tends to mask the differences in the scientific extra-mathematical status, in the credibility of the conclusions and in appropriate ways of checking assumptions and results.

Mathematisation can – and therein lies its success – make existing rationality transparent; mathematisation cannot introduce rationality to a system where it is absent (such as in an economy with practically unlimited and free flowing capital) or compensate for a deficit of knowledge (as in many ecological and environmental issues).

The ease of programmed calculation, with the added prestige of the information technology used for calculation and display, supports the proliferation of this kind of mathematical reasoning – however dubious it may be.

The fact that identical mathematical formulae may serve in models whose scientific-theoretical status, significance, mode of verification, and range of validity may be very different, renders the necessary distinctions between the character of mathematical models difficult. The binomial theorem $(a+b)(a-b) = a^2 - b^2$ expresses a property of distributive, commutative and associative number systems no longer valid, not even in a precise approximate sense, in computer arithmetic (where it has to be replaced by the completely different laws of interval arithmetic). Pythagoras' theorem $a^2 + b^2 = c^2$ for the side lengths of a right-angle triangle follows from the axioms of Euclidean geometry where it possesses a precise approximation in the cosine theorem $a^2 + b^2 - 2ab \cos \angle(a,b) = c^2$, whereas its generalisation to curved space requires completely different tools. The formula of gravitation, $F = Gm_1m_2/r^2$, however, expresses a law of nature. Here, the exponent 2 in the expression r^2 is exact. This depends conclusively upon the fact that space has three dimensions – precisely, not approximately. The same is true for Coulombs law $F = kq_1q_2/r^2$, at least in the limits of classical electrodynamics, whereas Ohm's law, $U = RI$, is just the linearisation of much more complex and not yet well understood relations, though it is very reliable within the ranges of temperature, voltage and current in everyday applications. The formula of risk, $R = PC$, with P standing for the probability of an event to occur and C for the consequence, imputes numerical values; in striking contrast to its wide application and high prestige in the debate about the acceptance of nuclear power stations and other hazardous installations, it is just a definition of doubtful relevance. Traditional teaching of mathematical modelling and applications tends to disguise these differences.

Models based on *ad hoc* assumptions are not necessarily less valuable or less reliable than models embedded in a theory universe – if the model formulation with its underlying *ad hoc* assumptions is appropriate for gathering and presentating empirical evidence and for evaluation of deviations and irregularities. As long as such a model represents empirically verifiable facts, it may even be outstanding and irreplaceable, just like a valid railway timetable or any other reliable chart. An application beyond its range of validity without sufficient empirical verification will, as a rule, be worthless and, if the previously acquired authenticity is mechanically imputed to the new situation, it will be misleading at best or even dangerous.

Theoretically substantiated models, such as Newton's celestial mechanics, are not necessarily more precise than *ad hoc* models; the coding of experience in the form of a theory, however, allows a more flexible use of the model, since its embedding in a theory universe permits a theoretical check of at least some of its assumptions – a theoretical assessment of the precision and of possible deviations of the model can be based on the model itself.

Models based on *ad hoc* assumptions are totally dependent on comparison with the latest empirical data, where access might be more restricted and better controlled than access to theoretical insight.

3. WARNING: MORE FRIGHTENING EXAMPLES

3.1 Inadequate technical mastery: heuristic value and production of simulated, not real, evidence

A quite different, and certainly not better, situation arises when the mathematical means used are outside the kernel of mathematics which is theoretically well founded and which is generally well understood by its users. Intricate inverse problems, for example, are raised in the analysis of geoelectrical measurements needed to separate regions of different specific resistance in the earth to get an impression of size and shape of the various geological layers. We still do not understand all relevant aspects of the transport of electrical charges in porous media nor the whole sophisticated mathematics of three-dimensional inverse problems. Nevertheless, by combining geological knowledge and efficient machine calculation one can obtain quite reasonable guestimates, for such things as the distribution of fresh water resources underground. Mathematical models as heuristic means, however limited in their truth content, can save many expensive drillings and will be adequately tested when wells are drilled. Recurrent failures will, moreover, support continued awareness of the weaknesses of the mathematical method.

The situation is again quite different if the foreseeable consequences of failure would not be acceptable. The same parties of the previous example, perhaps municipal authorities and consulting agencies, may now perform the same type of geoelectrical measurements and apply the same program package for determining the geological structure and possible liquid permeability of the earth in the search for safe depositories for toxic waste. The only difference is that the validity test is postponed to years or decades later when the drinking water, perhaps for a large population, may be contaminated. The identity of methods and procedures masks the wide difference of situation and encourages the indiscriminate use of non-validated mathematical methods in totally unacceptable contexts.

Similar problems arise in the highly popular field of Computational Fluid Dynamics (CFD) which also has only heuristic value in the absence of a satisfactory theory of the Navier-Stokes equation, of the discretisation of partial differential equations and of the truncation problems in machine arithmetic - and in the presence of various paradoxes of numerical analysis. See Abbott and Basco (1989) who recognise therefore only 'a small step from alchemy to CFD' and the thoughtful comments in Davis and Hersh (1988).

3.2 Blind and powerful: ruthless, model-based, digitalised control
Another type of mathematics application - and this is the worst case - is the support of devices or processes in cases where reliability is completely unattainable due to the coincidence of the absence of theoretical foundation in the application domain, as in Example 2.3, and the absence of mathematical insight as in Example 3.1.

Such is the case in the following examples: in attempts by nuclear engineers to model the growth of cracks in concrete or in metal alloys under high temperature, pressure and radiation; in the psychological modelling of the behaviour of operators in the control centre of nuclear energy stations under crisis pressure; in the design of steering automata for directionally unstable high-speed container ships. For most of these situations it is characteristic that the calculations may be acceptably descriptive under normal circumstances but worthless under special circumstances, as witnessed in Harrisburg, in Tchernobyl, in the Challenger disaster, in ship losses due to control failures or in the loss of a fully loaded new model Airbus during its maiden flight. In some cases the technology failed, in other cases system failure was due to human failure induced by misguided confidence in the technology - see Källström and Ottosson (1983), Feynman (1988), Horgan (1988) and Lenorowitz (1988).

A comprehensive analysis of the risks of the societal use of mathematical modelling at the borders of knowledge is presented in Booss-Bavnbek *et al* (1989). It turns out that the presently increasing use and search for mathematical models in industry is often related to the concept that model-based digital control permits operating closer to energy and/or material optimal and critical points. Earlier, the operators worked with a good deal of routine activity at lower process speed, or further from critical values. Moreover, analogue control elements, fuses and valves, excluded in practice the surpassing of critical values - often at the cost of a shorter or longer interruption of operations. Now occasional surpassing of a few critical values can be permitted, since the necessary counteraction usually is already anticipated by the model-based digital control. If, nevertheless, something goes wrong, there is still the human operator who has to find a way out of the now always new, unexperienced, surprising and disturbing situation. Subsequent 'human failures' are pre-programmed. Mathematisation and automation have not removed the human factor but formulated new dependencies on human activity and new challenges for qualification and democratic control.

4. WARNING: DO NOT UNDER-RATE THEORETICAL MATHEMATICS

Contrary to the advances in pure mathematics and to the best traditions in the teaching of pure mathematics - with their emphasis on borders of reasoning validity (proofs and counterexamples) and focus on complexity itself - teaching mathematical modelling seems biased to idealised, completely unrealistic, unrepresentative and misleading 'success' stories.

Owing to the pedagogically well-intended selection of examples, those chosen are generally very misleading. They are too simple, too smooth, giving the student a wrong impression of formal concepts matching reality. The students do not see the true problems in modelling, in simplification and generalisation, in the formulation, analytical treatment and numerical solution of equations, in the interpretation and control of the results - or they see them only in a systematically reduced form.

The choice of context is, moreover, often misleading. Population biology or economic modelling, and other fields which lend themselves to arbitrarily simple modelling, are preferred choices. They are easily grasped by the mathematics learner precisely due to the lack of mathematical tradition and quality standards such as those obtained, for example, in engineering, geodesics, and meteorology. In these fields, with well-established traditions of mathematical modelling, practical experience prohibits the use of simple or toy models. These more established subjects are more representative of real modelling of complex reality and therefore, alas, more difficult to teach.

5. WARNING: THE EROSION OF SCIENCES' INTEGRITY

As a general rule the emergence of new and powerful models indicates the presence of very sad problems in society.

- The struggle for military superiority – the main source of 'modelling advances'
- The perceived or the real need to act in an excessively interlaced environment of inscrutable complexity
- The vicious acceptance of the technological risks and the irrationalities of growth economy.

It indicates also the presence of severe methodological problems, the lack of theory and of critical data.

In this way society has been transformed into a laboratory for scientific experiments, where mathematical models play the role of an opiate, tranquilising the feeling of uncertainty.

Example. Unrealistic expectations are directed towards environmental modelling. Models are expected to provide foundations for all of the following.

- The interpretation of data at hand – often worthless data
- The design of experiments – usually by continuing or extending risks for large human populations
- The extrapolation of trends – usually ill-founded and irresponsible
- The evaluation of alternative options – usually confined to narrow technological and/or commercial frames.

A long experience of science and technology shows that, to some extent, models based on good theory (like celestial mechanics) can compensate for lack of data, and that models based on broad evidence (as in parts of chemistry) can compensate for lack of theory. However, models alone (such as high speed numerical simulations and advanced statistical models) can hardly compensate for the lack of both. The advertised new paradigm of scientific computation is of only a very limited value, and by no means the third pillar of modern science in addition to the old-fashioned theory and experiment. This is now seen, by some, as becoming as important as theory, observation and laboratory experiments and in fact better in that infeasible experiments can be done and questionable approximations and simplifications can be avoided, as praised, for example, by Hut and Sussman (1987).

Example. The questionable use of models in crisis management, the perversion of models from illuminating heuristic means to misleading foundations of decisions, often reflects nothing but the state of ignorance. This is not only a political problem of misuse, but also a methodological problem of the erosion of sciences' integrity. This goes for all complex models, no matter how well-intentioned they may be.

The methodological weakness of contemporary applied mathematics can best be understood as a reflection of the origins and history of the contexts of research situations. Ill-founded and irresponsible complexity modelling and complexity handling, equally accepted and demanded in military and dominant civilian circles, faithfully reflects their predominately military origin. A comprehensive treatment can be found in Booss-Bavnbek and Høyrup (1989). Also see Hoare (1981), Parnas (1985), Abrahams (1988) and Domke (1988).

6. CONSEQUENCES: A HIPPOCRATIC OATH FOR MATHEMATICAL SCIENTISTS

6.1 Hands off irresponsible modelling!

The fourth warning says tell the students to keep their hands off irresponsible modelling, supporting illusions of controllability where uncontrollable complexity is introduced in technology and society.

Teach the students to distinguish when the highly esteemed classical approaches in science and engineering are appropriate, and when not. The

- extremely selective choice of data,
- focussing only on single aspects,
- dramatic simplifications, extensions and generalisations,
- courageous formulation of hypotheses,
- partial modelling,
- ingenious inventions of only partially understood technical devices

have all played a positive role in history in overcoming poverty and backwardness. However, one may ask whether there is any necessity to continue along that path in wealthy countries, where a program of de industrialisation might be more relevant. Perhaps what we need is simple approaches wherever complexity can be avoided and, as a whole, more cautious approaches, involving also an analysis of the whole range of related social processes.

6.2 Constant, effective monitoring is needed

The first and second warning say that if, in spite of all this, we still want to go on teaching mathematical modelling, we must find ways to explain the open or hidden but decisive differences in the structure and status of mathematical models and calculations to our students and to the public. To avoid harm, it is necessary to orient towards *ambitiously high quality standards*. Even in situations where information hiding, for example the use of black boxes, seems appropriate in order not to drown in information, complete *accessibility* has to be guaranteed, and the use of black boxes should not be allowed to lead to abrogation of responsibility for the contents. Against the cancerous growth of risk, constant, effective *monitoring* is needed. The model-based technological changes have to be slowed down. The elaboration of theoretical foundations must be demanded for each single step, since fault tolerance is diminishing. As the consequences of mathematical modelling and applications are experienced in the community at large, it follows that all of the people potentially affected by design effects or side effects should be involved in the discussions.

Secrecy must be opposed wherever it arises, because it sabotages all hope of timely evaluation of effects.

6.3 Against mathematical support for risky decisions and technologies

I want to close with a story I read in an article by Abdus Salam (1989), famous for his work in gauge-theoretic physics.

"Nine hundred years ago, a great physician of Islam, Ali Asuli, living in Bukhara, wrote a medical pharmacopoeia which he divided into two parts: diseases of the rich and diseases of the poor. If Ali Asuli were alive and writing today about the afflictions wrought upon itself by mankind, I am sure he would divide his pharmacopoeia into the same two parts. One part of his book would speak of the affliction of possible nuclear annihilation inflicted on humanity by its richer half. The second part of his book would speak of the affliction which poor humanity suffers from - underdevelopment, undernourishment and famine. He would add that both these diseases spring from a common cause - excess of science and technology for the case of the rich, and a lack of science and technology for the case of the poor.

He might also add that the persistence of the second affliction of humankind - underdevelopment - was the harder to understand, considering that the remedies for it are readily available in that the world has enough resources, technical, scientific, and material, to eradicate poverty, disease and early death for the whole of humanity. It has only to eschew deployment of these resources towards aggravating the first affliction."

Perhaps the time is now to extend the Hippocratic Oath - first do no harm - to mathematicians, as proposed by Chandler Davis (1989).

In this sense, modelling, thinking and expressing our knowledge and our thoughts about models, must take a central place in the tetrahedron-shaped mathematics teaching universe as I see it, with the corners *mathematics, reality, information technology, democratic education.*

REFERENCES

Abbott MB and Basco DR. (1989). *Computational Fluid Dynamics*. Longman/Wiley, London.

Abrahams P. (188). Specifications and illusions. President's letter. *Comm ACM*, 31/5, 480–481.

Booss–Bavnbek B et al. (1989). Vurdering af matematisk teknologi – Technology Assessment – Technikfolgenabschätzung. IMFUFA–tekst no. 164*, Roskilde.

Booss–Bavnbek B and Høyrup J. (1989). On Mathematics and War. An Essay on the Implications, Past and Present, of the Military Involvement of the Mathematical Sciences for Their Development and Potentials. Series Filosofi og videnskabsteori på Roskilde Universitetscenter, Roskilde.

Davis C. (1989). A Hippocratic oath for mathematicians? In Keitel C (Ed). *Mathematics, Education and Society*, Science and Technology Education – Document Series, No 35. UNESCO, Paris.

Davis PJ and Hersh R. (1988). *Descartes' Dream. The World According to Mathematics.* Penguin Books, London.

Domke M. (1988). Einflußnahame von Politik, Militär und Industrie auf die Informatik am Beispiel Supercomputer. In Kitzing R, Linder–Kostka U and Obermaier R (Eds), *Schöne neue Computerwelt. Zur gesellschaftlichen Verantwortung der Informatiker.* Verlag für Ausbildung und Studium (VAS), Berlin (West), 136–163.

Feynman RP. (1988). An outsider's inside view of the Challenger inquiry. *Physics Today*, 2/88, 26–37.

Hoare CAR. (1981). Star wars of the seas. Do the lessons of the Iranian Airbus tragedy apply to SDI? *Scientific American*, September, 12–13.

Hut P and Sussman GJ. (1987). Advanced computing for science. *Scientific American*, 257/4, October, 137–144.

Jensen JH. (1980). Matematiske modeller – vejledning eller vildledning. I. *Naturkampen* **18**, 14–22.

Jensen JH. (1988). Matematiske modeller – vejledning eller vildledning. II. *GMMA – Tidsskrift for fysik* **72**, (Nov 1988), 17–31.

Källström CG and Ottosson P. (1983). The generation and control of roll motion of ships in close turns. In Volta E, (Ed) *Ship Operation Automation.* Proceedings of the 4th IFIP/IFAC Symposium (Genua 1982). North–Holland Publ. Comp., Amsterdam, 25–36.

Lenorowitz JM. (1988). A320 crash investigation centers on crew's judgement during flyby. *AW&ST*, 4 July, 28–29. A320 crash inquiry finds no aircraft technical faults. *AW&ST*, 8 August, 28–31.

Naur P. (1989). The place of strictly defined notation in human insight. Workshop on Programming Logic. Baastad (Sweden), May 21–26.

Parnas DL. (1985). Software aspects of strategic defense systems. *American Scientist* **73**, 432–440.

Salam A. (1989). Science, high technology and development. *Scientific World*, **33**, 2, 12–15.

CHAPTER 7

Application-orientated Mathematics Teaching : A Survey of the Theoretical Debate

G Kaiser-Messmer
Kassel University, FR Germany

SUMMARY

A detailed analysis of the theoretical discussion on applications and modelling in mathematics teaching which revived in the late sixties is described, which points out that the discussion has not been uniform and regular, rather it is possible to distinguish different trends with different emphasis. Within the international discussion, two trends can be discerned: approaches from the English language area, which place utilitarian or pragmatic goals for application-orientated mathematics teaching in the foreground, in contrast to approaches from the Romance language area emphasising the science of mathematics and humanistic educational ideals. Approaches from the German language express aspects from both strands of the discussion and develop integrating trends, emphasising a broad range of goals for application-orientated mathematics teaching.

1. INTRODUCTION

The consideration of applications and modelling examples in mathematics teaching has once again become more prominent since the beginning of the seventies, mainly as a reaction to the almost complete displacement of applications by the structure-oriented mathematics education discussion of the late sixties. The paper intends to give a survey of this extensive debate which has taken place in recent years, based on the discrimination of different trends within the international debate as well as the German–speaking debate. This description should provide a framework for a better understanding of the discussion, and prevent us getting 'drowned' in the large number of proposals. The description is of course influenced by my own point of view, and my subjective perception of the discussion. Thus, I do not claim the description given as a true, factual approach, but as a stimulation for personal reflection.

2. INTERNATIONAL DEBATE

Ideally it is possible to distinguish two trends in the international debate on applications and modelling in mathematics instruction

- the *pragmatic trend*: largely accepted in the English language area
- the *scientific–humanistic trend*: widely accepted in the Romance language area.

These two trends differ from each other mainly in the goals for mathematics teaching.

The *pragmatic trend* places utilitarian or pragmatic goals in the foreground, namely the ability of the students to use mathematics for the solution of real world problems. The main purpose of mathematics education should be the teaching of such mathematical methods needed for daily life, the professional area and intelligent citizenship. The students should have their own experiences in the solution of real problems by carrying through modelling processes, and this should give them insight into the broad applicability of mathematics. For the promotion of this aim a change of the school curriculum is required, such as incorporating new mathematical topics with many applications in daily life, vocational areas or the sciences, and removing mathematical topics without applications in non–mathematical areas. Put briefly, the mathematical topics taught in school should be those which are useful for society (see for example Pollak 1986, Burkhardt 1989, Hobbs and Burghes 1989).

In contrast to these ideas, the *scientific–humanistic trend* is more interested in the science of mathematics and humanistic educational ideas, and emphasises the ability of the students to establish relations between mathematics and the real world. The students should not learn applied mathematics, but learn how to apply mathematics, because mathematics is mainly seen as an activity, a method to tackle problems. Such an appropriate understanding of mathematics, or adequate attitude towards mathematics, is needed for becoming a human being. Freudenthal (1973) has expressed this point of view, which has also recently been specified by De Lange (1987) in his approach of mathematics for all. Therefore mathematics should be taught and learned in context, that means starting from the increasingly broadening life of the students and reaching, at least sometimes, mathematical strands as well. Recent approaches, developed by De Lange (1987) and Niss (1989), emphasise the development of a critical attitude as a main goal of mathematics teaching.

Furthermore, there exist significant differences between these two trends of the international debate concerning the *relevance* applications should have to mathematics teaching, and the ideas of *how* to consider *applications* and *modelling examples* in mathematics teaching. Although it is generally agreed that applications should be an indispensable part of mathematics teaching, and that mathematics instruction should not be reduced to applications and modelling, there are remarkable differences concerning the importance and status of the science of mathematics and its structure.

Thus, several exponents of the pragmatic trend require an *interdisciplinary approach*, or the inclusion of applications and modelling as an additional activity, realised in special islands of applications (for example, Burkhardt 1981). Other approaches go so far as to demand a fundamental change of the curriculum, which involves an *exclusive use* of *extra-mathematical teaching units* and the abandonment of the mathematical structure as the orienting guide of mathematics teaching (see, for example, Burghes in Spode Group, 1989).

In contrast to this, the scientific-humanistic trend calls for a teaching of *mathematics in context* which means developing mathematics from the real world, constantly connected with applications and *integrated* from the very beginning. The exponents of this trend reject reducing the comprehension of context to the real world and the lives of the students and favour a comprehension of context, which includes the inner world of mathematics as well. In contrast to the pragmatic approach the scientific-humanistic trend considers the science of mathematics and its structure as an indispensable orienting guide for mathematics teaching, which cannot be abandoned (see, for example, Freudenthal 1973).

The different comprehension of the ability to apply mathematics developed by the two trends of the debate leads to different demands on the *usage* of *mathematics* and the *mathematical activities* of the students.

Thus, the pragmatic trend emphasises the modelling or model-building *process* precisely, as the loop from real world via mathematics back to the real world, and corresponding activities of the students to solve real world problems. Descriptions of this process discriminate the phase of the recognition of the real world problem, the phase of the formulation of the mathematical problem, often defined as the mathematisation phase, the problem-solving phase, the interpretation of the mathematical solution in the real world situation and, if necessary, the improvement of the solution (see, for example, Burkhardt 1981 and Pollak 1979).

The scientific-humanistic trend stresses the *mathematisation* process and appropriate activities of the students, which means the ability to mathematise parts of the real world as well as parts of mathematics. Freudenthal (1973) describes these activities as the structuring of mathematical and non-mathematical subjects, or more recently (1987) as *vertical* and *horizontal* mathematisation, specifying modelling as mathematisation on the lowest level, namely the mathematisation of non-mathematical areas. De Lange (1987) has recently elaborated these ideas by discriminating two kinds of mathematisation

- applied mathematisation – as part of the problem-solving process,
- conceptual mathematisation – as a way to introduce new concepts.

These two kinds of mathematisation are based on each other, with the conceptual mathematisation as the first phase of the mathematisation process, followed by the applied mathematisation phase.

Thus the differences in terminology are related to different ways of thinking about the required mathematical activities of the students, even though the words modelling and mathematisation are sometimes used synonymously.

In summary, it can be stated that within the *pragmatic trend* two directions for the intended change of mathematics teaching can be distinguished.

- On the one hand a change of the mathematical areas taught is intended, including such mathematical areas with broad applicability in everyday life and professional areas (which I have called *content related procedure*), mainly proposed by Pollak (1986) and Hobbs & Burghes (1989).
- On the other hand the inclusion of special phases aiming to train such abilities needed by the students to solve problems of general importance or from daily life, whereby such problem-solving abilities include the abilities to apply mathematical methods, without being all of them (which I have called *ability related procedure*), mainly proposed by Burkhardt (1989).

The main aim of the *scientific-humanistic trend*, in contrast to the pragmatic trend, is a change of the mathematics taught towards a *mathematics in context*, including the teaching of an appropriate picture of the science of mathematics as a basic goal of mathematics instruction. Accordingly, the main emphasis of this approach lies in a change in the introduction of concepts and their treatment in such a way that the students are able to use them in their actual and future life.

In the last few years several changes and developments have taken place. On the one hand the approaches from the scientific-humanistic trend have adopted elements from the pragmatic trend, especially the ideas of applied and conceptual mathematisation, which emphasise the connection from mathematics back to the real world which, in the beginning of the discussion, had low importance. The idea of applied mathematisation has strong relations to the conception of modelling processes emphasised by the pragmatic trend. On the other hand, approaches from the pragmatic trend have developed new conceptions, which are almost exclusively restricted to utilitarian goals, which was not the case in the beginning of the debate.

These differences concerning the goals of mathematics education have to be seen in the light of *global differences* between the *Romance* and the *English-speaking educational debate.* Niss (1981) has pointed out that, since the end of the last century, educational goals referring to the *individual* and the *individual benefit* have mainly been required in English-speaking countries, in contrast to the Romance and German language areas, which emphasise goals which refer to *society* and the *social benefit*.

3. GERMAN-SPEAKING DEBATE

Ideally it is possible to distinguish three trends in the German-speaking debate on applications and modelling in mathematics teaching

- the *emancipatory trend*
- the *science-oriented trend*
- the *integrating trend.*

These trends mainly differ from each other by the *goals connected* with *applications* and *modelling* in mathematics education.

Thus, the *emancipatory trend* starts from the *interest in emancipation* as the main orientation, and claims chiefly utilitarian goals. Mathematics teaching should enable the students to act in a rational and autonomous manner in actual or future real life situations (see, for example, Volk 1979 and 1980, Damerow 1984, Keitel 1985).

The *science-oriented trend* places the *science of mathematics* and its *epistemology* in the foreground, and requires mainly methodological and epistemological experiences for the students. This trend emphasises the necessity to gain insight into the modelling and mathematisation processes (see for example Steiner 1976, Engel 1982, Steinbring 1980).

The *integrating trend* demands a *balanced relation* between *utilitarian, methodological/epistemological* and *mathematical goals.* This trend is strongly influenced by a proposal from Winter (1975) for general pedagogical goals for mathematics education (see Blum 1985, Fischer and Malle 1985, Schupp 1989, Wittmann 1981).

The German discussion was polarised from its beginning until the mid eighties, an important part of the theoretical background referring to the emancipatory and especially to the science-oriented trends. In the last few years conceptions belonging to the integrating trend have become more relevant, in the course of which some of the extreme positions have lost their importance, others changed their conceptions mainly by including utilitarian goals (for example, the Austrians Fischer and Malle, 1985). In summary, it is nowadays a *consensus*, largely accepted in the educational debate, that applications and modelling should perform the following goals (for details see Kaiser-Messmer, 1986).

- Utilitarian or pragmatic goals: promotion of abilities to master everyday life
- Methodological goals: teaching of modelling abilities and abilities to apply mathematics in real world situations
- Goals referring to mathematics: increase of the motivation of the students to do mathematics, promotion of the long-term retention and enhancement of the comprehension of mathematics
- General pedagogical goals like the promotion of creativity or problem solving abilities
- Science-oriented goals, that is to show a realistic picture of mathematics as a cultural and social phenomenon

These three strands of the debate also develop different thinking on the *relevance* of applications and modelling in mathematics teaching, and the way of considering them. The science-oriented and the emancipatory approaches represent the poles; whilst the science-oriented trend assigns only low importance to applications and modelling, the emancipatory trend abandons the mathematical structure as the orienting guide of mathematics teaching and demands an interdisciplinary teaching.

The integrating trend formulates an intermediate position between these two approaches, such as giving applications and modelling a high priority in mathematics teaching without accepting them to be the only criteria for the selection of mathematical content and the structure of mathematics teaching.

The thinking on the usage of mathematics and the mathematical activities of the students has changed considerably in the last few years. It is largely agreed, however, that modelling and mathematisation processes have to be considered in mathematics teaching, but an extensive discussion comparable to those of the English-language area has not taken place.

The debate has been influenced strongly by a scheme for mathematisation processes already developed in the mid-seventies by Steiner (1976). In this description Steiner tried to develop a synthesis between the conception of modelling processes from the pragmatic trend and the conception of mathematisation processes from the scientific-humanistic trend, leading to mathematical theory as well.

So far, it is largely accepted in the educational discussion that the modelling process and the teaching of modelling abilities have to be stressed, as against the teaching of standard models. In contrast to most of the pragmatic approaches, Fisher and Malle (1985) as well as Blum (1985) emphasise the necessity to *explicitly reflect* on the modelling process and the underlying epistemological questions in mathematics teaching.

Winter (1980) develops a description of the introduction of mathematical concepts embedded in modelling and mathematisation processes, that means in close connection with relevant real-world situations. With this proposal he links up with the tradition of the practical applied arithmetic, the so-called *Sachrechnen*, which is very specific for the German educational discussion and has influenced it since the last century.

Altogether, it can be stated that the German discussion considers and refers to the international debate, initially to both trends of the international debate but in the last few years more and more to ideas of the scientific-humanistic trend. However, I would like to point out that the developed approaches on modelling and applications are continuing some typical German traditions, especially the previously mentioned *Sachrechnen*. Furthermore, approaches from the integrating trend also link up with ideas from the *Meraner Reform*, a reform movement from the beginning of this century, which required a balanced relation between utilitarian, methodological and mathematical goals. The ideas of the emancipatory trend rely strongly on the reform movement of the reform pedagogy, and the so called *Arbeitsschulbewegung* emphasising the necessity to base all learning on the activities of the students and trying to close the gap between life and school.

I have tried to show, with the example of the German discussion, that on the one hand the discussion going on is very different in diverse countries, and a deeper understanding of the discussion in one country very often needs some knowledge of the specific traditions. However, on the other hand, there are a lot of common points, so if we are aware of these differences we should be able to understand and learn from each other.

REFERENCES

Blum W. (1985). Anwendungsorientierter Mathematikunterricht in der didaktischen Diskussion. *Mathematische Semesterberichte*, **32**, 195–232.

Burkhardt H. (1981). *The Real World and Mathematics.* Blackie, Glasgow.

Burkhardt H. (1989). Mathematical modelling in the curriculum. In W Blum *et al* (eds), *Applications and Modelling in Learning and Teaching Mathematics.* Ellis Horwood, Chichester, 1–11.

Damerow P. (1984). Mathematics for all – ideas, problems, implications. *Zentralblatt für Didaktik der Mathematik*, **16**, 81–85.

Engel A. (1982). Statistik auf der Schule: Ideen und Beispiele aus neuerer Zeit. *Mathematikunterricht*, **28**, 57–85.

Fischer R and Malle G. (1978). *Fachdidaktik Mathematik.* Klagenfurt: Universität für Bildungswissenschaften.

Fischer R and Malle G (with cooperation of Bürger H). (1985). *Mensch und Mathematik.* Bibliographisches Institut, Mannheim.

Freudenthal H. (1973). *Mathematics as an Educational Task.* Reidel, Dordrecht.

Freudenthal H. (1987). Theoriebildung zum Mathematikunterricht. *Zentralblatt für Didaktik der Mathematik*, **19**, 96–103.

Hobbs D, Burghes D. (1989). Enterprising mathematics: A cross–curricular modular course for 14–16 year olds. In W Blum *et al* (eds), *Applications and Modelling in Learning and Teaching Mathematics.* Ellis Horwood, Chichester, 159–165.

Kaiser–Messmer G. (1986). *Anwendungen im Mathematikunterricht.* Vol 1, Theoretische Konzeptionen. Franzbecker, Bad Salzdetfurth.

Keitel C. (1985). Mathematik für alle – ein Ziel; was sind die Ziele einer "Mathematik für alle"? Hans Freudenthal zum 80. Geburtstag. *Zentralblatt für Didaktik der Mathematik*, **16**, 1977–186.

De Lange J. (1987). *Mathematics – Insight and Meaning.* OW&OC, Utrecht.

Niss M. (1981). Goals as a reflection of the needs of society. In R Morris (ed), *Studies in mathematics education*, Vol 2. UNESCO, Paris, 1-21.

Niss M. (1989). Aims and scope of applications and modelling in mathematics curricula. In W Blum *et al* (eds), *Applications and Modelling in Learning and Teaching Matheamtics.* Ellis Horwood, Chichester, 22-31.

Pollak HO. (1979). The interaction between mathematics and other school subjects. In UNESCO (ed), *New Trends in Mathematics Teaching IV.* Paris, 232-248.

Pollak HO. (1986). The effects of technology on the mathematics curriculum. In M Carss (ed), *Proceedings of the Fifth International Congress on Mathematical Education.* Birkhäuser, Boston, 346-351.

Schupp H. (1989). Applied mathematics instruction in the lower secondary level - between traditional and new approaches. In W Blum *et al* (eds), *Applications and Modelling in Learning and Teaching Mathematics.* Ellis Horwood, Chichester, 37-46.

Spode Group. (1989). Introducing decision mathematics into the school syllabus. In W Blum *et al* (eds), *Applications and Modelling in Learning and Teaching Mathematics.* Ellis Horwood, Chichester, 207-212.

Steinbring H. (1980). *Zur Entwicklung des Wahrscheinlichkeitsbegriffs* - das Anwendungsproblem in der Wahrscheinlichkeitstheorie aus didaktischer Sicht. Institut für Didaktik der Mathematik, Bielefeld.

Steiner HG. (1976). Zur Methodik des mathematisierenden Unterrichts. In Dörfler W and Fisher R (eds), *Anwendungsorientierte Mathematik in der Sekundarstufe II.* Heyn, Klagenfurt, 211-245.

Volk D. (1979, 1980). Vol A : *Handlungsorientierende Unterrichtslehre am Beispiel Mathematikunterricht.* Vol B : *Zur Wissenschaftstheorie der Mathematik.* Päd-extra-Buchverlag, Bensheim.

Volk D. (1989). Mathematics classes and enlightenment. In W Blum *et al* (eds), *Applications and Modelling in Learning and Teaching Mathematics.* Ellis Horwood, Chichester, 187-191.

Winter H. (1975). Allgemeine Lernziele für den Mathematikunterricht? *Zentralblatt für Didaktik der Mathematik*, **7**, 106-116.

Winter H. (1980). Zur Durchdringung von Algebra und Sachrechnen in der Hauptschule. In H–J Vollrath (ed), *Sachrechnen.* Klett, Stuttgart, 80–123.

Wittmann E. (1981). *Grundfragen des Mathematikunterrichts.* Vieweg, Braunschweig.

CHAPTER 8

The Real World behind Real-world Problems - some findings of a micro-ethnographical study

H Jungwirth
University of Linz, Austria

SUMMARY

The findings are presented of an analysis of interactions in those phases of mathematical lessons in which extra-mathematical fields of applications are discussed. The aim is to find out from which perspectives teachers and students see the actual real-world topic, and how teachers organise the lesson. Firstly, the findings show, that in such phases framework ambiguities may exist for the students. Secondly, two modes of dealing with the non-mathematical knowledge could be reconstructed: one mode using the asking question method as is customary in the mathematics classroom, and one mode in which this knowledge is merely demonstrated.

1. INTRODUCTION

The lack of balance in New Maths, when it was introduced into schools, led to the neglect of the principles and objectives of mathematics teaching, which have gained even greater importance as a result. An important objective now is to make the aspect of applying mathematics a subject, even when it is argued about in detail in various ways (see Kaiser et al, 1982). It must be said that mathematics geared to application is often limited to solving highly pre-structured problems which are cut down to fit mathematical units which have just been taught or are about to be taught. This does not mean that such tasks are worthless in general. On the one hand, it is certainly legitimate to motivate the learning of mathematics in this way; on the other hand, the problems dealt with can certainly be important for everyday life and work. What must be established is that tasks such as these cannot, in themselves, meet all the aims of teaching applications and modelling. In what follows, I would like to present mathematics teaching in which more happens than the solution of such pre-structured examples of application. Extra-mathematical fields of examples of application are discussed too, and the students have to exploit the knowledge they have acquired outside their mathematics lessons and/or from their experiences in everyday life.

The relevant observations were made in the framework of a research project, the general aim of which was to find out whether and how interaction in the mathematics lesson was coloured by the fact that the students were boys or girls (Jungwirth, 1989). Over and above this, it was possible to gain a general insight into how students dealt with extra-mathematical questions in mathematics lessons, and reconstruct forms of interaction in these sections. (It was shown that, even in these sections, gender differences are present - see Jungwirth, 1989). Video and audio recordings of 38 lessons in 11 classes in Austrian grammar schools from the fifth to the twelveth grade are the basis for these data. In four classes, extra-mathematical questions were dealt with in connection with problems of application.

2. THE THEORETICAL BACKGROUND AND ITS METHODOLOGICAL IMPLICATIONS

The theoretical background for the analysis lies in micro-sociological approaches stemming from constructivist scientific theory: symbolical interactionism and ethnomethodology (Blumer 1980, Garfinkel 1980). On the basis of the approaches outlined, interaction in mathematics lessons has been successfully analysed by university mathematics teaching departments for some years now (Bauersfeld 1983, Krummheuer 1983, Voigt 1984, 1989, Bauersfeld, Krummheuer & Voigt 1988). The hypothesis that people give meaning to objects is essential in these approaches. There is such a thing as common knowledge, for example meaning and ideas, which are agreed upon by many people. In social interaction they tell each other how they see the objects in question, and in this way define the meaning which is to apply. From this point of view, such a negotiation of meaning takes place in the matheamtics classroom too - teachers and students interpret themes and teaching processes differently, and must therefore come to an agreement as to what knowledge should be the standard one.

As for the investigation of interaction processes, a consequence is to direct the analysis towards the reconstruction of the reality which was constructed by the participants. In this reconstruction, therefore, the interactions are seen from the participants' point of view. Over and above this, regularities in acting and patterns in the interactions are shown that cannot be understood from the interpretations of the participants alone. Their interpretations are accessible for the researcher because, as I mentioned above, the participants tell each other how they see the objects in question.

The findings of the interaction analysis are those reconstructions that are gained from comparing the interpretation of several interaction sequences – in the case of interest here, the sequences in which extra-mathematical fields are thematised. Hypotheses made from particular sequences are applied to the other comparable sequences at the researcher's disposal. In this process the hypotheses are developed into the resulting interpretations. The modes of interpreting are in accordance with the standards of qualitative research (Erickson, 1986). Thus the 'stories' presented in the following passages are sequences involved in this process. In this context they should serve the purpose of enabling the reader to follow the given interpretations.

The most important concept for the analysis of dealing with extra-mathematical subjects was shown to be 'framework' (Goffman, 1980). It describes the customary pattern of interpretation, the background thinking which organises experiences and guides behaviour. Modes of communication in mathematics lessons were analysed according to this concept. In algebra lessons a different framework for teachers and students was reconstructed, and it could be shown to what extent teachers and students adjust to each other other (Krummheuer, 1982 and 1983).

3. FRAMEWORK AMBIGUITIES IN THE RECOURSE TO EXTRA-MATHEMATICAL KNOWLEDGE

These phenomena are not only evident in algebra lessons. They appear in connection with the thematising of real life aspects of problems too. Uncertainty as to the framework used is even greater here because, in addition to an (inner) mathematical framework, real life ones can be used. Everyday life is an area where people are supposed always to agree about meanings; teachers, however, may assume all the more that students think in the way they themselves do.

The following scene illustrates such a problem of 'different possibilities of interpretation with regard to framework' (Krummheuer, 1982). The questioning situation is not organised so that an unambiguous intra- or extra-mathematical framework can be assumed for all concerned. It took place in a seventh grade, which was learning about addition and subtraction of terms.

Teacher: Let's read through the first example. Gerda, please begin.
Gerda: In the course of collecting for a good cause, 'a' schillings were collected. The government contributed the same amount. How much is the total amount which can be handed over? Use as simple a term as possible.
Teacher: Yes, that will do. It's quite clear, it's in the text.
Gerda: a + a = 2a
Teacher: Yes, good. How can we prove that now, and what does 'a' stand for?
Gerda: For schillings.
Teacher: For any amount. Right. Let's have an example. Anyone got an example?
Gerda: Yes, 10 schillings. [Shrugs her shoulders]
Teacher: 10 schillings? Do you think that's a realistic suggestion, a collection which comes to 10 schillings is doubled by the government? No, I don't think so. [Laughter]
Gerda: Thousand?
Teacher: Do you think that's right? [Gerda shrugs her shoulders again] Yes? [Looks at Paul]
Paul: A million?
Teacher: That seems a bit much.
Gerald: 100 000.
Teacher: Yes.

Up to the point where the question arises as to how to check the correctness of the solution given, (9), the problem has a mathematical framework. In this situation, concerned with the question of a numerical example, one can act with a mathematical background, but with a real life component too. Gerda's shrug of her shoulders at the answer, (14), shows the uncertainty she felt. Her answer "10 schillings" can be understood as the expression of a mathematical framework, but also in relation to an everyday situation – Gerda would give 10 schillings to this fund. The teacher is thinking about the problem with her background knowledge about such public fund-raising (15–17). She takes it for granted, and therefore nearly makes fun of Gerda's answer. Gerda and other pupils then try to guess amounts that might be accepted by the teacher (18, 21, 23). Part of the uncertainty felt by students may be due to the fact that they often find that the knowledge they possess, which has a logic of its own and which they have acquired outside school, is not relevant to the mathematics lesson (see Voigt, 1984).

General non-mathematical knowledge is treated in various ways in mathematics teaching. One form of interaction can be called 'the eliciting information through asking questions method'; another may be termed 'demonstrating knowledge' (Goffman 1980, p79).

4. GAINING GENERAL KNOWLEDGE THROUGH ASKING QUESTIONS

In general, interaction in constituting mathematical knowledge shows a structure of asking questions. This is also seen in gaining general, other than mathematical, knowledge. By general knowledge, we mean knowledge which is based, on the one hand, on the everyday experience of the students outside school; on the other hand, we mean scientific knowledge gained at school in subjects other than mathematics.

On the surface, interaction takes place without friction, but below the surface differences of interpretation exist. The teacher and the students achieve interaction by means of well-tried routines. On the part of the teacher this takes the form of provisional queries, critical follow-up questions and, finally, suggestive questions. The students use the trial-and-error strategy, and the routine 'speech reduction' in answering (ambiguous) questions (see Voigt, 1984).

An important difference in mathematics teaching lies in the fact that the teacher's general knowledge also plays an important part in developing other question sequences. There is the danger here that the personal views of the teacher, and his or her own explanation patterns, will be judged to be valid, and thus given the status of truth, which is of course not necessarily the case.

The following scene may illustrate this technique of taking subjects from extra-mathematical fields; it took place in a tenth grade. The problem of applying a geometrical sequence is calculated at the board with the help of the students. "How much does the value of money go down in one year if the prices go up by 6%? How much of its original value has a certain sum of money got after 5, 10 or 20 years if prices rise steadily?"

After the problem has been solved, the teacher talks about the social implications of depreciation.

Teacher: What happens when money loses its value – its purchasing power? This has consequences – what are they? What happens when people can buy less and less?
Vera: People get poorer.
Teacher: Yes, first of all, people get poorer. What about the economy? What does it live on? [Laughter, murmuring] Yes?
Boy: On purchasing power.
Teacher: Yes, on purchasing power, certainly. So how will we try to offset it? What happens each year?

Boy: Inflation [Tentatively]
Teacher: Yes, we have inflation. It's very slight in Austria at the moment, I believe.
Vera: 1.8%
Teacher: Yes, under that. It changes all the time. I don't know exactly, but in any case it's a lot under 2% at present, isn't it? We have had much higher inflation rates. What is the position now? How can I offset inflation rates when the purchasing power gets less? It's an absolutely natural cycle. How can I offset it?
Chris: Through increasing wages.
Teacher: Yes, through increasing wages. What if I want a real wage increase – that is, if I receive more wages, what must happen to the wage increase? What must it lie above?
Adolf: It must lie above the inflation rate.
Teacher: Yes, above the inflation rate. What else must lie above the inflation rate? Let's think about what we did just now. Yes? Who knows? [Hans points to Otto]
Boys: Interest
Teacher: Whose interest?
Otto: The banks' interest.

The use of the term 'loss of purchasing power' (1–2), and its interpretation into everyday language, shows the changeover to a framework which can be described as that of everyday economics. This is based on certain parts of economic theory, but also on the experience of the everyday world concerned with how people really live. Vera's answer, "People get poorer" (5), points to an everyday way of looking at things, but at the same time shows that she is thinking about people's actual living conditions. The teacher, however, keeps to his own way of thinking. Step by step he develops his argument that counter-measures to inflation lie in wage increases, and the rate of interest being above the inflation rate (22–33). The real-life component in his background knowledge makes the teacher give his point of view absolute authority. Just one perspective of economic processes and relationships appear to be the sole possible one. What is here taken for granted – "it is an absolutely natural cycle" (21) – is not obvious when considered objectively.

5. DEMONSTRATING THE GENERAL KNOWLEDGE AT YOUR DISPOSAL

Subjects drawn from general knowledge can be used in a second way in mathematics lessons. General knowledge is here so-called everyday knowledge, for which experience outside school is relevant. At first, the teacher asks explicitly how the problem they have been talking about can occur in everyday life. This is in fact an open question, because the teacher, as the further course of the interaction shows, has no special answer in mind to which he or she guides the conversation. It is, in this phase, aimed rather at collecting several possible answers, that is, several points of view or solutions to the problem. The students name possibilities in everyday life with reference to their background knowledge. Their behaviour can be described as 'making a show of opinions'. They offer suggestions which are not necessarily their own opinions. According to the routine of 'finely adjusted evaluation' (Voigt, 1984), the teacher comments on the contributions of the students and says whether they are plausible, practical, and so on, without, however, channelling them to a predetermined point. This evaluation is the only form of direct comment on the contributions. The students, amongst themselves, either ignore their utterances or at least do not refer to them explicitly. Seen as an analysis of a framework, it is a demonstration of the availability of general knowledge. Recourse to this knowledge serves other aims than those otherwise related to it; it is not representing a real problem from everyday experience which has to be solved. Goffman calls this transformation of a framework a 'demonstration', "the performance of an action outside its normal functional context, to enable someone to see its exact development" (Goffman 1980, p79).

The following scene from a lesson in an eighth grade illustrates this way of using general knowledge. The following inequality problem was set as homework.

"Mrs Sweet wants to get some strawberries from a strawberry field. In the one nearest to her home the strawberries cost 16 schillings a kilo, in another one further away they cost 14.50. It costs 30 schillings to get the second one, that is, 12 schillings more than to the first one. Ask a question, and answer it by solving an inequality".

After discussing the solution, the teacher speaks about the everyday situation.

Teacher: Do you think price is the only thing I would be thinking about when deciding which field to go to? What happens in real life?
Simon: Maybe the second field has much nicer strawberries. They are much sweeter and taste better.
Teacher: Yes.
Simon: Or the first has smaller ones which don't taste so good.
Teacher: Yes, quite possible. Martin?
Martin: If it's a long way away, and I come back with the strawberries and the road is rough or something, the

strawberries could go bad or something. [Laughter].

Teacher: Well, they wouldn't be very good keepers, would they? [Looks at Hans]

Hans: Well, if the first field was right next to the motorway, or next to the main road, and the second was in the middle of a wood

Teacher: Which would you choose, Hans?

Hans: The one in the wood.

Teacher: Have you got something against motorways?

Hans: Well, I'm thinking about the exhaust fumes – they're not very good for the strawberries. I think it would be worth the 4.50.

Teacher: Mhm. Gerrit what do you think?

Gerrit: Well if you were in a hurry, you'd probably go to the nearest one

Teacher: Yes, time is also an important factor. The other field is really too far away, too out of the way, or perhaps there's no direct bus or train route there. So time plays an important part. You have to consider a lot of things. What you like doing. No, I'll go to the nearest field after all perhaps I'll meet some people I know. I haven't been to the other one so often.

The teacher's remark "What happens in real life?" (3), expresses explicitly the change of perspective. We are no longer talking about a mathematical problem. The background thinking necessary to answer the teacher's question is general knowledge about an everyday situation – buying strawberries. The students are asked to list criteria for the choice of where to buy them according to their general knowledge. Several pupils take part: Simon brings up the criteria of taste (4–8), Martin talks about the problem of transporting the perishable fruit (10–12), Hans mentions pollution (15–23) and Gerrit the time needed for the journey (25–26). The teacher agrees that there is a problem of time, and perhaps one of meeting friends (27–33).

6. CONCLUDING REMARKS

The aim of the teacher, to introduce aspects of the question which are not directly mathematical into the mathematics lessons, is certainly welcome. However, in evaluating their realisation, some doubts occur. In the one case the teacher implicitly rejects the students' orientations. Her/his view comes to stay. Pre-existing framework differences between teacher and students are kept up, without being thematised. In the other case, the opinions of the students are not channelled towards the sole valid one. The mere demonstration of everyday knowledge, however, leads to the question, what the students really do learn in this mode of dealing with extra-mathematical fields. In my opinion, there is the danger that students will consider such references to real life as arbitrary or odd, and that this will lessen their interest in the subject. So, a new approach to the problem of real-life contexts in mathematics lessons seems to be necessary. A first step towards this aim is that teachers examine their classroom practice, and become aware of the patterns in the teaching process.

REFERENCES

Bauersfeld H. (1983). Subjektive Erfahrungsbereiche als Grundlage einer Interaktionstheorie des Mathematiklernens und – lehrens. In Bauersfeld H *et al* (eds), (1983), *Lernen und Lehren von Mathematik*, Aulis, 1–56.

Bauersfeld H, Krummheuer G and Voigt J. (1988). Interactional theory of learning and teaching mathematics and related microethnographical studies. In Steiner HG and Vermandel A (eds), *Foundations and methodology of the discipline mathematics education*, Antwerpen, 174–188.

Blumer H. (1980). Der methodologische Standort des Symbolischen Interaktionismus. In Arbeitsgruppe Bielefelder Soziologen (eds): *Alltagswissen, Interaktion und gesellschaftliche Wirklichkeit,* Opladen, 80–146.

Erickson F. (1986). Qualitative methods in research of teaching. In Wittrock MC (ed), (1986), *Handbook of research on teaching*, Macmillan, New York, 119–161.

Garfinkel H. (1980). Das Alltagswissen über soziale und innerhalb sozialer Strukturen. In Arbeitsgruppe Bielefelder Soziologen (eds): *Alltagswissen, Interaktion und gesellschaftliche Wirklichkeit*, Opladen, 189–213.

Goffman E. (1980). *Rahmenanalyse. Ein Versuch über die Organisation von Alltagserfahrungen.* Frankfurt.

Jungwirth H. (1989). Die geschlechtliche Dimension der *Interaktionsstrukturen im Mathematikunterricht und ihre Folgen.* Projekt – Abschlußbericht. Linz.

Kaiser G, Blum W and Schober M. (1982). *Dokumentation ausgewählter Literatur zum anwendungsorientierten Mathematikunterricht.* Fachinformationszentrum Energie, Physik, Mathematik, Karlsruhe.

Krummheuer G. (1982). Rahmenanalyse zum Unterricht einer achten Klasse über "Termumformungen". In Bauersfeld H *et al* (eds), (1982), *IDM-Reihe Untersuchungen zum Mathematikunterricht*. Band 5 Analysen zum Unterrichtshandeln, Köln, 41-103.

Krummheuer G. (1983). Algebraische Termumformungen in der Sekundarstufe I. Abschlußbericht eines Forschungsprojekts, *Materialien und Studien Band* 31, IDM Bielefeld.

Voigt J. (1984). *Interaktionsmuster und Routinen im Mathematikunterricht*. Weinheim und Basel.

Voigt J. (1989). The Social Constitution of the Mathematics Province - A Microethnographical Study in Classroom Interaction. In *The Quarterley Newsletter of the Laboratory of Comparative Human Cognition*, Vol 11, **1** and **2**, 27-35.

CHAPTER 9

Technology Transfer – a Didactical Problem?

J Maass and W Schlöglmann
University of Linz, Austria

ABSTRACT

In many countries, the number of those studying mathematics in order to become teachers is decreasing. This is one reason for us to consider if there might be new directions in the teaching of mathematics in the university, and new fields of working for mathematics education. If students become fewer, the question as to jobs and money available for departments of mathematics education becomes ever more urgent. In West Germany, ministers wanting to carry out economy measures and to set up new priorities in their work, have announced that teaching positions becoming free should be given up to other subjects, such as medicine or information technology (see Bauersfeld, 1988).

Apart from our natural, insider interest in the long-term preservation and growth of our profession, we are mainly concerned about the social developments which focus our attention on new tasks for university mathematics and mathematics education departments. The new technologies, which are altering our lives so radically, are essentially mathematical technologies. Computer hardware and software would be unthinkable without the mathematical theories on which they are based. For this reason, we believe that it will be ever more necessary to understand the mathematical pre-conditions of the new technologies in order to apply them in the most meaningful way, and to understand the effects of their application in all areas of society. We should also aim to extend the number of people with whom we are involved in working out the processes used in learning mathematics. Which scientific discipline, if not the department of mathematics education at the university, is better qualified to contribute well-substantiated statements in order to lead to a more general and better understanding of mathematics in the population as a whole, and especially for users in industry and in the administration?

That is why, in this paper, we shall be suggesting that the university departments of mathematics education adopt new priorities in their work – let *us* try to find a new and better orientation for our work!

1. TECHNOLOGY TRANSFER: VARYING ATTITUDES TO A FASHIONABLE TERM

When using trendy words of this kind, there is always the danger that we may lose sight of their real meaning. It does not help to have a careful definition (which can limit the meaning to what the author has in mind) if it cannot influence the common usage of the term. That is why such terms are normally avoided in scientific discussion. However, since we are nevertheless concerned here with the relationship of technology transfer to departments of mathematics education at the university, we shall try to bear in mind the situation above as well as the tendency towards financial and other support being granted increasingly according to non-scientific criteria and in circles outside the university.

2. TECHNOLOGY TRANSFER: AN OUTLINE OF THE PROBLEM

The term technology transfer was originally used for export of technology to the so-called Third World countries. In this connection, the essential problems of such a transfer are which (adapted, environmentally and socially compatible) technologies can be transferred, and in which way? What part do the social and economic contexts of development and use play, and what are the differences in communication between developers and users? If the transfer of knowledge from the university to other sub-systems of society is also called technology transfer, similar questions arise. When the most common transfer route is taken – that is, from university to industry – the problem in each particular case is how to achieve what is called in the language of system theory 'the transfer of meaning' – there must be a conversion from the scientific communication code of 'truth' into the code of the economic system 'money'. The systems, which are differentiated according to their various functions, realise from the variety of the total complexity of the world around them the information which they can process in each individual case according to the rules which they have institutionalised in their inner structure. The typical method of processing in science, the specific way of reducing complexity, is the differentiation between true and not-true statements, and thus leads to an increase in scientific knowledge as such. If such knowledge is not further processed it is, in itself, not necessarily usable outside the scientific system. A mathematical theory, such as linear optimisation or graph theory, does not in itself help the economy; it is only when mathematical theories are converted into a computer program controlling optimality, and thus profitability – the production of, say, cabinets in a factory – that the transfer can be said to be successful. It is essential that mathematics, as a sub-system of the science system (Maass, 1988), can achieve a successful transfer, not only by working out new theories and algorithms, but by working them out in such a way that they can usefully be brought to bear on the system of application.

3. TECHNOLOGY TRANSFER: FOUR WAYS

In what follows it may be useful to differentiate between the following methods of transfer.

- **Education**, in the form of a course of study, is the classical method by which knowledge from the university reaches systems of application. Related questions as to the meaningfulness and possibilities of orientation towards practical application, for example, have long been under discussion in the mathematics education community (Maass and Schulz-Reese, 1989).

- **Further education**, in particular the scientific training of users of mathematics in industry, such as engineers and scientists, is a relatively new field. Didactically thought-through experience in this field, as well as in the training in vocational education, is not yet available.

- **Black boxes** enable a transfer of knowledge (Maass and Schlöglmann, 1988), even when the users have no idea about what is inside them. When you operate a light switch, you do not need to know anything about physics or technology. When you use a software package for linear optimisation, you do not need to prove, or even understand, the simplex algorithm involved.

- **Cooperation**, for example in the form of the university and the user working out a project together, is always useful and necessary when a problem is to be solved with the help of scientific knowledge not at the disposal of the user but which, because of its complexity and variability, is not soluble through using an existing black box, or where it can only be solved by constructing a black box together.

4. MATHEMATICS AS TECHNOLOGY?

The importance and the function of university mathematics education departments for the first two methods of transfer of mathematical knowledge are obvious. Before we can discuss their possible role with respect to the other two, we must first go into the question of whether mathematics can be thought of as a transferable technology at all.

According to a definition by Beckmann, technology is the *theory* of the processing of raw materials, or the *science* of the transformation of raw materials into finished products and the *method* of their processing. When rules for processing raw materials are described in mathematical terms – for example, as a ratio of the various elements included in a mixture, or as physical or chemical laws set out in formulae which establish the properties of substances and the fundamental principles of their ability to change, and which also necessitate a numerical solution of a system of equations in controlling a production plant – then mathematical knowledge is obviously technology in the sense of this definition. As this concerns only one aspect of mathematics (or indeed of its application), and as the use of the term technology suggested in the dictionary does not include one essential component – that is, its being inseparable from its social context – we need to try once again to answer the question as to whether mathematics is to be regarded as technology, or if it is at least technologically efficient. That is why, in the course of a conference on this subject in September 1988 in Strobl (Wolfgangsee, Austria), university teachers of mathematics, mathematicians from the university and industry, scientists, engineers, and philosphers discussed, in interdisciplinary cooperation, different aspects of the question Mathematics as Technology? and its consequences for interaction between mathematics, new technologies, education and further education (Maass and Schlöglmann, 1989).

We can only outline a few selected examples of their findings here. From the point of view of philosophy, the question of the relationship between mathematics and technology first comes up against the problem that mathematics is not an empirical science. So, for Ruben, the "posing of the problem of 'mathematics as technology' is a pre-condition, and we mean by 'technology' not only the theory of techniques, but also the totality of theories of real and possible production processes and the cooperation between mathematics and each specific technology" (see Maass and Schlöglmann, 1989). Hülsmann (1985) believes that this division in the preconditions of the technological formation of society emphasises too much the formal aspects and does not stress the formative aspects enough. "If mathematics is thought of as *mathesis universalis*, as Leibniz has it, and thus has the identity of scientific knowledge in mind, for example logic, arithmetic, geometry and mechanics, the formal aspect is by no means empty of content, but it is rather a possible approach to a reality which as such represents a structural connection, a unity, a world. It seems important to me here that in its technological form, each individual science should not be understood and determined on its own. This *mathesis universalis* is concrete and real. This is shown by the fact that it is no longer a matter of the operation of a single science, but that the operation of knowledge must be understood as technological totality, as it realises its technological form in all the disciplines" (Hülsmann 1985, p124).

A second result of the discussion in the working group Consequences for School at the Strobl conference, was that a consensus was reached according to the technological aspect, for example, "laying stress on the *technical-instrumental* aspects of mathematics, including the *consideration* and weighing-up of the conditions and implications of these aspects" should be an integral component of the mathematics curriculum. This can at present only be formulated and hoped for as an aim as, although a corresponding formulation of aims is to be found in the preamble to almost all syllabuses, the usual mathematics teaching in practice suffers from great weight being laid on method and calculation and, in the upper grades, teaching is largely orientated to the teaching normally at university. As possible steps towards change, subjects such as methods and contents, meaningful use of the computer, orientation to extra-mathematical application, open, and even non-numerical problems, and so on, were discussed.

The working group Consequences for University Studies discussed courses of study initiated or planned over the last few years such as Technomathematics (Kaiserslautern, BRD) or Industrial Mathematics (Linz, Austria) and, corresponding partial study programmes or lectures within the framework of 'normal' courses of study terminating in a diploma, as well as mathematics for teaching.

In new courses of study, technical sectors are explicitly planned. However, even in traditional courses of study, problem seminars in which students solve practical problems in industry can provide essential steps towards training geared to real-life situations.

5. THE CONTRIBUTION OF MATHEMATICS EDUCATION DEPARTMENTS AT THE UNIVERSITY TO THE TECHNOLOGY TRANSFER

No complete answer can be given here to the question of new challenges to mathematics education departments, with participation in constructing black boxes and in industrial projects. Yet it is clear that these departments can make a positive contribution to these two methods of transfer too, by bringing to bear their knowledge about, and experience with, transfer and learning processes. Common transfer problems, such as the sensible organisation of a technology or, more concretely, of a user interface in a black box, can be solved with the help of typical didactical knowledge.

In addition, we draw attention to the result of a discussion a few years ago in Western Germany about polyvalence and flexibility in the training of teachers. In view of the growing number of unemployed teachers, certain politicians in the educational field, as well as managers, demanded that the curriculum be altered. Pedagogics and specialised didactics were to be dropped, so that there was more time for attaining specialised qualifications. According to these proposals, only those graduates who actually teach in a school should learn didactics and pedagogics at the university (Maass and Rahmann, 1985). This proposal proved not to be very practical for a number of reasons. One of these reasons is especially important for our present considerations. Research carried out by the Ministry of Labour's Institute for Research into the Labour Market and Employment (Havers *et al*, 1983) showed that teachers who could not find work in a school were very often given employment in industry because of their pedagogical and didactic abilities. Their new employers appreciated their non-specialised qualifications, which enable them to get on well with people, as something very valuable for their companies.

We suggest that precisely this ability – compared in general with the 'typical mathematician' – is a feature of most teachers working in mathematics education departments. People who have above-average linguistic and communicative competence can, when cooperating with industrial mathematicians, be helpful in crucial situations such as the following.

> Before the beginning of a project, a contact has to be established and it must be made clear to prospective customers what the possibilities and limits of using mathematical methods would be.
>
> In the course of a project, provisional results and additional information necessary to the project processing have to be given.
>
> At the end of a project, the results have to be presented in a form understandable to the mathematical layman, for example the manager responsible.

As our research into the acceptance of mathematical technology has shown, all three points are by no means irrelevant to success.

6. NEW TASKS FOR UNIVERSITY MATHEMATICS EDUCATION DEPARTMENTS IN RESEARCH AND TRAINING

The task of doing research into the transfer of mathematical knowledge as a whole appears to us even more important than the possibility to work in cooperation with industrial mathematicians. Such research demands not only specialised knowledge of mathematics and its application, but also an awareness of the methods of social science and empirical methods, as well as contact with industrial mathematics. This combination cannot be expected of either mathematicians or of social scientists. Even if not all people employed in mathematics education departments have the relevant abilities and experience at their command at present, we believe that our branch of science is the most likely way to be able to fulfil all the necessary conditions.

Another research project is in further education in mathematics. The question here is essentially 'how do adults learn mathematics?'. In view of the growing importance of further education, mathematics teaching is called upon to develop specific methodological concepts for the various levels and to evaluate their application.

In view of the tasks of university mathematics departments in education, certain consequences result. At present there are insufficient didactically sound possibilities for education and further education for those who are, or would like to be, active in vocational adult education.

Courses for industrial mathematicians who want to learn something about the didactic components of their work would also be conceivable.

In view of the training of future teachers, the university can gain a great deal if, through personal contacts in industry, it can get a deeper insight into the real life application of mathematics.

REFERENCES

Bauersfeld H. (1988). Quo Vadis? Zu den Perspektiven der Fachdidaktik. In *mathematica didactica* 2, 3–24.

Havers N, Parmentier K and Stooss F. (1983). *Alternative Einsatzfelder für Lehrer? Eine Bestandsaufnahme zur aktuellen Diskussion.* Beiträge zur Arbeitsmarkt- und Berufsforschung 73, Institut für Arbeitsmarkt- und Berufsforschung der Bundesanstalt für Arbeit. Nürnberg.

Hülsmann H. (1985). *Die technologische Formation.* Berlin.

Maass J. (1988). *Mathematik als soziales System.* Weinheim.

Maass J and Schlöglmann W (eds). (1989). *Mathematik als Technologie?* Wechselwirkungen zwischen Mathematik, Neuen Technologien, Aus- und Weiterbildung. Weinheim.

Maass J and Schlöglmann W. (1988). The mathematical world in the black box – significance of the black box as a medium of mathematizing. In *Cybernetics and Systems, An International Journal*, 19, 295–309.

Maass J, and Schulz-Reese M. (1989). Wissenschaftliche mathematische Weiterbildung als Technologietransfer. In *Mathematik als Technologie?* Wechselwirkungen zwischen Mathematik, Neuen Technologien, Aus- und Weiterbildung.

Maass J and Rahmann G. (1985). Polyvalenz und Flexibilität – Neue Perspektiven für die Lehrerausbildung? In Schaper R (ed), (1985), *Hochschuldidaktik der Mathematik*, Alsbach.

CHAPTER 10

Future Trends in Mathematical Modelling and Applications

MS Arora
University of Bahrain, Isa Town, Bahrain
and
A Rogerson
MISP, Hawthorn, Australia

SUMMARY

A working definition of mathematical modelling is given. A few remarks on historical development are made. Despite the tremendous changes in the mathematics curricula in the last three decades of the present century, why is there so much dissatisfaction with the mathematics graduates that we produce? The present state of the art in mathematical modelling is examined. The impact of the computer and information technology is studied and future trends in mathematical modelling and its applications are outlined, these trends being essentially general in nature.

1. INTRODUCTION

The word 'model' is often used in different meanings in different contexts. In common parlance, it usually means 'a small object, built to scale, to represent some existing, or yet to be produced, object', such as scale models of an airplane or a building. All the different usages, however, have one underlying feature in common, namely approximation to some *'real'* situation.

What then is a 'mathematical model'? It is the representation or transformation of a real situation into mathematical terms, in order to understand more precisely, analyse and possibly predict therefrom? Finally, 'mathematical modelling' is the art and exercise of building and working with mathematical models. This, of course, is no simple task.

We do not intend here to detail the methodology of mathematical modelling. There is an abundance of literature on the subject, for example Bender (1978), Cross and Moscardini (1985), Maki and Thompson (1973), Meyer (1985), Saati and Alexander (1981).

Certainly, mathematical models have been around since mathematics itself. We remind the reader of the blind shepherd who sat at the mouth of the cave in the morning and put a pebble in his pouch for each sheep that he let out of the cave and, in the evening, took a pebble for each sheep that he let in the cave.

Of importance in the historical perspective is the unfolding of the last part of the seventeenth century which saw the epoch-making invention of calculus by Isaac Newton and Leibniz, working independently. In subsequent years, researchers employed this exciting discovery of calculus and other new-found ideas in the creation of mathematical models to study natural phenomena, for example, the gravitational theory by Isaac Newton, the theories of electromagnetic waves by Maxwell and relativity by Einstein.

The technological revolution brought on by the invention of computers, various social and economic factors, and the wars of the twentieth century, has resulted in a division of mathematics into pure and applied. Loosely speaking, pure mathematics is mathematics that is studied for its own sake, while applied mathematics is concerned with immediate practical uses of mathematics.

The twentieth century also saw the birth of the so-called 'modern mathematics' or the 'new mathematics' movement, spanning approximately two decades of the post-sputnik era, with its pros and cons and all-encompassing debates: the back-to-basics cry of the seventies, the explosion of information technology in the eighties, and its impact on educational curricula in general and mathematics curricula in particular. They are documented in research literature all over the world and we shall not take time to recapitulate or discuss them here. What concerns us is, and we quote from Arora (to be published), "... despite these tremendous and far-reaching changes in mathematics education, why is there a growing dissatisfaction all around the world with the mathematical competency of the high school graduates that we produce? *Why are the literates from school often so mathematically illiterate?*" Any serious thinker in mathematics education today, anywhere, in any culture, is confronted with these fundamental questions.

One of the outcomes - ironically, an indirect one - of the revision of the mathematics curricula in many countries has been the development of a strong interest in applications of mathematics to real-life and useful situations. This emphasis on establishing a link between the real-world and mathematics has led to the introduction of mathematical modelling as an important component of mathematics curricula, both at school and university levels. Admittedly, the modelling component is still relatively small, particularly at the school level. All the same, it provides an important extension to the range of application problems available. Several important issues are involved in introducing modelling as an integral part of the mathematics curriculum at different levels of school and higher education. They are discussed in Burkhardt (1989).

2. THE PRESENT STATE OF THE ART IN MATHEMATICAL MODELLING AND APPLICATIONS

Until the 1920s, mathematics at the secondary stage consisted mainly of teaching arithmetic, algebra, Euclidean geometry and possibly some eighteenth centry calculus. University courses in mathematics were built on these prerequisites and this pattern was followed closely all over the world.

The impact of the profound and far reaching changes in the mathematics of the nineteenth century - the introduction and insistence on rigour in mathematical thinking, the unfolding of the new areas of research subsequent to Cantor's work, the division of mathematics into pure and applied, the heralding of the post-sputnik era, the birth of modern mathematics and other developments - generated a new, rich legacy of mathematics which became reflected in the revision of curricula, both at school and university levels. Probability, statistics, sets, logic, transformation geometry, and so on, filtered down to school mathematics. Schools of pure and applied mathematics came into existence and distinct undergraduate and graduate degrees in the two fields evolved. Traditional courses in statics, dynamics, hydrostatics, attraction and potential gave way to rigourous courses in analysis and abstract algebra, which became the norm.

Traditionally, applications in mathematics have been either examples of applied mathematics, from the fields of Newtonian mechanics of particles, or irrelevant problems of the type, "... a train leaves station A at 10.00 hours with a speed of 80 km per hour and simultaneously another train from the opposite direction leaves station B at ..." or "... if John can do a job in 6 hours and Tom can do the same job in 9 hours ...". This is beginning to change. Real-life situation models are being developed and introduced in the classroom. Emphasis in the classroom is shifting from problem-doing to problem-solving. The National Council of Teachers of Mathematics (NCTM) in the USA, in its Agenda for Action (1980) highlights in Recommendation 1 that "Problem solving should be the focus of school mathematics in the 1980s". In recommending actions, it asks that "Mathematics teachers should create classroom environments in which problem solving can flourish" and that "Appropriate curricular materials to teach problem solving should be developed for all grade levels".

The development of the electronic computer has made possible many new fields of applications of mathematics. The increasing availability of computers in the classroom has had a tremendous impact on the expectations of the computational skills required of all citizens and the type of application problems that can be introduced. In fact, the very meaning of the word solution has undergone a radical change and, with the availability of these new technologies, a more pragmatic, creative and open-ended approach to problem solving can now be adopted in the classroom. Without doubt, the access to computers and computer power, at low cost, has made possible the teaching of mathematical modelling and applications to real-life problems in the classroom and this will increasingly become so. The NCTM's Agenda for Action, in its Recommendation 3, states that "Mathematics programs must take full advantage of the power of calculators and computers at all grade levels".

This is not to say that mathematical modelling and applications have permeated and become an integral part of the school and the university level curricula. Far from it. However, the fact remains that, in the past fifteen years or so - partly as a reaction to the excessive formalism of the modern or new mathematics, and partly as a result of continuing developments in the applications of mathematics to other disciplines, especially geography, biology and the social and economic sciences - mathematical modelling and applications have come to be recognised as necessary and desirable to stimulate and revitalise the teaching and learning of mathematics at all levels. The relevance of teaching mathematical modelling is being emphasised by various mathematical educationists and voiced in national and international conferences and forums. Some of the projects and conferences that deserve mention are the Unified Sciences and Mathematics in Elementary Schools (USMES) project in the USA, the Spode Group in the UK, the International Congresses on Mathematics Education (ICME), the Bielefeld Meeting in 1978, the ICTMA-3 Conference in 1987 and the ICTMA-4 Conference in 1989.

3. FUTURE TRENDS IN MATHEMATICAL MODELLING AND APPLICATIONS

So, how do we view the future of mathematical modelling and its applications? We believe that more and more emphasis will be placed on problem solving and modelling in mathematics, to bring out its utility in day-to-day life. To quote Henry Pollak in his opening lecture in the Exeter Conference on Teaching and Applying Mathematical Modelling in

1983, "Society provides the time to teach mathematics in our schools every year. Why? Not because mathematics is beautiful - which it is - or because it provides great training for the mind, but because it is so useful". We submit that mathematics is only useful to the extent to which it can be applied to a particular problem. The increasing relevance of mathematics and mathematics education to society and culture will be reflected in problem solving in mathematics. Modelling exercises and their applications will become more *thematic* and will relate more closely to the society and to particular cultural contexts.

We foresee changes in mathematical curricula to include, in an integrated manner, applications of mathematics to real-life or real-world problems and, at the same time, maintain a balance between applications and fundamental concepts. This will require development of curricular materials, stressing problem solving and concept development simultaneously. It will also place greater demands on teachers in that they alone can create an environment in the classroom in which students, faced initially with an unresolved situation or problem, are encouraged to engage in lateral or divergent thinking, using this approach throughout the modelling exercise; not being afraid to take risks or to try out new ideas or to explore apparently unconnected stages in what we call the modelling search for solutions. This, no doubt, raises some very pertinent issues in the training of the teacher to which we need to address ourselves.

Curriculum development in mathematics will take for granted and utilise the availability of calculators, computers and other computational aids. Electronic computers will become more powerful, faster, smaller and cheaper, and will influence the life pattern of every citizen. This will demand the design and development of new hardware, and new educational software which, in turn, will create versatile teaching and learning environments. New expectations in terms of computational skills required by society will be identified and this, in turn, will affect the nature of modelling situations that can be introduced in the classroom. However, we are of the view that the gap between the developing and the developed countries in computer and information technology will persist. The gap might narrow somewhat, but we cannot see it disappearing.

We envisage a distinct and conscious move away from teaching mathematical modelling as a subject towards developing skills and confidence in the student to model real-life situations - to doing mathematical modelling. Students must be encouraged to define the problems and objectives, to approach them with open minds, to recall and apply the mathematical concepts that they have learned, to transfer skills to new situations, to experiment and to reason.

We have only sketched above some general or nonspecific future trends in mathematical modelling. They point to a general direction as we foresee it. Specific examples of expectations and problems cannot be given within the scope of this paper. We have stated, for instance, that modelling exercises in the curriculum will become more thematic and culture oriented. Certainly, we cannot begin to furnish even a partial list of these themes. Similarly, we have pointed to the fact that there will be problems with the teachers unless we are cognisant of the issues in their training.

REFERENCES

An Agenda For Action. (1980). *Recommendations for School Mathematics of the 1980s.* The National Council of Teachers of Mathematics Inc, Virginia.

Arora MS (Ed). Mathematics Education into 21st Century. To be published.

Bender EA. (1978). *An Introduction to Mathematical Modelling.* John Wiley & Sons Inc, New York.

Burkhardt H. (1989). Mathematical Modelling in the Curriculum. *Applications and Modelling in Learning and Teaching Mathematics.* Blum W *et al* (Eds). Ellis Horwood, Chichester.

Cross M and Moscardini AO. (1985). *Learning the Art of Mathematical Modelling.* Ellis Horwood, Chichester.

Maki DP and Thompson M. (1973). *Mathematical Models and Applications.* Prentice-Hall Inc, New Jersey.

Meyer WJ. (1985). *Concepts of Mathematical Modelling.* McGraw Hill Book Co Inc, New York.

Saati T and Alexander JM. (1981). *Thinking with Models. Mathematical Models in the Physical, Biological and Social Sciences.* Pergamon Press, New York.

CHAPTER 11

The Importance of the Teaching of Mathematical Modelling in Bangladesh

H Banu
University of Dhaka, Bangladesh

SUMMARY

Students and teachers should be sufficiently inspired to undertake investigation of large-scale national problems through mathematical techniques and mathematical modelling. Mathematics is an efficient tool to assist in resolving complex problems such as population growth, flood, storms, epidemics and so on, which affect the day to day life of inhabitants of a country. Emphasis is placed on the need for a comprehensive teaching programme in applicable mathematics to tackle the most challenging problems of the developing countries.

1. INTRODUCTION

Mathematical modelling is an attempt to describe some part of the real world in mathematical terms. Mathematics, a man-made universe, has been created to solve problems of day-to-day life, of business, industry and commerce. The essence of mathematics teaching is mathematising, mathematical modelling and understanding situations mathematically.

Mathematical modelling can avoid or reduce the need for costly, undesirable or impossible experiments in the real world. Instead of dealing with a tower or a river or a human body, we have to deal with mathematical equations. When we deal with the problems of optimisation, there is no alternative to mathematical modelling in this case.

Underdeveloped countries like Bangladesh have suffered from enormous deterioration in the standards of mathematics instruction at all levels. They do not have properly trained teachers or mathematicians who can accept the challenge of the problems of science, technology and society waiting for mathematical modelling and mathematical solutions.

2. APPLICATIONS OF MATHEMATICAL MODELLING

Mathematical models are applied in describing the environment, including models for growth of population, for interaction of species, for quantitative description of pollution in air and water, and also for epidemics and storm prediction. The coastal region of Bangladesh is frequently exposed to severe cyclones which cause heavy loss of life. A storm surge in the Bay of Bengal during a December 1981 cyclone has been simulated with the help of a mathematical model based on numerical solution of shallow-water equations in two-dimensional cases. Accurate determination of variation in surge height with respect to time and distance is necessary to provide information for the planning and design of engineering works such as embankments, cross dams, ports and bank protection.

Mathematical models are now widely used in water resources planning in many developing countries. In Bangladesh a few modelling projects have also been initiated for flood forecasting, flood control, drainage and irrigation schemes. A mathematical study using the model of the Danish Hydraulic Institute was made to assess the impact of flood control polders upon flood level of the Atrai river in Bangladesh. However, due to the limitations of flood time hydromorphological data, the models are not very satisfactory.

3. COMPREHENSIVE TEACHING PROGRAMMES

The main objectives of teaching mathematics in developing countries like Bangladesh are as follows.

- To increase the students' ability and skills in mental calculations and estimations and applying the rules of calculations to the practical problems faced in daily life.
- To encourage the students to develop mathematical models depending on the requirements of the country, by making complete use of the local resources.

Students in our country indirectly learn and apply mathematical modelling, but this has not been introduced as a subject so far. They use geometrical models, structures and graphs depending on mathematical problems. It is hoped that the government will order the inclusion of computing in the curriculum of the upper secondary level from 1990. It is therefore hoped that students can work on simple discrete mathematical modelling projects, solving equations numerically on the computer.

Although there are no formal courses in mathematical modelling either at the undergraduate level or at the postgraduate level in the universities of Bangladesh, we have already offered a course on mathematical modelling in our University of Dhaka at the end of the second year undergraduate classes. Also projects are offered on mathematical modelling at the postgraduate research level. This article proposes a comprehensive teaching plan to carry out modelling-aided teaching of mathematics at different levels in the educational institutions of developing countries like Bangladesh. An attempt has been made to envisage a modified approach to the teaching and utility of applicable mathematics and mathematical modelling, with special relevance to the growth of applied sciences in developing countries. The teaching of mathematics may be split up in the following ways at different educational levels.

- At school or secondary level the first plan is the direct teaching of basic topics like sets, algebra, geometry and so on. The course should start with the quantitative study of the field problems (not imported but obtained locally) involving mathematical modelling. Also concepts in real analysis through periodicity, continuity and discreteness in various problems should be brought in.

- At college or undergraduate level, the application-oriented mathematics courses should appear as a part of each discipline. Mathematical modelling and computer applications have become part of research in many new areas of science and the humanities.

A computer at this stage will be of tremendous use in teaching the final year of undergraduate level. The student can take courses on mathematical modelling and numerical analysis and can undertake simple projects on biosciences and so on.

- At the postgraduate level, apart from the traditional courses there should be an advanced course on mathematical modelling and applications, followed by a dissertation on the practical problems of the country.

In the university of engineering and technology there are projects on mathematical models, solving problems like flood control, storm prediction and other important local problems of the country. Mathematical modelling is primarily a research activity. The best method of learning and appreciating mathematics is via research, that is by the students investigating problems which are new for them, though they may not be new to others. Every student and every teacher has to be research oriented and the most important component of a student's training should be mathematical modelling of the new and unfamiliar situation of the country.

4. CONCLUSIONS

The experience of mathematical modelling should start in the primary classes and continue throughout the school, college and university education. We should investigate what types of mathematical modelling are in progress in neighbouring countries, in other developing countries having problems of a similar kind. By comparison of the results we might develop much better models to solve the problems of developing countries such as Bangladesh.

REFERENCES

Bangladesh Water Development Board. (1987). Flood in Bangladesh.

Daa PK and Sinha MC. (1974). Storm Surges in the Bay of Bengal. *Quart JR Mat Soc*, **100**.

Danish Hydraulic Institute Report. (1988). Development of a Surface Water Simulation Model for the Second Coastal Embankment Rehabilitation Project. Bangladesh.

Haberman R. (1977). *Mathematical Models*. Prentice-Hall, New Jersey.

Storm Warning Centre. Depression and Cyclonic Storms of the Bay of Bengal in 1981. BND Report, Agargaon, Dhaka.

CHAPTER 12

The Role of Specification Languages in Model Formulation

A Norcliffe
Sheffield City Polytechnic, UK

SUMMARY

Mathematically-based specification languages, such as Z, now offer the software engineer a real alternative to English as a specification medium. In this paper we investigate what languages, such as Z, can offer the mathematical modeller, and report on our experiences of introducing mathematicians to Z in modelling classes at Sheffield City Polytechnic.

1. INTRODUCTION

Mathematical modelling has undergone a revolution with the advent of computing. Not only is there the raw computing power of the super computer, to assist with numerical work, and enhanced computer graphics, to see what computations mean, but there are now sophisticated computer algebra packages to help with analytical work. With the availability of all this computing power, the ability to formulate a problem properly is becoming increasingly important. Unfortunately, most mathematicians still tend to concentrate on developing methods of solution, once problems have been formulated, instead of developing these vital skills of problem formulation.

Software engineers, to their credit, have realised the central importance of specification in problem solving and languages, such as Z, now exist to assist in the production of correct coding. At present, however, Z tends only to be used by software engineers and programmers. Few mathematicians, if any, really know about such languages, or of their potential for developing good problem formulation skills.

In this paper we aim to show that specification languages have a valuable role to play in mathematical modelling work in developing these skills, and in helping to put the model formulation stage on a more rigorous footing. We also wish to report, briefly, on our experience of using Z in modelling classes at Sheffield City Polytechnic, where Z has been used in a wider context than just software engineering.

2. WHAT IS Z?

Z is a notation based on set theory and predicate logic. It was developed by the Programming Research Group in Oxford and is fast becoming an industry-standard specification language. At the heart of Z is the structure known as a *schema,* in which the mathematics of what is being specified is set out.

The best way of explaining Z is to see it in action. Suppose we wish to develop a simple software system that is to be part of a security system for a building, and suppose the system must be able to tell us who the valid users of the building are, and who is currently in or out of the building at any given time. A schema that specifies the state of the building can be written in Z as follows.

```
┌─State──────────────────────────
│ users, in, out: ℙ(Person)
├────────────────────
│ in ∩ out = { }
│ in ∪ out = users
└────────────────────────────────
```

All schemas have a name, ours is called State. The section above the middle line is called the signature and introduces the relevant variables and identifiers. Thus users, in and out are introduced as subsets of the set Person and are the state variables in our system.

The text below the middle line is the predicate. Each line is conjoined with the rest of the predicate using the logical AND connective. The predicate specifies how variables, introduced in the signature, are related. In our system we require that no person is both in and out of the building at the same time, and that those inside, together with those outside, make up the valid users. This predicate forms a data invariant, and any operation that might change the current state of the building must not violate this invariant. When a valid user, for example, is checked out of the building the state of the building changes. The check out operation, that preserves the data invariant, can be written as a schema in Z as follows.

```
┌─CheckOut──────────────────
│ Δ State
│ user? : Person
├─────────────────
│ user? ∈ in
│ in' = in \ {user?}
│ out' = out ∪ {user?}
└───────────────────────────
```

The Δ notation that appears introduces all the variables into CheckOut that are in State, together with decorated versions of these. Decorated state variables have a dash after their names. The undecorated and decorated variables represent the before and after states of the CheckOut operation. The input to Checkout is user?. Inputs to operations carry a query mark after their name; outputs carry an exclamation mark.

For CheckOut to work, user? must have previously checked into the building, hence the requirement user? ∈ in. When user? is checked out then user? is taken out of the subset in and added to out. This is specified via the difference and union operations indicated in the schema.

Other operations that change the state of the system involve checking in, adding, and removing users. As above, these can be specified as schemas. Similarly, schemas that simply supply information about who is in or out of the building can be written, and in a piecemeal way the full system can be built.

3. WHAT CAN Z OFFER THE MATHEMATICAL MODELLER?

If we look at the Z we have written, we see that we have built a precise and clear model of what the software is required to do. Everything we have written down is amenable to formal reasoning. We can be critical about what we have written and use Z as a basis for communication. We have formulated in a rigorous and structured way a mathematical model that can then usefully provide the solution to our problem – that is, the coding for the system. This approach to software development is well documented and holds out the promise of better software quality, and improved productivity in the software industry.

So, let us now look at a quote from an influential report (McLone 1973) relating to the training of mathematicians.

> "Good at solving problems, not so good at formulating them, the graduate has a reasonable knowledge of literature and technique, he has some ingenuity and is capable of seeking out further knowledge. On the other hand, the graduate is not good at planning his work,

nor of making a critical evaluation of it when completed; and in any event he has to keep his work to himself as he has apparently little idea of how to communicate it to others"

Today, on our degree courses, we would argue that these criticisms now receive more attention than they used to. But if we are honest, we have to admit that we still tend to neglect vital issues of model formulation and specification in our teaching. Perhaps specification languages do have an important role to play in the training of mathematicians.

In the next section we give some examples of the use of Z, in a wider context than just software engineering, to show that Z has a role to play in developing problem formulation skills. We have deliberately kept the examples simple; specifying more complicated systems should present few additional problems.

4. WIDER USES OF Z: SPECIFYING PHYSICAL SYSTEMS

Let us look at the problem of particle dynamics. The state of a particle of mass m might be specified in Z as follows.

```
┌─State──────────────────────
│ m : M
│ a: T ⇸ LT⁻²
│ v: T ⇸ LT⁻¹
│ r: T ⇸ L
├────────────────────────────
│ ∀t : T | t ≥ t₀ •
│ d/dt v(t) = a(t)
│ d/dt r(t) = v(t)
└────────────────────────────
```

$$m : M$$
$$\underset{\sim}{a} : T \nrightarrow \underset{\sim}{L}T^{-2}$$
$$\underset{\sim}{v} : T \nrightarrow \underset{\sim}{L}T^{-1}$$
$$\underset{\sim}{r} : T \nrightarrow \underset{\sim}{L}$$
$$\forall t : T \mid t \geq t_0 \bullet$$
$$\frac{d}{dt}\underset{\sim}{v}(t) = \underset{\sim}{a}(t)$$
$$\frac{d}{dt}\underset{\sim}{r}(t) = \underset{\sim}{v}(t)$$

In our schema M is just the set of masses, and m is being declared a member of this set. The particle's acceleration, $\underset{\sim}{a}$, its velocity, $\underset{\sim}{v}$, and its position, $\underset{\sim}{r}$, are all being modelled as functions of time. T is the set of possible times, $\underset{\sim}{L}$ the set of vector lengths, and so on. In the predicate we are defining a data invariant that must hold, whatever happens, for all times t after the particle came into existence at time t_0.

If a force $\underset{\sim}{F}$? is now applied, at time t?, to the particle, then the following ApplyForce schema will specify the ensuing motion. The state of the particle remains unaltered for times t less than t? but, for times greater than or equal to t?, Newton's second law modifies the acceleration accordingly.

```
┌─ApplyForce──────────────────────────────────
│ Δ State
│ F? : T ⇸ MLT⁻²
│ t? : T
├──────────────────────────────
│ t? > t0
│ dom(F?) = { t | t ∈ T ∧ t > t? }
│ ∀t : T | t ≥ t0•
│ (t<t?) ⇒ a'(t) = a(t) ∧ m' = m
│          v'(t) = v(t) ∧ r'(t) = r(t)
│ (t≥t?) ⇒ a'(t) = F?(t)/m' ∧ m' = m
│          v'(t?) = v(t?) ∧ r'(t?) = r(t?)
└─────────────────────────────────────────────
```

Various other operations could be specified, such as a collision with another particle, but we will leave this to the interested reader to pursue.

Indeed, many physical systems can easily be formulated in Z. A system consisting of liquid in a storage tank, for example, can be described in a schema, and then the operations of filling the tank, and emptying it, specified. Other interesting dynamical systems can also be considered and a whole host of steady-state problems, such as heat mass transfer problems, exist that lend themselves naturally to being specified in Z. Unfortunately it is not possible to consider more problems here. Suffice it to say that the list is as long as we wish to make it.

5. COMMENTS

Positive features which Z promotes, that are vital in modelling work, can be seen to be the following.

- In schemas all variables have to be clearly specified. How often, in modelling work, do we fail to do this properly?

- In the predicate the precise connection between these variables is set out along with assumptions being made. In modelling it is easy to make assumptions and then overlook them. In Z we cannot do this because the schema would simply not work. Schemas can be systematically checked for consistency, using precondition analysis, which then automatically brings to light any overlooked assumptions.

- Having to write a specification forces us to separate out the formulation and solution stages of a problem. Too often we write down equations and procede to solve them before realising that key constaints have been violated.

- Writing schemas helps to break the problem into manageable parts and, in turn, this structures our thoughts about how the problem should then be solved.

- Finally, schemas provide a definite focus of interest, which helps concentrate the mind when communicating ideas to peer mathematicians or to the lay person.

6. EXPERIENCES WITH Z AT SHEFFIELD CITY POLYTECHNIC

Z is taught on a range of courses at the Polytechnic. Degree and postgraduate students in Computer Studies receive formal instruction in Z because of its importance in software engineering. Students on the Computing Mathematics degree also learn Z for the same reason but, additionally, they study Z in modelling classes for the specific purpose of developing key problem formulation skills .

In modelling workshops students tackle a range of problems from specifying software systems for banking, library use, security systems, electronic mail, and suchlike, to describing how vending machines and other physical systems work, to formulating logic puzzles for solution in PROLOG. Students work in groups and give presentations where they defend the specifications they have written or the models they have formulated. Word-processed reports, aimed at the lay person in some instances, are produced and assessed. By and large they enjoy this activity. They see it as different, creative and exciting, and a welcome change from mundane problem solving exercises that, all too frequently, they are given.

Observations, based on our experience of operating these kinds of workshops in the past few years, are the following.

- Students are happiest creating and writing specifications. Reasoning, formally, about what they have written, is difficult for them. Proof ideas tend to come only slowly.

- Students prefer to specify systems where they are in charge of defining the rules. They are happiest modelling libraries, banks etc, as opposed to physical systems where additional knowledge of physics or mechanics is needed.

- Their ability to communicate about specifications seems to be enhanced. The structure of the specification seems to help structure the reports that are produced.

- It is surprising how little discrete mathematics is needed before students can begin to write Z specifications. There is no reason why first year students, in their second or third terms, should not make a start on specifying simple systems such as vending machines or particle dynamics, based only on a knowledge of sets, functions, and logic.

- Finally, as we have said, students enjoy writing specifications, working in groups and using each other as sounding boards. They all agree that Z helps focus the mind on the essential features of the problem and helps develop a disciplined approach to problem formulation.

In conclusion, then, we feel that Z does have a valuable role to play in modelling and the training of mathematicians, and should not remain an approach to problem formulation to which only software engineers are exposed.

REFERENCES

McLone R. (1973). The Training of Mathematicians. SSRC Report.

CHAPTER 13

The Role of Applications in a Curriculum Project for School Mathematics

P Abrantes
University of Lisbon, Portugal

SUMMARY

Applications of mathematics were extensively used in two seventh grade classes in relation to a variety of students' activities, including practical and project work. Attention paid to extra-mathematical aspects and to the process of using mathematics in real situations seems to contribute to increased motivation and also to appreciation of mathematics and the understanding of its role. For this purpose, conditions related to time, assessment methods, and the teacher's role, tend to be highly beneficial.

1. INTRODUCTION

The MAT789 Project is developing an experimental curriculum for school mathematics in grades 7–9 (ages 12–15) working with two classes throughout the period 1988–91, and with two other classes throughout 1989–92. The Project Team is composed of two researchers and three mathematics teachers from the two secondary schools involved. The Project also has an Advisory Committee and institutional support from the University. Although independent, it is officially authorised by the Ministry of Education which is a necessary condition for its development, since Portugal has a centralised school structure with rigid national curricula and programmes for all the disciplines at elementary and secondary levels.

Grades 7–9 correspond to a critical school level in Portugal because many students actually leave the school as a result of successive failures in one of those grades - and mathematics strongly contributes to that situation - and it was recently decided to extend general and compulsory education from grade 6 to grade 9.

2. PRINCIPLES AND GOALS OF THE EXPERIMENTAL CURRICULUM

The MAT789 Project sees school mathematics as a positive, personal and cultural experience, centred in processes and not in content topics, relevant in itself and not only because of its utility for future studies. The richness and variety of learning experiences and students' activities (as recommended in several recent documents, for example NCTM, 1989), the consideration of individual and social interests and motivations, and the need to develop adequate methods of assessment and evaluation, are major and permanent concerns.

The experimental curriculum emphasises problem solving and the relations between mathematics and reality. The use of manipulative materials, calculators and computers are also important aspects of its guidelines.

The main educational goals of the curriculum are the development of positive attitudes towards mathematics and the contribution to a better appreciation of mathematics and its role for individuals and for society. These goals try to integrate capacities, attitudes and habits: curiosity, sense of criticism, pleasure in formulating and solving problems, habits of organising, communicating and discussing reasoning processes, creativity and intellectual self-confidence.

3. THEMES AND ACTIVITIES

Concerning grade 7 (where the experiment has been taking place since September 1988) we could say, mathematically speaking, that content came from areas such as number, combinatorics, statistics, graphs, functions and geometry. However, problem situations were always not only the starting point but actually the core of the study.

Various kinds of activities were organised inside and outside the classroom, providing opportunities both for individual and team work. The activities generally involved problem solving, exploration and discussion of situations, practical work or project work.

The main written products of students' work are comments or reports about situations or projects - with descriptions, representations, conclusions. These products were used to evaluate students' progress, together with two-stage written tests similar to those developed by the Hewet Project (De Lange, 1987).

The schedule below gives an overview of the themes, activities and projects developed throughout the school year.

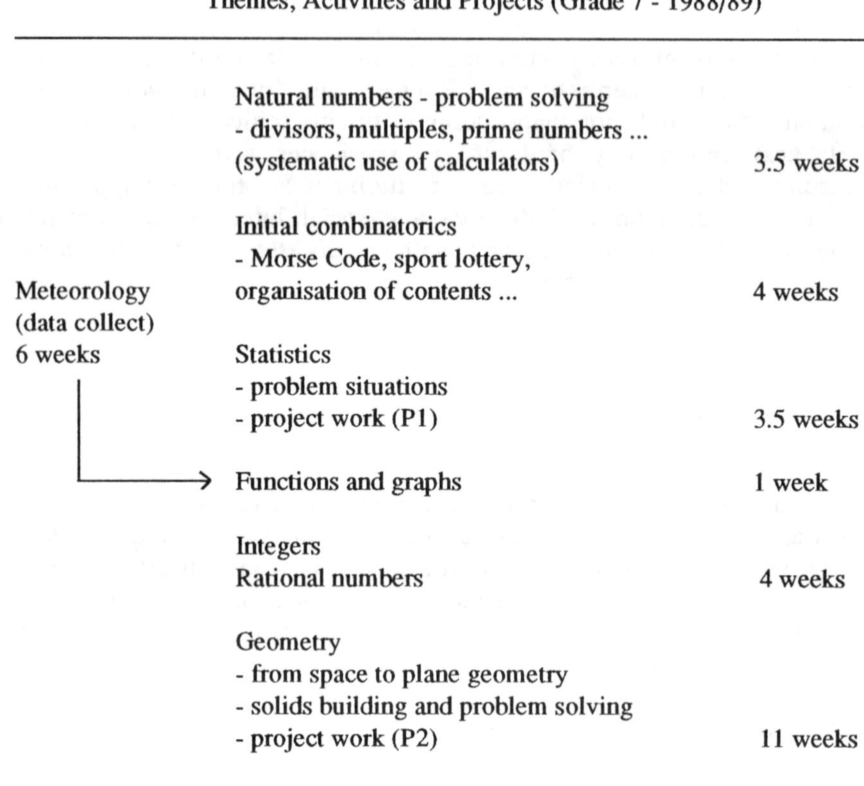

Themes, Activities and Projects (Grade 7 - 1988/89)

	Theme	Duration
	Natural numbers - problem solving - divisors, multiples, prime numbers ... (systematic use of calculators)	3.5 weeks
Meteorology (data collect) 6 weeks	Initial combinatorics - Morse Code, sport lottery, organisation of contents ...	4 weeks
	Statistics - problem situations - project work (P1)	3.5 weeks
→	Functions and graphs	1 week
	Integers Rational numbers	4 weeks
	Geometry - from space to plane geometry - solids building and problem solving - project work (P2)	11 weeks

4. ACTIVITIES INVOLVING APPLICATIONS AND MODELLING

The units Combinatorics, Statistics, Functions and also the last part of Geometry, consisted entirely of application-oriented activities (which were also present in other units, for example in the use of some properties of the natural numbers or in the introduction of the integers). This is quite an unusual situation in Portugal.

Applications as well as modelling, especially the stages corresponding to mathematisation (as defined by Niss, 1989), were in fact a major source of activities in the experimental curriculum. Working with applications and modelling was considered a necessary way to develop appreciation of mathematics, understanding of its role in society and competence, confidence and sense of criticism about its use.

Here are examples of activities developed in 1988/89 by the students of the two seventh grade classes.

(1) Study of communication systems with arms and flags, and study of the Morse code, including both mathematical and historical aspects. Beginning with concrete experiences and with the support of written materials, the activities included the use of computers to code/decode words in Morse.

(2) Study of the Portuguese lottery system based on the scores of football matches (The *Toto*), namely the ways of filling in the official bulletins with the 'multiples' systems. The so-called 'economic systems', as well as processes of determining what can be earned, became too difficult for the students and therefore they were limited to slight references and left to one of the next school years.

(3) Planning calendars for different kinds of sports competitions (championships with one or two rounds, and so on).

(4) Comments on the systems of registration marks used to identify vehicles in several countries (car number plates), their evolution and perspectives. This issue was included as an open question in a two-stage test (with much information and calling for an opinion on the Portuguese situation) and was retaken three months later in the form of a comment about a letter published in a magazine (focussing on the same problem and containing some errors).

(5) Reading and graphing meteorological data involving temperatures and precipitation. This activity lasted for about six weeks, and involved practical work mainly outside the classroom.

(6) Exploration of simulated and simplified problematic situations (workers' salaries, scores on students' tests) requiring the use, and criticism of the misuse of concepts and procedures from elementary descriptive statistics.

(7) Investigation of the evolution of the number of children born in the last three generations of students' families. This study was organised as a three-week project developed inside and outside the classroom, and included the presentation of final written reports.

(8) Elaboration of a proposal to adapt ground existing in the school for the practice of several sports. The work was organised as a four-week project and included measuring, searching and consulting information sources and people, preparing a formal proposal (a text and maps in adequate scales), and finally meeting with the school

administration.

5. PRACTICAL WORK AND PROJECT WORK

Considering both the kind of problems studied and the nature of students' work, two of the above examples merit a more elaborate description.

The activity with meteorological data – example (5) – gave the students an opportunity to work with real instruments and data and to have a contact with the way professionals work in a particular domain. Over six weeks, small groups of students had different tasks: collecting data about temperatures (minimum and maximum at given times) or precipitation, using the meteorological station existing in the school; finding related information about Lisbon and other places in the world by consulting daily newspapers, and so on. In the meantime, a visit to the Meteorological Office of Lisbon was organised: students could see instruments, hear explanations and ask questions. Finally, the work came into the classroom: data was presented in graphs and discussions took place.

During the period of practical work, many problems emerged concerning the use of the instruments or the meaning of the information (How to read the minimum temperature in a day? Why precipitation is usually given in millimetres?). Students always showed considerable motivation and an increasing sense of responsibility (working in the breaks between lessons, keeping the key of the station, and suchlike). It is true that many of their results were not plausible but, in the final stage of the work, comparing data and discussing why we could not believe in some values was at least as important as learning how to draw cartesian graphs.

The study concerning the evolution of the number of children born in the last three generations of students' families – example (7) – gave the students their first experience of project work in mathematics. The objective of the study was discussed in the classroom as well as appropriate methods of collecting relevant data and developing the work. Obviously the next step was individual, and involved students' families. After that phase, students worked in the classroom, in small groups, trying to use their previous knowledge about statistics – see example (6) – to represent numeric data properly and to draw conclusions. Individual reports would then be written as homework. How to organise such a report had to be stressed in the classroom and, during two weeks, students often asked for the teacher's opinions about some particular points of the report. In the meantime, official data about the evolution of the number of children born in Portugal and other countries was available for discussion. At the end, some reports were of high quality, with a lot of creativity both in form and content, and showed a high personal commitment from many students. Moreover, there is evidence that, in some cases, the theme of the work was discussed not only at school but also at home.

6. THE ROLE OF AND APPROACHES TO APPLICATIONS

It has already been mentioned that, because of Portuguese tradition, applications are not usually given much time and attention in mathematics lessons. However, the innovative role of applications in the experimental curriculum is related primarily to the independent objectives of the corresponding work. In fact, applications were not used to provide a kind of external motivation to introduce mathematical topics or some pleasant exercises to practise previously learned techniques. Instead, the main arguments for their inclusion in the curriculum were (using the terms of Niss, 1989) to develop critical potential towards the use and misuse of mathematics in extra-mathematical contexts, to prepare students to practise applications and modelling, and to establish a balanced picture of mathematics which must include all its aspects, applications being one of them.

Taking into account the diverse reasons for their inclusion in the curriculum, the role of applications in the different activities varied. The following perspectives can be identified.

(a) Understanding/discussing
- (i) things that really exist or existed – [1], [3]
- (ii) daily life situations often misunderstood – [2], [4]

(b) Acquiring practical knowledge
- (i) by working like professionals – [5]
- (ii) from simulated situation – [6]

(c) Acting in new situations
- (i) investigating real (simplified) situations, applying concepts and procedures – [7]
- (ii) developing an original and actually useful project (with openly non-mathematical purposes) – [8]

The approaches used to develop the eight activities were already suggested, together with the short description of each. The table below attempts to summarise this.

	problem situations	practical work	project work
simulated materials	[3]a [6]a		
real (simplified) materials	[1]a		[7]c
real materials	[2]a [4]c	[5]b	[8]c

Letters *a*, *b* and *c* refer to the nature of the main things students should produce in each situation:

a – answers to problems and questions
b – practical tasks
c – larger written productions (comments, reports, and so on)

7. SOME PRELIMINARY RESULTS

A large range of instruments was used to evaluate the work: questionnaires to students and parents, registration of events, produced materials and tests, and individual interviews of the 50 students at the end of the year.

The emphasis on applications seemed to fit well with the main goals of the curriculum although it should be noted that we are dealing with long-term goals.

Working a lot with applications had a very favourable response from most students. At the same time, some of them showed a considerable (and maybe unexpected) capacity for reflecting on the style of the discipline as a whole. In particular (as stated by Pollak, 1979), the importance given to applications did not undervalue understanding in relation to techniques, on the contrary it seemed to reinforce problem-solving and reasoning aspects.

> "This is a different mathematics ... one has to think a lot to come to a conclusion, one explores ..."
> "There was always a problem in everything, but it was not solved in a straightforward way as last year ... there were other things, it was more involved ..."
> "We think that the question is only to write but we begin and we see that there is more to say"

There was not a single student opposing the mix of pure with applied mathematics and some even made curious associations.

> "I liked most the *Toto* work and the Geometry because of the *mentalities* we had to use"

The attention paid to extra-mathematical aspects of real problems was surprising for the students in the beginning ("Read temperatures in mathematics ... isn't it geography?") but it often became quite motivating. However, the deepest innovative point goes beyond motivation. Students should be inside the process of using mathematics in a real situation, therefore we tried not to jump over any stage of that process. This may have contributed to changing some students' beliefs: computations may be exact but the answers to a lot of questions are not a matter of certainty. (What kind of values are acceptable? What is to be done if a student is not sure about the values he/she brought to the classroom? How are we to obtain the real dimensions of the ground and what is the appropriate precision for them?)

In this sense, practical work and project work proved to be essential (not replaceable) educational approaches. However, our experience suggests that a combination of these and other approaches (namely simplified applicational activities in the classroom) may be a realistic and useful guideline.

The work with applications as described here cannot be separated from some important conditions not usual in our ordinary curricula. Firstly, the time that was given to students to work in the various situations. Secondly, the assessment and testing methods quite different in nature from the usual written tests and examinations. Thirdly, the role of the teachers as persons who organise activities and materials and give suggestions, instead of teaching how to do and checking whether or not answers are correct.

A last remark should be made about the inclusion in the mathematics curriculum of activities and projects involving applications. No doubt some of them could have been developed elsewhere, in another discipline or in a general area. However, the effects of combining them with other aspects of mathematics are not to be neglected, namely the contribution to a better understanding and appreciation of mathematics.

REFERENCES

De Lange J. (1987). *Mathematics Insight and Meaning.* OW & OC, Utrecht.

NCTM. (1989). *Curriculum and evaluation standards for school mathematics*. NCTM, Reston.

Niss M. (1989). Aims and scope of applications and modelling in mathematics curricula. In Blum W *et al* (eds). (1989). *Applications and Modelling in Learning and Teaching Mathematics.* Ellis Horwood, Chichester.

Pollak HO. (1979). The interaction between mathematics and other school subjects. In *New Trends in Mathematics Teaching IV*. Paris, UNESCO, 232–248.

CHAPTER 14

Mathematical Modelling For All Abilities

M Swan
Shell Centre for Mathematical Education, University of Nottingham, UK

SUMMARY

Students are rarely offered opportunities to deploy their mathematical knowledge in anything other than an imitative way on problems that are at best illustrative and artificial. This paper describes two rich learning environments which encourage students to pose and tackle their own problems in realistic contexts. Both technical and strategic skills are developed. The integration of mathematical techniques into real problem solving and assessment methods are discussed.

1. THE NUMERACY THROUGH PROBLEM SOLVING PROJECT

The Shell Centre for Mathematical Education and the Joint Matriculation Board have, since 1984, collaborated to develop classroom activities and an associated assessment scheme for 14 to 16 year old students under the title Numeracy Through Problem Solving. In this context, the term numerate is used in the sense described in the Cockcroft report (1982, para 39).

> "We would wish the word 'numerate' to imply the possession of two attributes. The first is an 'at-homeness' with numbers and an ability to make use of mathematical skills which enable an individual to cope with the practical mathematical demands of his everyday life. The second is an ability to have some appreciation and understanding of information which is presented in mathematical terms, for instance in graphs, charts or tables ... Taken together, these imply that a numerate person should be expected to be able

> to appreciate and understand some of the ways that mathematics can be used as a means of communication. Our concern is that those who set out to make pupils 'numerate' should pay attention to the wider aspects of numeracy and not be content merely to develop the skills of computation."

The recurring emphasis is on developing *understanding, appreciation* and *confidence* in order that mathematics may be *useable* in the everyday lives of students.

Traditionally, mathematics is presented in a topic-focussed manner. We structure our curriculum around topics, (matrices and vectors), illustrating the power of these ideas with a few applications drawn from the real world (figure 1). The danger is that topics become isolated from each other and mathematics becomes a fragmented and virtually unuseable subject. (One may imagine trying to teach carpentry in a similar manner. "This week we will learn how to saw pieces of wood correctly. Next week we will learn how to make neat dovetail joints. The following week we will learn how to use the lathe ...". Imagine the absurdity of this approach if the students are never encouraged to put their skills together to *make* something!) It is essential that times are set aside when *students* are required to *select* appropriate strategies and techniques from their existing repertoire and *integrate* these skills in the tackling of real problems. The periods of skill *acquisition* must be interspersed with periods of skill *deployment*, where the problem becomes the focus of attention, not the mathematics (figure 2).

Indeed, some of the skills that students deploy may be drawn from areas of the curriculum other than mathematics. Although this sometimes worries the mathematics teacher, it is essential to allow this if we want our students to be able to use mathematics autonomously in life outside the classroom. Often there is some dishonesty on the teacher's part. The students are told that the problem is the focus, but the teacher has in mind the real goal - a specific piece of mathematics that he or she wants the students to learn. Students are thus discouraged from attempting solutions which involve original or 'inappropriate' approaches which perhaps involve less mathematics. A genuine problem-centred approach requires considerable skill and courage on the part of the teacher.

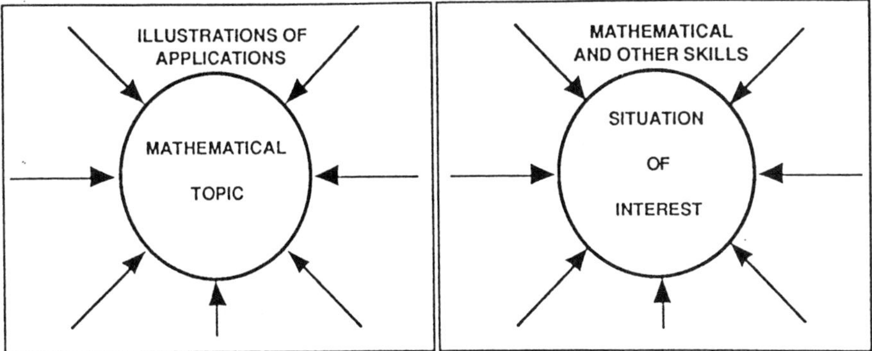

Figure 1
The traditional curriculum

Figure 2
Real Problem Solving

The Numeracy project has enabled many teachers to introduce a real problem-centred element into their mathematics lessons for the first time. The materials have been developed on a modular basis, with each module being designed to occupy between 10 and 20 hours. The modules provide themes within which the students take responsibility for planning, organising and designing. Students are able to choose which areas of mathematics to use and they live with the consequences of their own decisions. The teacher moves away from the more traditional explaining and task-setting roles towards those of an advisor, facilitator and encourager.

Five modules have been published.

- **Design a Board Game**, in which each group of students designs and produces an original board game which can then be played and evaluated by other members in the class. (Geometrical and probabilistic ideas predominate)

- **Produce a Quiz Show**, in which students devise, schedule, run and evaluate their own classroom quizzes. (Statistical and geometrical work)

- **Plan a Trip**, in which students plan and undertake a class trip. (Maps, timetables, costings, scheduling, surveys and everyday arithmetic)

- **Be a Paper Engineer**, in which students design, make and evaluate three-dimensional paper products such as pop-up cards and gift boxes. (Three-dimensional geometry and algebra)

Be a Shrewd Chooser, in which students conduct consumer research and produce consumer reports which inform people on how to make better choices. (Experimental design, surveys and statistics).

The early development of the classroom materials, the assessment scheme and the teacher support materials have been described in previous ICTMA papers (see Binns et al 1987, 1989; Gillespie et al 1987, 1989). This paper focusses on the most recent work of the project concerned with the two modules Be a Paper Engineer and Be a Shrewd Chooser. In particular, these two modules illustrate how the teachers are provided with a secure structure within which to work, while the students are still allowed to retain ownership of the problems being tackled.

2. THE BE A PAPER ENGINEER MODULE

Geometry at secondary level is almost always presented as a static, two-dimensional subject. On the few occasions when we leave this 'flatland' it is usually only to construct a few polyhedra or to perform some sterile, abstract, technical exercise; students are rarely given the opportunity to explore a rich, three-dimensional environment. There are, however, some indications that this situation is improving. Recently, in England, the National Criteria have specified that all examinations leading to a GCSE award in mathematics must assess practical work. Some examination boards are now encouraging students to present extended pieces of coursework which involve a practical geometry task, for example the designing of a cardboard box which will hold five tennis balls. (Even these tasks sometimes give the feeling that mathematics is being dragged in in an artificial way, for example when students are asked by their teachers to calculate the volume of air in the box after the tennis balls have been introduced.)

In the Be a Paper Engineer module we offer students the opportunity to design and make a product from paper or thin card and then produce a kit containing full instructions so that someone else can recreate it. They may then, perhaps, set up a small business enterprise based on the products. The design process is arranged into four stages.

1. Looking at examples

Before designing an original product, it is sensible to look at a few examples of existing products to stimulate ideas and become familiar with the techniques involved. In the module package, we provide a collection of 32 pop-up cards and gift boxes which may be photocopied and given to groups of students. After making these, students are asked to find more examples from home, and then reflect on structural differences using a classification game. Figure 3 illustrates the kind of products we provide.

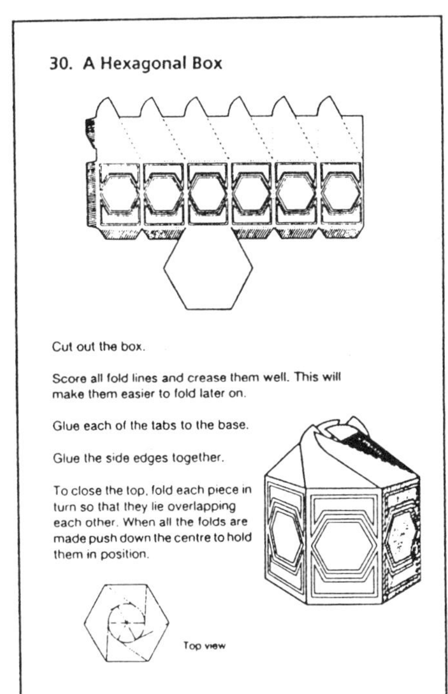

Figure 3 : Two examples from Stage 1

2. Exploring techniques

In this stage students explore and develop, in more depth, some of the techniques that have already been introduced. They may, for example, discover what happens when they change the positions and angles of both cuts and fold lines when making pop-up cards, and devise theorems which must hold true if the products are to function properly. Students are expected to keep full written records of all their discoveries, including failures.

3. Making an original prototype

In groups, students now brainstorm ideas for their own products, then work individually to prepare rough prototypes. This process involves a combination of the techniques and theorems developed in stage two, and trial and improvement, where ideas are successively refined until satisfactory results are achieved.

4. Going into production

Students now attempt to draw accurate templates for their products, accompanied by full instructions which enable other people to recreate them. These kits are now photocopied and tested by other students in the class.

During these four stages, students call on a variety of mathematical and technical skills. These are likely to include many of the following.

- Understanding and using angles and symmetry
- Estimating and measuring lengths and angles
- Following instructions presented in words and diagrams
- Making and testing conjectures, explaining and proving
- Visualising
- Creating three-dimensional objects from two-dimensional representations
- Drawing two-dimensional representations of three-dimensional objects
- Designing, making and using levers and linkages
- Writing clear, concise and complete instructions using diagrams or photographs where appropriate
- Perspective drawing

The level at which such skills are deployed depends both on the nature of the product being made and on the ability of the student. Clearly, however, there is a challenge here for students at every level of ability.

3. THE BE A SHREWD CHOOSER MODULE

Students frequently face and make consumer decisions. Such decisions are often made on impulse, with little appreciation of the many factors which could be taken into account to help the student make a better choice. In mathematics lessons we often simplify the situation to comparisons of, for example, prices per unit weight, a model that rarely applies to real situations. In this module students reflect on how people *really* make consumer decisions, and produce consumer reports which may be used to inform better choices. The material is again presented in four stages.

1. Learning from experience

Students listen to a radio show, recorded on audiotape, which contains a number of interviews with people who have just purchased different items, and an interview with two students who have produced a report on choosing orange drinks. This is supported by a written copy of the report which students can discuss critically. These activities enable students to consider the factors and methods involved in decision-making, and the processes and difficulties involved in writing a consumer report.

2. Preparing the research

In this stage students begin to plan their consumer research based on items of their own choice. This choice has to be restricted, to ensure that a rich variety of classroom activities can take place, using the following criteria.

- People in the class must have some experience of choosing the item
- The item is cheap enough for samples to be brought to school
- It is possible to carry out tests/experiments on the item to measure its quality

Suitable items are confectionery, breakfast cereals or other foodstuffs, soft drinks, batteries, writing instruments and other small, frequently purchased items. The students then have to list their research aims and methods, and prepare any tables and questionnaires that will be required for data collection.

3. Carrying out the research

Students now carry out the research they have planned. These are mainly surveys and experiments to discover, for example, whether more expensive products perform or taste any better than cheaper products. Students also need to choose appropriate ways of presenting data, draw conclusions from their data, and prepare a report. These activities give students the experience of using important statistical concepts which may later be applied to other situations, for example, sampling techniques, graphical representation, graphical interpretation and measures of central tendency. Students may have experienced all of these, but are unlikely to have been asked when and where to use them.

4. Presenting and evaluating the reports

In this final stage all the written reports are circulated around the other groups in the class, and any group wishing to make an oral presentation does so. The reports are then evaluated by the class, and each group is given an opportunity to improve its own report, taking these comments into account.

During this module, students are likely to call on a variety of mathematical and technical skills, including the following.

- Devising questionnaires and conducting interviews
- Designing and carrying out experiments
- Analysing and presenting data in various ways
- Selecting and using sources of information
- Interpreting data and presenting clear and reasoned recommendations
- Handling money, other everyday measures, percentages, statistics, graphical representation, ratio and proportion in real-life contexts

4. INTEGRATING THE MATHEMATICS WITHIN THE MODELLING

While students are working on a particular module theme, their main objective is to produce an attractive paper product or an interesting consumer report, not to develop particular mathematical techniques. The mathematics is used as a tool and is not seen as an end in itself.

Students are only likely to deploy those skills with which they are already confident, and may resist any attempts to teach them new techniques while they are involved in their projects. The teacher, however, may wish to use the many opportunities offered by the module to motivate the learning of specific mathematical techniques in a more explicit way. How can this be achieved, without destroying the essential flow of activities contained in the modules?

Occasionally, mathematical activities are initiated by the student. A student may, for example, become aware of the need to acquire a particular skill in order to make a design work. "How can I be sure that my pop-up card will open and close without creasing or tearing in the wrong place? What lengths/angles have to be equal? Do I have to make the card first or is there a way of calculating it?" This kind of situation can lead to an invaluable learning experience, because the student *wants* and *needs* to know. Such opportunities occur rather unpredictably, however, and it is not always possible to spend time helping one individual when there is a large class to supervise. We have found that one possible solution is to ask the student to describe the problem to the whole class, and invite help and advice from other students.

Teacher-initiated work on mathematical techniques relating to the module theme may occur before, during or after working on the module.

Before:

This timing has the advantage that the student will, if all goes well, have relevant techniques polished and available, but it may seem artificial to learn new techniques before seeing a need for them. Students may also tend to assume that the module is merely a vehicle for practising these techniques, rather than to develop autonomy in problem solving.

During:

Some teachers have suggested that, during the course of a module, one lesson a week is devoted to developing techniques. This enables the teacher to respond to needs as they arise, but the work on the module can tend to drag on over many weeks and lead to boredom.

After:

The experience of working on a module may motivate students and enable them to see the value of techniques when they are taught. Students may still not be able to use these techniques autonomously, unless they are given further opportunities to apply them in other real problem-solving contexts.

The integration of mathematical technique into modelling activities is clearly an area that needs further research. Whatever is decided, we must try to preserve the student's strategic control of their work; it is too easy to allow them to revert to the imitative role that the traditional curriculum encourages.

5. SOME ASSESSMENT ISSUES

The scheme has been designed to assess strategic skills, for example the ability to measure accurately. The procedures are designed to assess whether or not a student can satisfy a number of criteria. In figure 4, we list some of these criteria as they appear in the two modules under discussion. (Gillespie *et al,* 1989, gives details about criteria and other assessment issues relating to the Plan a Trip module).

(i) follow instructions,
(ii) cut, fold and glue accurately to assemble a 3-dimensional product,
(iii) recognise structural features of a design,
(iv) make a 3-dimensional object from a 2-dimensional representation,
(v) draw a 2-dimensional representation of a 3-dimensional product,
(vi) give a reasoned explanation for design features,
(vii) identify and correct design faults,
(viii) develop an existing idea for a paper product,
(ix) generate possibilities for a design with original features,
(x) draw a design to an acceptable degree of accuracy,
(xi) construct a prototype with original features,
(xii) devise instructions to enable someone else to make the product.

Be a Paper Engineer

(i) identify important factors and methods involved in decision-making
(ii) obtain and interpret information from oral interviews
(iii) obtain and interpret information from tables and graphs
(iv) identify possible research aims
(v) select appropriate research methods
(vi) devise suitable methods for the collection and organisation of data
(vii) present a summary of research data in a clear, organised way
(viii) draw sensible conclusions from a collection of research data
(ix) take an active part in compiling their own reports
(x) evaluate a report and suggest improvements to it.

Be a Shrewd Chooser

Figure 4 The module criteria

Such criteria assist in providing a useful profile of relative strengths and weaknesses, but they have little absolute meaning without specifying the context, the frequency of success, the amount of help that was given, the distance of transfer from the student's previous experience, and the mode of response (in writing or orally). As the students work through the modules in the series, they are likely to demonstrate similar strategic skills in a variety of different contexts. This enables the teacher to make more general statements about the students' progress.

Each module is accompanied by a collection of short *Basic* level assessment tasks, some of which are completed in the normal course of the work and some of which are administered at the end of the appropriate stage in the module. These are intended to be accessible to the vast majority of students. They are supplemented by two written examination papers at *Standard* and *Extension* levels, which about 80% and 40% of students respectively should be able to pass (although these are not norm references). One (or both) of these papers is sat during the term following that in which the student completed the module. On the basis of these assessments, a student may be awarded a short *Statement of Achievement* for each module. This lists the criteria that a student has satisfied. If a student has been successful in three or more modules then he or she may be awarded a *Certificate in Numeracy Through Problem Solving* which gathers together the entire collection of criteria that have been satisfied, in a more generalisable form. This assessment scheme has proved very popular with the schools that have been involved.

REFERENCES

Binns BS, Burkhardt H, Gillespie J and Swan MB. (1989). Mathematical Modelling in the School Classroom. In Blum W *et al* (eds), *Applications and Modelling in Learning and Teaching Mathematics.* Ellis Horwood, Chichester.

Binns BS, Burkhardt H and Gillespie J. (1987). Bottom–up Numeracy. In Berry JS *et al* (eds), *Mathematical Modelling Courses.* Ellis Horwood, Chichester.

Cockcroft Report. (1982). *Mathematics Counts.* HMSO.

Gillespie J, Binns BS, Burkhardt H and Swan MB. (1989). Assessment of Mathematical Modelling. In Blum W *et al* (eds), *Applications and Modelling in Learning and Teaching Mathematics.* Ellis Horwood, Chichester.

Shell Centre for Mathematical Education, Joint Matriculation Board. (1987–9). Design a Board Game, Produce a Quiz Show, Plan a Trip, Be a Paper Engineer, Be a Shrewd Chooser. Longman, UK.

CHAPTER 15

Enterprising Mathematics : A Context-based Course with Context-based Assessment

R Francis and D Hobbs
University of Exeter, UK

SUMMARY

This paper outlines a novel assessment scheme which has been devised for a context-based GCSE course currently being piloted in a group of secondary schools in England. This course aims to provide materials and appropriate assessment to implement the teaching styles advocated in the Cockcroft report and to satisfy the National Criteria for GCSE.

1. INTRODUCTION

At the Centre for Innovation in Mathematics Teaching, Exeter University, a modular mathematics course has been developed for 15 and 16 year old students, leading to a General Certificate of Secondary Education (GCSE) qualification. (GCSE examinations, based on national criteria, were introduced in 1988 in England and Wales to form a common system of examinations at the end of the compulsory period of schooling, replacing the former GCE and CSE examinations). A feature of the course is that it is context-based. Details of the philosophy behind the course are given in Hobbs (1987, 1988) and Hobbs and Burghes (1989).

The *Enterprising Mathematics Course*, as it is called, is organised according to application and interest topics into five modules, presented at three ability levels (see figure 1). Each module is approximately one term in length and is taught in two blocks.

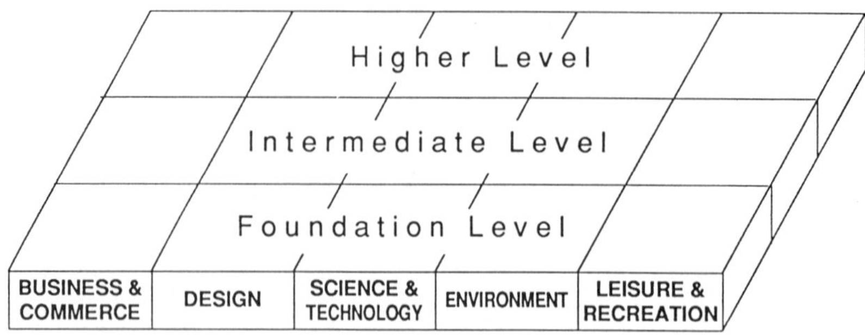

Figure 1

The course has been approved as a Mathematics GCSE by the School Examinations and Assessment Council, for a pilot scheme starting in September 1989. This paper outlines the assessment arrangements.

2. ASSESSMENT

The course has been designed to satisfy the aims, objectives and mathematical content of the GCSE National Criteria for Mathematics (DES, 1985). In particular, attention has been given to objectives 16 and 17:

- respond orally to questions about mathematics, discuss mathematical ideas and carry out mental calculations
- carry out practical and investigational work and undertake extended pieces of work.

All GCSE schemes in mathematics are required to give between 20% and 50% of the total assessment to aspects such as these, which cannot be assessed in formal, written examinations. The contextual nature of Enterprising Mathematics enables these requirements to be met in a natural way, and influenced us to allocate the maximum allowable amount of coursework to those aspects. The other 50% of the total assessment is allocated to written achievement tests.

A second important feature of the assessment scheme is that, like the course, it is modular. This avoids having one long examination at the end, and enables feedback to be given to students during the course. It also has the advantage that students do not have to do all the modules at the same level.

Achievement Tests

The *Achievement Tests* at the end of each module are based on contexts. For example, a specimen test for the *Business & Commerce* module relates to a filling station. Information about the context is provided a week or so in advance (see Appendix 1). It is used by the teacher as a basis for some lessons to familiarise students with the context and with problems arising from it. To assist the teacher an activity sheet is provided, including questions such as the following.

- What is a lease?
- What weight of petrol does a full tanker carry?
- Use some form of graph to show how the monthly sales of two different fuels compare throughout the year.
- Explain the variations in the amount of 4-star petrol sold during a day.
- How would you change a petrol consumption in mpg (miles per gallon) into litres per 100 kilometres?

In the achievement test the information sheet is provided and the test consists of questions based on it. The specimen test at Intermediate Level is included in Appendix 2.

One of the five written tests also includes an *aural test*, containing questions which are largely set in context.

Achievement tests are marked according to the usual method/accuracy schemes. Students are then awarded a level number according to eight level descriptors such as

Level 1: Shows evidence of Knowledge, Skills and Applications (as defined in the Assessment Objectives) on some of the Foundation Level subject content.

Level 4: Shows a very good mastery of Knowledge, Skills and Applications on all of the Foundation Level subject content, particularly in solving problems. Can solve some two-stage problems involving the Foundation Level subject content.

Level 8: Shows a very good mastery of Knowledge, Skills and Applications on all of the Higher Level subject content. Has a very good facility in Problem Solving (as defined in the Assessment Objectives), in particular in making conjectures, forming generalisations, and expressing them algebraically, and providing by logical deduction generalised statements involving algebraic and spatial ideas.

Coursework

For coursework, students submit a portfolio appropriate to the module. It is intended to include a variety of styles of work (for example case studies, surveys, investigations, practical work). Normally at least two coursework items would be expected per module. These items are not prescribed, but suggestions are included in the course material. Figure 2 shows a page from a Higher Level topic unit in the Design module which gives ideas for coursework items.

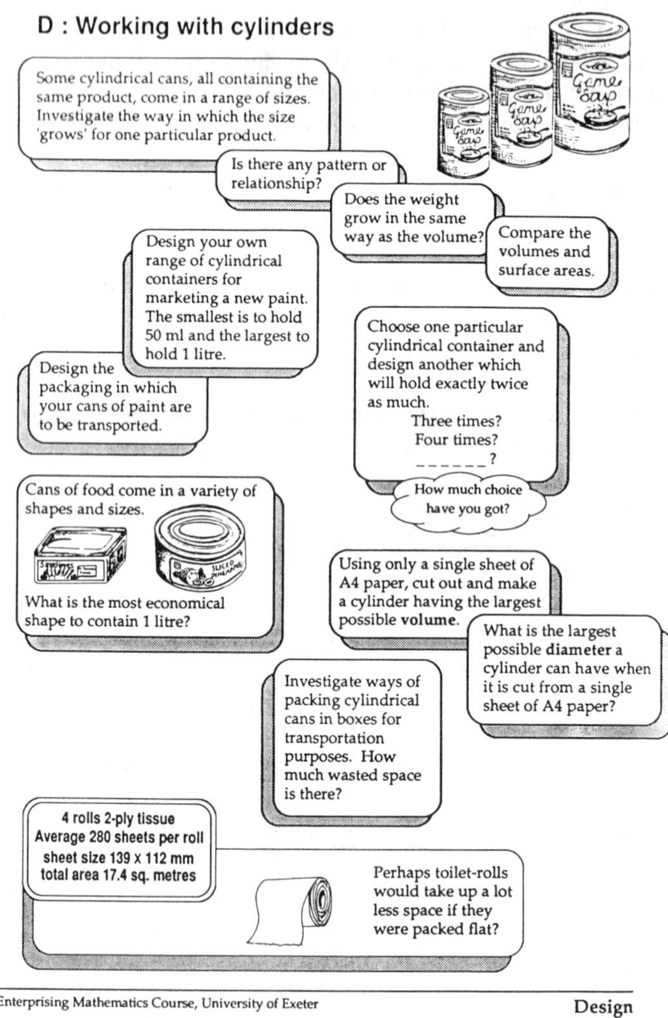

Figure 2

During a module teachers comment on potential portfolio items under the headings identifying and planning, implementing, evaluating and communicating. These comments are given to students as *formative* feedback. When the module is completed teachers determine a *summative* assessment based on the comments sheet according to eight level descriptors, such as the following.

Level 1: Can select materials and mathematics to use for a task.
Can carry out simple mathematical tasks accurately and record results in a simple form.
Can estimate and check results; consider whether it is a simple answer.
Can talk about own work and ask questions; record work done and results obtained.

Level 4: Can design a task and select appropriate mathematics and resources; check there is sufficient information and obtain any that is missing.
Can make and test simple statements; use oral, written, visual or concrete forms to record and present findings.
Can use examples to test statements or definitions; make and test simple statements.
Can describe the outcomes of the work done in the context of the original task.

Level 8: Can justify the selected route.
Can use symbolisation with confidence; construct a proof, including proof by contradiction.
Can use symbolisation with confidence; recognise and use necessary and sufficient conditions.
Can invent own mathematical structures, language and notation where necessary.

The four statements at each level relate to the headings above. In addition, a level of achievement for *oral* contributions during the module is recorded according to level descriptors such as the following.

Level 1: Can make responses to mathematical questions asked by the teacher.

Level 4: Can contribute mathematical ideas.

Level 8: Can discuss and criticise mathematical arguments proposed by others.

A summary level of achievement is then recorded, giving equal weight to the five components: identifying and planning, implementing, evaluating, communicating and oral work.

Achieving a GCSE grade
For each module the levels of achievement attained in both the achievement test and the coursework are combined to produce a module result. The module results are combined to produce a GCSE grade in the range A (high) to G (low).

3. CONCLUSION

Enterprising Mathematics differs from many other courses in the emphasis which it puts on contexts. A submission to the Cockcroft report suggested that mathematics lessons in secondary schools are very often not about anything (1982, para 462). The Enterprising Mathematics course attempts to show that mathematics *is* about something. Furthermore, to match this philosophy, an assessment scheme has been devised to encourage the *doing* of mathematics during the course and to test it in context.

REFERENCES

Cockcroft WH. (1982). *Mathematics Counts.* HMSO, London.

Department of Education and Science. (1985). *GCSE : The National Criteria.* HMSO, London.

Hobbs D. (1987). Mathematics – An Alternative Approach. *Teaching Mathematics and its Applications*, Vol 6, **4**, 160–163.

Hobbs D. (1988). Enterprising Mathematics Course. *Teaching Mathematics and its Applications,* Vol 7, **2**, 53–70.

Hobbs D and Burghes D. (1989). Enterprising Mathematics : A Cross–curricular Modular Course for 14–16 year olds. In Blum et al (Eds), *Applications and Modelling in Learning and Teaching Mathematics.* Ellis Horwood, Chichester.

Appendix 1

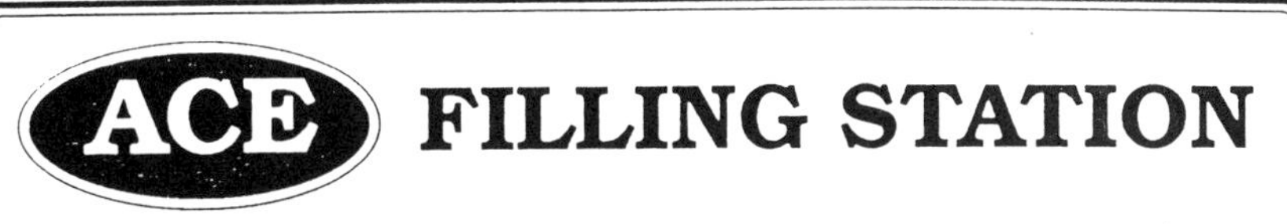

Jon and Sara run a petrol and diesel Filling Station. They do not own it, but lease it from the **ACE** Petroleum Company.

These charges had to be paid in 1989:

Leasing Charge Per Year	
Land	£ 325
Equipment	£1650
Buildings	£ 875

It is a condition of their lease that they **must** buy **ALL** their petrol and diesel-fuel from that company.

There is a small shop attached to the filling-station where they sell various things needed by motorists (replacement lamp bulbs, oil, sparking-plugs, batteries etc.) as well as a whole range of other goods (toys, sweets, torches etc.).

These goods they may buy from anywhere.

They employ one assistant who helps in operating the till for petrol sales as well as running the shop. This is Kim, who is paid on the basis of a 40-hour week, but who normally works more than that. A typical pay-slip looks like:

Name of Employee - Kim NEWINGTON

Week beginning:	**06-11-89**	**Deductions**	
Basic Pay (40 hours)	84.00	Income Tax	20.11
Overtime (15 hours)	47.25	National Insurance	12.10
Uniform Allowance	0.50		
Travel Allowance	2.70		
		Total Deductions	32.21
Total GROSS Pay	134.45	**Total NET Pay**	102.24

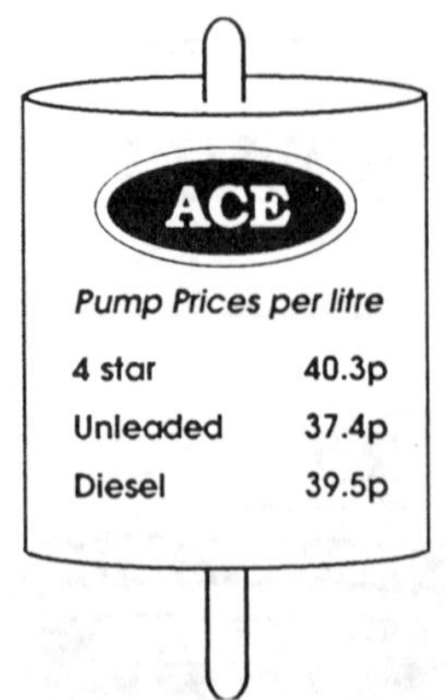

Tanker Capacity

The tankers which deliver the fuel each have a capacity of 30 000 litres.

Each tanker has six compartments, each of which can hold up to 5000 litres of fuel.

Compartments may hold 4-star petrol, unleaded petrol and diesel in any combination.

A compartment may be filled with petrol, but weight restrictions mean that it may contain no more than 4000 litres of diesel.

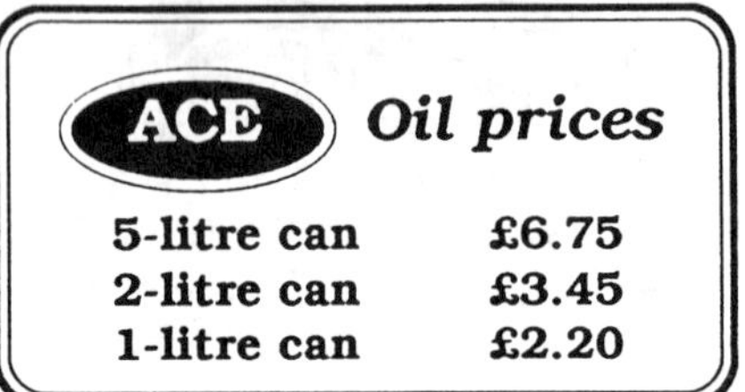

Filling-station Capacity

At the filling-station the fuel is stored in tanks buried in the ground. The capacities of these tanks are:

Petrol :	*4 star*	80,000 litres
	Unleaded	60,000 litres
Diesel		20,000 litres

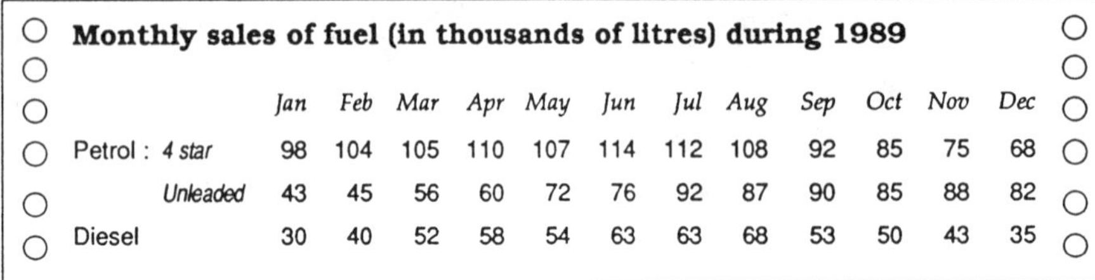

Monthly sales of fuel (in thousands of litres) during 1989

		Jan	*Feb*	*Mar*	*Apr*	*May*	*Jun*	*Jul*	*Aug*	*Sep*	*Oct*	*Nov*	*Dec*
Petrol :	*4 star*	98	104	105	110	107	114	112	108	92	85	75	68
	Unleaded	43	45	56	60	72	76	92	87	90	85	88	82
Diesel		30	40	52	58	54	63	63	68	53	50	43	35

The chart below shows the amount of 4 star petrol (in litres) sold on one particular week-day in September, 1989, from 8 a.m. to 8 p.m.

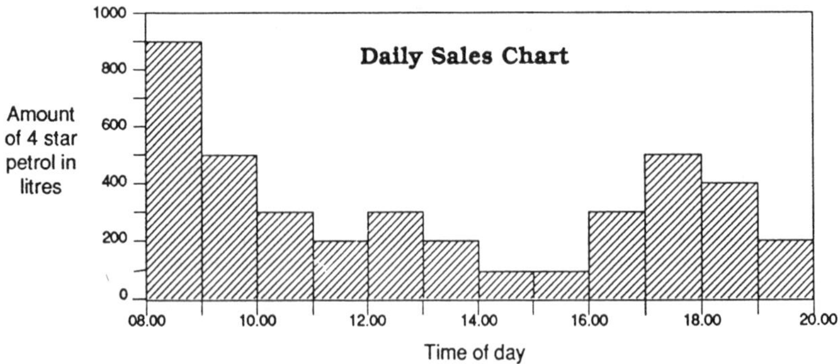

Appendix 2

SOUTHERN EXAMINING GROUP

General Certificate of Secondary Education

MATHEMATICS

Enterprising Mathematics

Business & Commerce Intermediate Level, Written
Time allowed: 45 minutes

SPECIMEN PAPER

1. Tessa's car has a petrol tank which is roughly in the shape of a cuboid

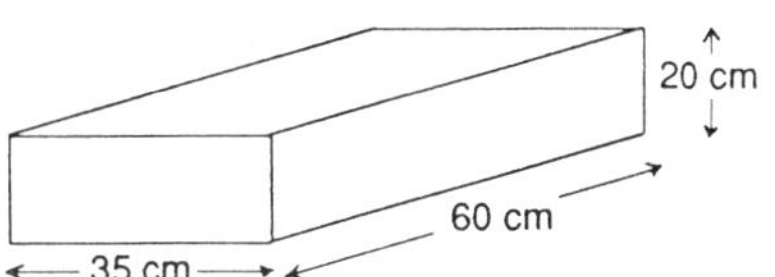

(a) What volume of petrol would the tank hold when full ?
Give your answer in litres. *(4 marks)*

(b) Tessa uses the **ACE** Filling Station to "fill up".
How much **more** does a tank full of 4-star cost than a tank full of unleaded petrol ?
Give your answer to the nearest penny. *(4 marks)*

2. The table shows the amount of unleaded petrol (in litres) sold on Friday, 13th October, 1989, in hourly intervals from 8 a.m. to 8 p.m.

Time interval	*Litres*	*Time interval*	*Litres*
0800 - 0900	810	1400 - 1500	70
0900 - 1000	470	1500 - 1600	60
1000 - 1100	250	1600 - 1700	210
1100 - 1200	190	1700 - 1800	530
1200 - 1300	270	1800 - 1900	420
1300 - 1400	180	1900 - 2000	170

(a) Calculate the mean (average) number of litres of unleaded petrol sold per hour on Friday, 13th October, 1989. *(3 marks)*

(b) What percentage of the day's sales of unleaded petrol were taken outside the period 9 a.m. to 5 p.m. ? *(4 marks)*

3. When Jon and Sara first opened for business they needed to fill their fuel tanks, which were previously empty.

The **ACE** Petroleum Company charges garages the following for their fuel:

Petrol	*Unleaded:*	£80	per 1000 litres
	4-star:	£100	per 1000 litres
Diesel		£90	per 1000 litres

*You will need to refer to the **Tanker Capacity** and **Filling-station Capacity** panels.*

(a) What would it cost Jon and Sara to fill all three tanks with fuel ? *(4 marks)*

(b) Imagine that you are the delivery manager at the **ACE** Petroleum Company.

You have to arrange a delivery schedule of tankers to fill the tanks at Jon and Sara's Filling Station.

Find the minimum number of tankers required, detailing the types and quantities of fuel to be carried in each tanker. *(6 marks)*

4.

Kim gets a pay-rise!

Her basic pay goes up to £2.30 per hour.

(a) What will Kim's basic pay be for a 40-hour week ? *(2 marks)*

Name of Employee - Kim NEWINGTON

Week beginning:	**12-02-90**	**Deductions**	
Basic Pay (40 hours)	....	Income Tax	....
Overtime (20 hours)	....	National Insurance	14.78
Uniform Allowance	0.50		
Travel Allowance	2.70		
		Total Deductions	....
Total GROSS Pay	....	**Total NET Pay**	....

Overtime is paid at "time-and-a-half"

(b) What will her gross pay be for the week beginning 12-02-90 when she works 20 hours overtime ? *(3 marks)*

Kim can earn £54.00 per week "tax free" (her tax allowance) and pays 25% of the rest of her gross pay in tax.

(c) How much tax did she pay that week ? *Give your answer to the nearest penny.*

(3 marks)

(d) Complete Kim's pay-slip for the week beginning 12-02-90. *(2 marks)*

5. Hannah and Jason are planning a trip in their car, a Ford Fiesta, which uses **unleaded** petrol.

Starting from Exeter, they are going to Sheffield, stopping off at Bristol and Birmingham on the way.

(a) Use the mileage chart below to estimate the total distance for the journey. *(3 marks)*

	Birmingham	Bristol	Cardiff	Exeter	London	Manchester	Nottingham	Plymouth	Sheffield
Bristol	88								
Cardiff	108	47							
Exeter	162	84	121						
London	120	120	155	200					
Manchester	89	172	192	246	204				
Nottingham	54	149	169	223	131	72			
Plymouth	203	125	162	45	241	287	264		
Sheffield	87	182	202	256	169	37	44	297	
York	130	225	245	299	212	71	87	340	61

On a "long run" the Ford Fiesta has a petrol consumption of 50 m.p.g. (miles per gallon).

(b) (i) How many **litres** of petrol would they expect to use on their journey ?
Give your answer to the nearest litre. *(4 marks)*

(ii) How much would this cost at the **ACE** Filling Station ?
Give your answer to the nearest penny. *(3 marks)*

6. Jon and Sara want to offer a 24-hour breakdown recovery service.

They decide to buy a second-hand breakdown truck, costing £15 000.

CREDIT PURCHASE TERMS	
Deposits	25 %
Annual interest rate	15 %
Arrangement fee	£5

They cannot afford to pay cash, so will have to buy it on Credit Purchase.

(a) How much will the deposit be ? *(2 marks)*

(b) They are going to pay the loan back over one year.
What will be the **total** cost of buying the break-down truck on Credit Purchase ? *(4 marks)*

(c) BUT! If they had paid in cash they would have been given a 10% discount.
How much **more** is the break-down truck costing them by buying on Credit Purchase ? *(4 marks)*

CHAPTER 16

Introducing Discrete Graphs to 12 year olds

G van den Heuvel and H Krabbendam
National Institute for Curriculum Development, Arnhem, The Netherlands

SUMMARY

This article asks the question 'Which type of graph is most suitable as a start to the subject modelling with a graph?'. Two examples of worksheets are used to discuss this problem and we shall then put forward arguments for our choice. However, before dealing with the question mentioned above, we shall provide some background information, in context with our work as members of a curriculum innovation project in The Netherlands. We shall outline briefly the state of the art in the project in the field of the introduction of topics from discrete mathematics.

1. CURRICULUM INNOVATION IN THE NETHERLANDS

In The Netherlands we have primary education for 4 to 12 year olds, followed by secondary education (12 to 16/18 year olds) which is subdivided into three streams.

There is no school leaving exam at the end of the primary school and also no formal curriculum. When teaching arithmetic/mathematics there is an increasing use of realistic methods in approximately 50% of primary schools.

In secondary education, however, there are several school leaving exams. The examination syllabus for upper secondary education (pre-college education) has recently been revised. Mathematics has been subdivided into two separate subjects, one best categorised as realistic mathematics –

aimed at applications - the other more theoretical in content - a preparation for further education in the sciences.

Thus, the 12-16 syllabus - which dates back to 1968 and is rather formal and analytical, based as it is on set theoretical principles - now finds itself between two modern syllabi. It is obvious that adjustments are badly needed. The Minister of Education and Science has commissioned a group of experts to develop a new curriculum for this age group on the basis of elaborate experiments. The curriculum should be ready in 1992, in order that implementation can commence from 1993 onwards. The project will be undertaken by SLO-Enschede and the OW&OC department of Utrecht University.

2. CHANGES IN THE 12-16 CURRICULUM

Present day society is increasingly confronted with problem areas such as

- problems of traffic, trade and infrastructure (from a mathematical point of view, all variations on network problems);
- problems of decisionmaking about important social topics such as energy, the environment, operations research;
- questions about information science, using block diagrams and structuring datafiles.

These are all examples of areas in which we have seen important developments in the past decades. Mathematically speaking, these examples all have their roots in discrete mathematics. We can also look at the skills people need at present and in the future, as in past decades we have seen changes in these skills.

In their book Descartes' Dream (1986), Davis and Hersh - as a sequel to a list of intellectual aspects of mathematisation - present the following list of skills specific to computerised society.

- **Algorithmic thinking**. Programs, recipes. The ability to consider a whole as split up in small parts, which are easy to solve and to survey.
- **Modular thinking**. The formation of useful algorithms in sub-areas; assessing and using them in appropriate places.
- **State thinking**. Thinking, working in modalities. Knowing the state, the mode of things.
- **Systems thinking**. The formation of a large structure from many parts organising component parts.
- **Meta-thinking**. The creation of programs whose input is a program and whose output is a program.

In education we can translate these skills. In the programme, attention should be paid to such general problem areas as finding structures in problems, turning a situation into a diagram, making a model, assessing a model. It should be aimed, in particular, at situations in which structures play a role within finite sets.

The conclusion we draw from these considerations is that society increasingly demands skills which can either be deduced directly from discrete mathematics or exemplified by topics from discrete mathematics. (The latter may well prove to be the better choice.) In the present programme for 12–16 year olds these skills play hardly any role, and this is the reason why we have opted to include elements of discrete mathematics as a substantial new part of the 12–16 programme.

In selecting the content for the new 12–16 syllabus we have considered two strategies.

1. Several new developments made during the past few years should be implemented in the new 12–16 syllabus.
2. Several completely new items (for 12–16 year olds) should be introduced into the syllabus.

We shall be dealing with the second point, focussing on discrete graphs, in the present paper.

3. DISCRETE MATHEMATICS FOR 12–16 YEAR OLDS

In programming discrete mathematics in the new curriculum, we will need to make selections from the many possibilities which this rich area has to offer. Some of the selections we made in the experiment are the following.

- Our point of departure is the important means of visualisation in this area, the graph. Modelling situations in graphs, and the reverse procedure, play an important role.
- We select a few topics for elaboration, in this instance network problems. The thought behind this is that we prefer 'a little more about a little less to a little bit of everything'.
- Next to creating a model, reasoning and algorithms are the key areas.
- As to the latter, we prefer to understand algorithms rather than execute them, because a computer can be used to do this.
- Pupils should be able to express in words why the algorithm works, and also why it is economical.
- In order to be able to understand the algorithm properly some sort of support is needed, for example combination counting.

- We also need to pay attention to the rationale for models and counting. Here we think of ordering, but the idea of predicting from models also comes to mind.

The items we have pointed out above form the provisional points of departure, and the provisional selections we have made in the field of discrete mathematics. These form the basis for experimentation, and are likely to be favourable developments.

Having presented a survey of the framework within which we work, we now move on to the specific question we were faced with in our development work.

4. A QUESTION FOR DEBATE: WHICH GRAPH COMES FIRST?

In programming the new curriculum, modelling by means of a graph plays an important role. We considered two approaches for the introduction of the new graph. Firstly, introducing the concept of the graph by means of a geographical situation – for example an underground network – or, secondly, starting from a more abstract model, unrelated to a geographical context – such as a tournament graph.

The question we asked ourselves was 'What are the advantages and disadvantages of these two introductions to the concept of a graph, and which of the two approaches should be emphasised at the introduction?' We present examples to illustrate both approaches.

5. STARTING FROM A GEOGRAPHICAL SITUATION

In thc cxamplc Project Week in London the point of departure is a context the pupils can easily imagine. They have never experienced the phenomenon of the underground, yet the model, so to speak, presents itself. The pupil first constructs his or her own underground network, then abstracts it from the map, followed by a confrontation with the 'real' underground graph. In this way important aspects of the representation form of the graph are shown – the characterisation of data in points and lines, the discrete character of the subject, the arbitrariness of the exact representation of the lines in the model. It is rarely possible for either a real or a fictitious underground traveller to find out the real geographical course of the underground lines, and in fact this does not really matter. The line indicates that there is a connection between the extremities it connects, and that is enough. Moreover, in the final questions, a start is made with an important subsequent area, namely the problems of routing.

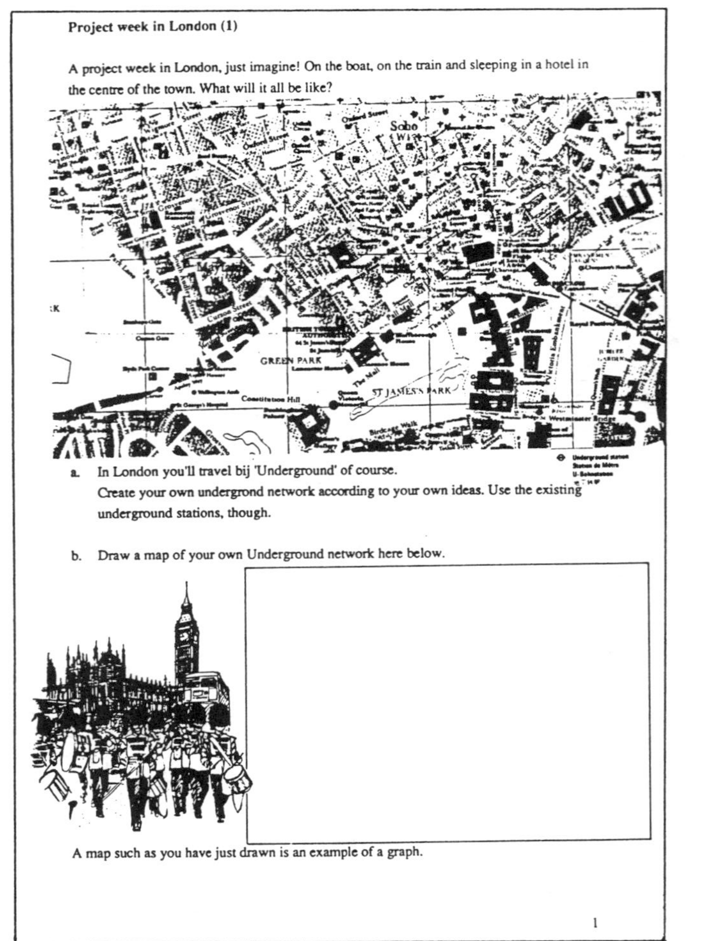

Project week in London (1)

A project week in London, just imagine! On the boat, on the train and sleeping in a hotel in the centre of the town. What will it all be like?

a. In London you'll travel bij 'Underground' of course. Create your own undergrond network according to your own ideas. Use the existing underground stations, though.

b. Draw a map of your own Underground network here below.

A map such as you have just drawn is an example of a graph.

1

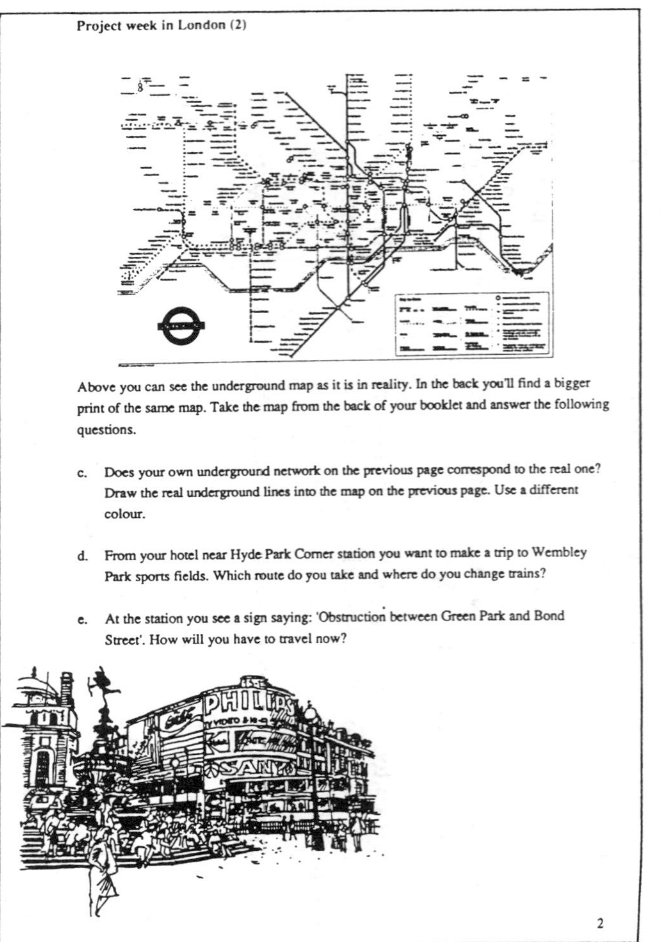

Project week in London (2)

Above you can see the underground map as it is in reality. In the back you'll find a bigger print of the same map. Take the map from the back of your booklet and answer the following questions.

c. Does your own underground network on the previous page correspond to the real one? Draw the real underground lines into the map on the previous page. Use a different colour.

d. From your hotel near Hyde Park Corner station you want to make a trip to Wembley Park sports fields. Which route do you take and where do you change trains?

e. At the station you see a sign saying: 'Obstruction between Green Park and Bond Street'. How will you travel now?

2

Worksheets used by pupils

However, some comments need to be made. In spite of the arguments presented above, one can hardly expect a pupil to think of one of the following solutions to question b in the worksheet.

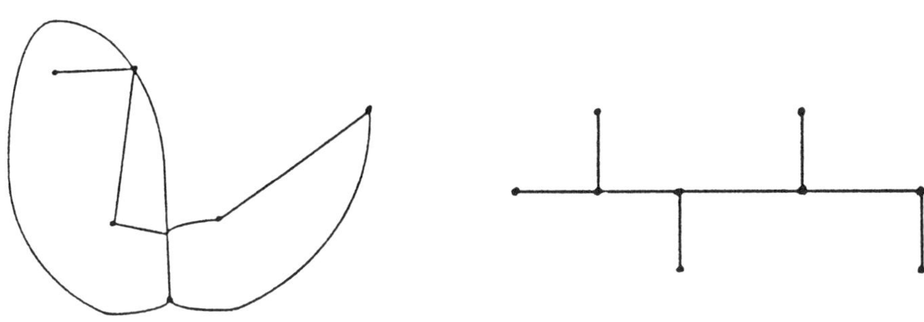

Yet it is only in answers like these that the basic features of the graph come across well. Likewise, the strength, and also the complexity, of the concept of the graph lies exactly in the way points and lines are manipulated for the purposes chosen. It is not sufficiently clear if pupils are only aware of the differences between the actual map and the model of the graph.

6. STARTING FROM A MORE ABSTRACT EXAMPLE

The next example is Tournaments. A volleyball tournament is a contest which a pupil can easily imagine, although even a volleyball tournament is tricky because not everyone knows the scoring of a volleyball game. There is sometimes a misunderstanding of the system used, for example half a competition and a knock-out tournament. However, these problems are not insurmountable. Judging from the speed with which the initial questions were solved, the model is very easy for the pupils to understand. They did not know how to represent the tournament results, but now that this system has been presented to them, they can handle it easily.

Besides, such matters as the arbitrary position of the points (for convenience's sake in something like a circle, but the sequence is irrelevant) and the fact that the lines only characterise a relation, are evident in this example. The problem presents an elegant kind of mathematics: 5 is greater than 4 and 4 greater than 3, so 5 greater than 3, but ... A beats B and B beats C, and so nothing, no transitivity in this world of graphs!

Tournaments

In the tournament graph below you can see the score of a volley ball tournament after the third round.

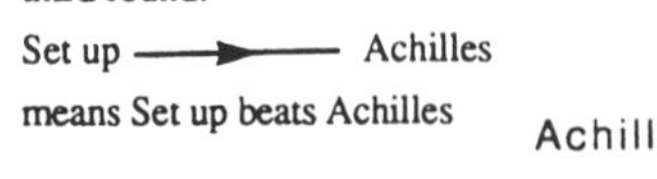

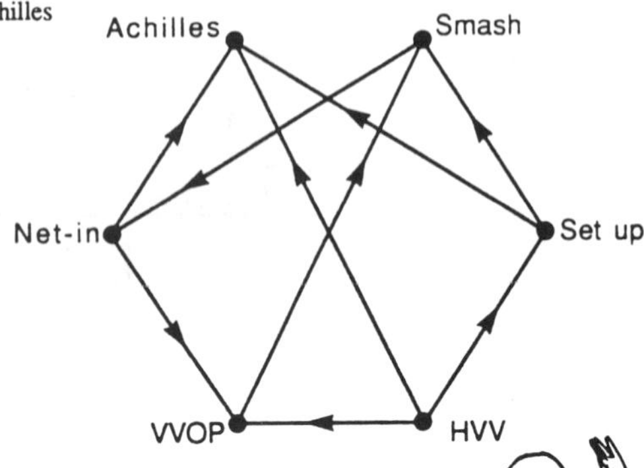

a. Who won the match Smash - VVOP?

b. Who are Net-In's remaining opponents?

c. What is the score after three rounds. Mind the order!

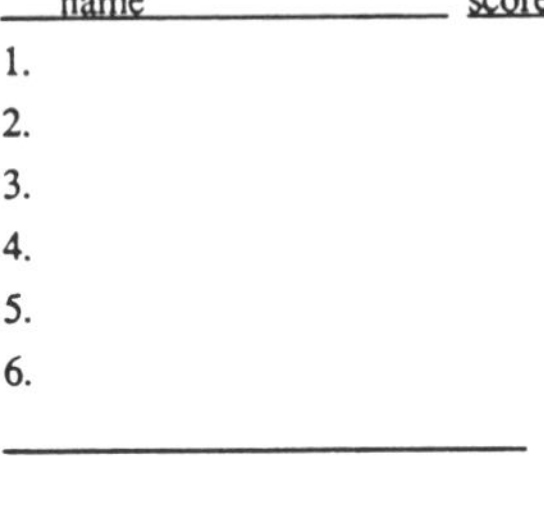

name	score
1.	
2.	
3.	
4.	
5.	
6.	

d. In the newspaper you will always find a score, never a tournament graph. Don't you think a graph would be much better?

e. Who do you believe will win the match Set up - Net-in and why?

f. How many matches will be played in total in this tournament?

g. How many would that be if there were 10 participating teams?

Tournament worksheet

Combinatorial counting is also brought up here in a functional manner, an enjoyable and, for many pupils, a challenging activity which is essential for discrete mathematics. In short, a worksheet full of perspectives, practical feasibility and many merits. All those subtle misunderstandings that can crop up with the map and the graph simply do not occur here.

However, there are also a few objections. An illustrative example is the answer to the question "Which of the two is more satisfactory in a newspaper, a score or a graph?" The majority of the pupils will almost blindly opt for the score, as a strange graph in a newspaper is simply not done. Obviously, there is little affinity with the model. The pupil understands what the teacher means, but it is a concept from a world different to that of the pupil, and this also makes other questions problematic. The pupil will gain some understanding of the arbitrary position of the points, but one wonders whether such matters are considered for a model that has not been created by the pupils themselves. Thus the initiative to pose questions and to reflect lies with the teacher who has also presented that special model. In Project Week in London, the representation of the situations raised questions and problems almost spontaneously but, in this case, they tended to remain somewhat hidden. What should pupils do with the worksheet on tournaments, with all its rich contents, when they have left the mathematics lesson? Will they make a link later to a shortest path question, also an important problem?

7. ONCE MORE: WHICH GRAPH COMES FIRST?

In this article, we have given two examples in our discussion of this question. If one opts for a particular introduction many of the following aspects appear, either implicitly or explicitly.

- Which are the content elements and structures that we should focus on? Do we leave more room for self-modelling and mathematisation, or do we opt for more emphasis on the abstract properties of the graph? One's vision on what mathematics is all about plays an important role here.
- What is the best way for pupils to learn a concept, in this case the graph? Should the point of departure be a more or less concrete situation leading to a gradual abstraction, or is the direct confrontation with abstracted structures of the subject to be preferred? For both opinions, arguments from educational psychology can be found.
- What kind of knowledge do we find important for the pupil? Do we strive after an amount of knowledge (with more links in daily life and consequently with better chances for the pupil to apply that knowledge in daily life), or is it more important to impart

knowledge about the subject area to the pupils, which will be of use to them in their future (school) career? This gives rise to all sorts of social considerations.

Our choice is an introduction from a geographical context. Thus we have taken a stand in the discussion about the dilemmas mentioned above. However, we also believe that mathematics should appear as an activity. The activity is, in our view, at least as important as its product. The latter, however, is still central in the present situation in schools.

To end our contribution we shall present another example, which will show how one can come to further abstraction by taking a geographical context as a point of departure.

8. GOING NORTH: ABSTRACTION FROM A GEOGRAPHICAL CONTEXT

The example Going North is an example of an assignment at a later stage of the programme. When we travel from our country (here, from the city of Arnhem) to Scandinavia (here, to the city of Oslo), the question which is the best way – the shortest, the cheapest, and so on – presents itself. The worksheets are just such examples, as they introduce the shortest path question, which is later elaborated algorithmically and with computer support. The experiences with the worksheets, limited though they are, have been encouraging. The pupils follow the steps taken with relative ease, and also realise that a more stylised model offers increased perspectives and enables a better and more sensible solution to be achieved.

Going North

Marieke and Yvonne live in Arnhem. This summer they are going to Oslo, perhaps even farther, with their partents by car. In any case part of the journey will have to be by boat. They can travel in many different ways. Let's find out which is the cleverest way of travelling.

a. What should they all know in order to determine the cleverest way? How can they find out?

You don't have to sort out all the material yourselves. We'll start with a map and a list of ferry connections. They have been taken from a travel agency's guide book.

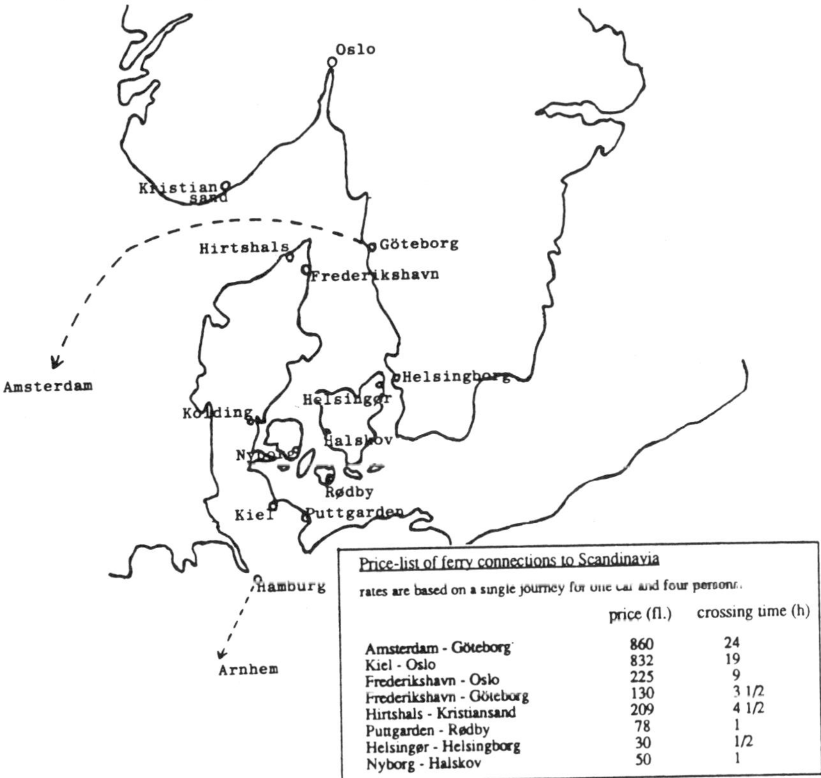

Price-list of ferry connections to Scandinavia

rates are based on a single journey for one car and four persons.

	price (fl.)	crossing time (h)
Amsterdam - Göteborg	860	24
Kiel - Oslo	832	19
Frederikshavn - Oslo	225	9
Frederikshavn - Göteborg	130	3 1/2
Hirtshals - Kristiansand	209	4 1/2
Puttgarden - Rødby	78	1
Helsingør - Helsingborg	30	1/2
Nyborg - Halskov	50	1

b. Which countries are on the map? Have you ever been there yourself?

c. Draw all the ferry connections into the map in dotted lines (work neatly please!)

Going North worksheet

For the journey by land Marieke and Yvonne use the distance chart below.

d. Draw the routes by land into the map in dotted lines.

e. What do you believe is the cleverest route? Why do you believe that?

Distance chart by land

distances based on travel by motorway

	distance (km)
Arnhem - Amsterdam	100
Arnhem - Hamburg	453
Hamburg - Kiel	87
Hamburg - Kolding	245
Hamburg - Puttgarden	145
Kolding - Frederikshavn	261
Kolding - Hirtshals	264
Kolding - Nyborg	94
Kristiansand - Oslo	339
Halskov - Helsingør	157
Rødby - Helsingør	203
Helsingborg- Göteborg	223
Göteborg - Oslo	321

There is rather a big price difference in the different boat trips and there is also a difference in the number of kilometers when travelling by land. Now Marieke and Yvonne will try to find out which is the cheapest and the fastest way.

f. What should they know in order to find out? And how should they tackle this?

Their car, a -Piep-424, costs all things considered 42 cent per kilometer. On such a long journey their average speed will be 90 km/h.

Going North worksheet continued

The central question here is: "What is the cheapest way to go from Arnhem to Oslo?". After collecting the prices of the different routes the pupil has to model the geographical context into a discrete graph. This graph is simplified in several stages with the following final result.

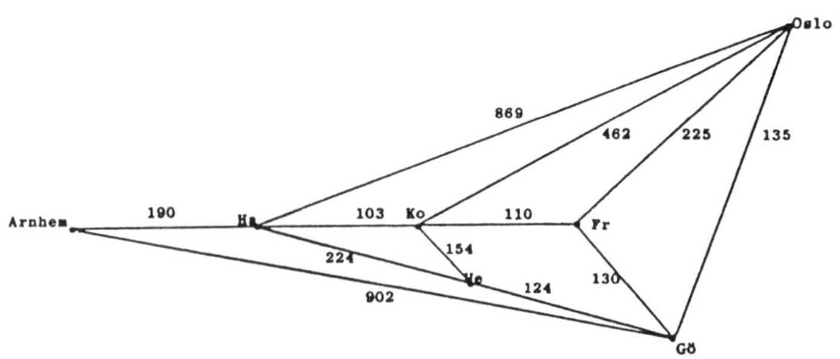

After this, the problem is solved. So we see a functional way to come from a geographical situation to an abstract model.

We notice that the essential properties of modelling occur and the features of the phenomenon of the graph become clearer. However, the link with reality always remains within reach and, at the same time, there is a clear rationale for the activities performed which makes it easier to solve the problem. The abstraction appears almost spontaneously from the problem stated. A possibility full of perspectives!

REFERENCES

Davis PJ and Hersh R. (1986). *Descartes' Dream.* Harcourt, Brace, Janovich.

CHAPTER 17

Mathematical Modelling in Schools – The Northern Ireland Further Mathematics Project

SK Houston
University of Ulster, Jordanstown, Northern Ireland

SUMMARY

The Northern Ireland Further Mathematics Project team initiated the Mode 2 Further Mathematics A-level course, offered by a number of schools in the province and assessed by the Northern Ireland Schools Examinations Council (NISEC). The course encourages mathematical investigation and mathematical modelling and has two novel assessment features – a major project and a mathematical comprehension paper.

This paper outlines the scheme briefly and gives details of the project work and the comprehension paper.

1. INTRODUCTION

At the beginning of the 1980s there was growing dissatisfaction with the way in which mathematics was taught, learned and assessed. It was recognised that there was little problem-solving activity or development of strategic skills (ie general strategies for dealing with different types of situation) or interpersonal skills. The Cockcroft Report (1982) crystallised this dissatisfaction and recommended that "Mathematics teaching at all levels should include opportunities for ... practical work ... problem solving and investigational work" (paragraph 243). It was suggested that the principal reason for this state of affairs was the examination system. Holcombe (1982) wrote "... the examination system at present actively discourages these innovations". While examining boards might take exception to the use of the word 'actively', arguing that examinations reflect what happens in schools and that examination boards can only lead where there is a consensus among teachers to follow, many teachers have said that the nature of the A-level examinations influences classroom activities to a considerable extent.

Consequently, in October 1982, an ad hoc group of teachers from secondary and grammar schools and tertiary institutions in Northern Ireland started discussing how to change examinations and hence classroom practice. In adopting these tactics, we were following the example of the British Government, who, according to Kingdom and Stobart (1988) "... set out to use the schools examination system as a means of influencing what went on in schools".

The involvement of school teachers throughout the development of this scheme was crucial to its success.

The group decided to concentrate initially on Further Mathematics for several reasons. Firstly, Cockcroft (1982, paragraph 561) pointed out that "... it is as important for students in the sixth form as for pupils of all other ages to develop problem-solving techniques, to pursue independent investigations and to discuss and communicate their ideas". Secondly, Further Mathematics is no longer generally a prerequisite for higher education and so changing it would not adversely affect pupils' opportunities for higher education. Thirdly, the number of pupils was small, therefore providing an opportunity for an easily controlled pilot scheme (about 2000 candidates per year sit NISEC's A-level GCE Mathematics examination, with about 200 also taking Further Mathematics). Fourthly, pupils already have a repertoire of skills and knowledge gained through a study of A-level. Fifthly, the experience gained by teachers at this highest level would eventually percolate down through the school, a process which was accelerated somewhat by the introduction of GCSE.

Accordingly, a proposal for a Mode 2 Further Mathematics examination was developed. This was accepted by NISEC in 1985 and approved by the Secondary Examinations Council later that year. Five of the nine schools originally associated with the project started teaching this scheme in 1986, and entered a total of 31 candidates for the first examination in 1988, and 19 candidates for the 1989 examination.

2. THE COURSE

Houston and Fitzpatrick presented details of the course to ICTMA-2 in Exeter, 1985. A full description is given in their paper (1987) which appeared in the conference proceedings. An account of the development of the course is given by Greer and McCartney (1989), and an initial appraisal of its operation was written by Houston (1989). Resource booklets for pupils and teachers were prepared (Nicholson 1986, McCartney 1986, Greer 1986) and extensive in-service training of teachers and preparation of pupils was carried out (see, for example, Nicholson 1989).

The aims and assessment objectives of the course are recorded elsewhere (see, for example, Houston 1989). Two aims in particular are important to the rest of this paper. They are

aim (ii) active involvement in investigative work in mathematics, and its applications

aim (vi) development of skills in the communication of mathematical ideas.

These aims are achieved particularly through the innovative features of the course – the project and the mathematical comprehension paper.

The candidate's total performance is assessed by these methods, and by two other written papers which are of more traditional form.

Paper I is on Pure Mathematics and paper II is on Applied Mathematics. Each of these papers contributes 27% to the total. Paper III contributes 26% and is the Mathematical Comprehension Paper mentioned above. A discussion of this paper follows below in section 4 and a critique of the 1988 paper is given in an article by McCartney (1989). The fourth element of the assessment, the individual pupil project, carries 20% of the total mark allocation. A discussion of the project follows in section 3 and a fuller report on the operation of the project is given in an article by Fitzpatrick and Greer (1989).

The syllabus contains much of the material usually found in A-level or Further Mathematics. Thus group theory, complex numbers, oscillations and sampling theory are among the topics taught. There are also a number of interesting new topics such as number theory, codes, population dynamics and Markov processes. The course aims to give students a grounding in the general processes of pure mathematical enquiry and of mathematical modelling, and the syllabus is taught so as to reflect these general processes. Further details of the content are published in Fitzpatrick and Houston (1987) and Houston (1989) and a complete syllabus may be obtained from NISEC. The syllabus content has been revised for the 1990 examinations to take account of changes in A-level. The aims and objectives of the course are unchanged.

The course seems to have been a success, both from the point of view of the teachers – who enjoyed teaching it – and from the point of view of the pupils – who all obtained examination grades at least as good as expected by their teachers. The 31 candidates for the 1988 examination all gained a grade D or better (on a 5-point range, A to E, of pass grades) with 17 candidates obtaining grade A, the top grade. One candidate went directly into employment and the others went to university. Most of them went to read an engineering subject or mathematics or physics, the others did medicine or a business related subject. The 19 candidates for the 1989 examination similarly all gained a grade D or better, with eight candidates obtaining grade A.

3. THE PROJECT

The individual pupil project carries 20% of the total mark allocation. Candidates carry out a project, under supervision of a teacher, and write a report on it. It may be a pure mathematics investigation or a 'real world' problem-solving exercise involving mathematical modelling. The submission date is 15 March in the year of the examination. The project is assessed internally by the supervisory teacher and is subject to external moderation by NISEC. In practice, in the first two years of operation, two teachers in each school assessed each project and returned an agreed mark. A moderator appointed by NISEC visited each school and interviewed each candidate. This helped to confirm that the work was the pupil's own and to achieve standardisation between schools. It was feasible to do this given the small number taking the examination; this level of moderation might not be possible with a large number of candidates. Pupils were firstly introduced to investigative and modelling activities, through group project work at two-day workshops (Nicholson 1989) - one workshop was devoted to investigations and the other to modelling. After an introductory discussion, and perhaps a video, pupils worked in small groups on a problem. At the end of the workshop, the group would give an oral presentation of their work, using an overhead projector, and would submit a written report afterwards. Pupils were also given practice at individual project work in their schools.

When it came to selecting a project for assessment in the examination, the choice of topic was left very much to the pupils themselves. Some made false starts on a project, gave up, and selected another topic. Some of them came forward with their own ideas for the project, while others selected a topic from a list provided in the resource book written by the project team (Greer 1986, 1988). After some practice on trial mini projects, pupils were quite enthusiastic about choosing their own topics, which were often related to another interest, for example, the economics of running a car, or the tennis serve. One pupil said "My project was chosen simply because it involved my hobby. There is a great sense of achievement in completing a project of some length which has been designed all by oneself". Another pupil said "It was good to experience the trials and tribulations of semi-research work and will auger well for university". Projects with the following titles were carried out in 1988 and 1989.

1988 Project Titles

The number 1729
*Integer sided triangles
Ship stabilising systems
*Fibonacci sequences
The greatest runner
Serving in tennis
Terminal velocity and drag
The trajectory of a golf ball
Folding and cutting shapes
Cube routes
Straight lines joining points
Chess and mathematics
Renting or buying a TV
The size of a golf ball and distance driven
A stochastic analysis of a tennis match
Order of stamps in a folded pile
Windscreen wipers
Space to park a car
Optimum distance for a place kicker to retreat in order to convert a try in rugby
Traffic ramps
Investigation of the Family Derby machine
How often should you change your car?
*Fuel consumption in a car
*A matrix investigation AB = B
*Roller coasters

*These project reports have been reproduced in full to accompany the revised guide to project work produced by the Further Mathematics Project team to assist pupils and teachers (Greer 1988).

1989 Project Titles

Predicting football results
Criticism of an interaction process
Rook tours
Abundant and deficient numbers
Network analysis
Interdimensional congruences (area = perimeter, etc)
Accuracy of approximations
Spirolaterals
Neat powers of complex numbers
Radiocarbon dating

Travelling salesman problem
Stochastic processes in the movement of pop records in the charts
Designs for windscreen wipers
Opening bids in bridge
Use of gears in cycling
Optimal sail sheeting angle
When should you change your car?
Rearrangements of 4 digit numbers
Comparison of record holders in athletics

The pupil guide to project work discusses both investigations in pure mathematics and investigations in applied mathematics (ie mathematical modelling).

Pupils are given advice on how to go about doing their project. It is suggested that their reports should contain about 3 000 words, that they should consult with their supervisor regularly, that they should plan their work, set goals and monitor progress. The proper use of sources of information and resources needed to do the job is discussed.

Generally, pupils started work on their project before the end of the first term of the second year of the course, with the intention of completing the work and submitting their reports by the following 15 March. Class time was made available for project work.

Project assessment guidelines have been produced with categories that apply to both types of investigation. These have previously been published by Fitzpatrick and Houston (1987). There are five categories.

Products (25%)
Processes (25%)
Evaluation (12.5%)
Use of sources and resources (12.5%)
Clarity of communication (25%)

Examiners were encouraged to use the whole range of marks, and scores between 40% and 100% were awarded.

4. THE COMPREHENSION PAPER

There are two sections in this paper, each with about 10 short questions to be answered in two hours.

Section 1 relates to a mathematical modelling article. A copy of this article is given to pupils six weeks before the date of the examination. Section 2 relates to a pure mathematics article which is relatively short and is given to pupils for the first time in the examination. The questions attempt to test understanding and lucid thinking, and the ability to verify mathematical assertions and to revise and interpret mathematical models. The mathematical knowledge required did not go beyond that contained in the core material of the syllabus.

The arrival of the modelling article six weeks before the examination catalysed great activity. Pupils worked together finding out as much as they possibly could about the topic. They (and their teachers) discussed the likely questions that might be asked. In 1988, one pupil wrote a programme to illustrate the model and made this available to other pupils and the other schools in the scheme. The modelling article used in the 1988 paper was about the modelling of Drug Therapy (Burghes et al 1982, 124–129). The article gives some background information and experimental data from which the pupils are led to a decaying exponential model. The objective of administering successive doses at regular intervals is to maintain the drug concentration in the bloodstream between certain limits. Pupils are led to work out a dosing strategy.

In the examination, pupils were asked to solve the differential equation and to verify that given values of the decay constant and the integration constant were consistent with the experimental data. They were asked to comment on the interpretation of various mathematical terms. The last two questions invited the pupils to work out a different dosing strategy from the one discussed in the article, and to discuss the advantages and disadvantages of each.

The pure mathematics article was the article by Larsen (1987) on Pell's equation $x^2 - 2y^2 = \pm 1$, where x and y are positive integers.

A critical analysis of the 1988 paper has been written by McCartney (1989).

The modelling article used in the 1989 paper was the article by Quadling (1987) called 'What the eye doesn't see, ...'. It dealt with the design of a door mirror on a car and it led pupils to calculate the blind spot in a car driver's field of vision. The question paper asks pupils, among other things, to discuss the simplifying modelling assumptions made in the article, and to formulate advice for buyers and drivers of cars using the predictions of the model.

The pure mathematical article used in 1989 was called 'Confusing quadratics' and was written by Bradley (1987). It deals with integer-sided right angled triangles.

Copies of these examination papers may be purchased from NISEC.

5. CONCLUSION

While the number of pupils taking this examination is small, we believe that the course and the assessment have achieved their objectives. It has been hard work for all concerned, especially the teachers, but it seems to have been an enjoyable experience for both pupils and teachers.

As a pilot scheme it has served its purpose well. Teachers and examiners have gained valuable experience in the novel forms of teaching and assessment. It is unlikely that this Mode 2 scheme will continue to have a separate existence for many years. We envisage revisions of A-level Mathematics and Further Mathematics which will encompass project work, and at least some elements of assessment which deal with comprehension. Discussion type questions are here to stay!

This paper has been written by SK Houston on behalf of the Project team.

REFERENCES

Bradley C. (1987). *Math Gazette*, **71**, 37–40.

Burghes DN, Huntley I and McDonald J. (1982). Applying Mathematics. Ellis Horwood, Chichester.

Cockcroft W. (1982). *Mathematics Counts. HMSO*, London.

Fitzpatrick M and Greer B. (1989). In preparation.

Fitzpatrick M and Houston SK. (1987). In Berry JS *et al* (eds), *Mathematical Modelling Courses*, 188–208. Ellis Horwood, Chichester.

Greer B. (1986). Investigations in Pure and Applied Mathematics. Unpublished.

Greer B. (1988). Investigations in Pure and Applied Mathematics, 2nd edition. Unpublished.

Greer B and McCartney JR. (1989). In Greer B and Mulhern G (eds), *New Directions in Mathematical Education*, Routledge, London.

Holcombe M. (1982). *Bull IMA*, **18**, 12–17.

Houston SK. (1989). *Teaching Maths and its Applications,* **8**, 115–121.

Kingdom M and Stobart G. (1988). *GCSE Examined.* Falmer Press, London.

Larsen ME. (1987). *Math Gazette*, **71**, 261–5.

McCartney JR. (1986). Processes in Pure Mathematics. Unpublished.

McCartney JR. (1989). *Teaching Mathematics and its Applications,* **8**. In press.

Nicholson A. (1986). Processes in Applied Mathematics. Unpublished.

Nicholson A. (1989). Sixth Form Mathematical Workshops. Submitted to *Teaching Mathematics and its Applications.*

Quadling D. (1987). *Math Gazette*, **71**, 198–201.

CHAPTER 18

Mathematical Models and Modelling. Examples from Danish Upper Secondary Teaching

B Hirsberg and K Hermann
The Ministry of Education, Denmark

SUMMARY

In the Danish Upper Secondary non-vocational education (the Gymnasium) a new curriculum comprising mathematical models and modelling came into force in August 1988.

This paper gives a short presentation of the new mathematics curriculum and syllabus, and shows how mathematical models and modelling are implemented.

Since computers play an increasingly large role in teaching and learning mathematics, the paper also includes comments on the impact of computers in the new curriculum.

1. INTRODUCTION

In recent years, many questions about the content of the mathematics curriculum have been subject to discussion as new demands and challenges have been met. For curriculum planners it is difficult to decide what to do, because trends in curriculum development are not unambiguous as they were, for instance, in the 1960s.

In August 1988 a new structure for the Danish Gymnasium (high school) came into force and, simultaneously, new curricula were worked out. One of the decisions made was to incorporate mathematical models and modelling into the curriculum. This decision, together with others on the content of the mathematics curriculum and syllabus, was the result of a long process comprising discussions among teachers and in-service educators, and the testing over a five-year period of an experimental curriculum published by the Minstry of Education.

In the following we will present the implementation of mathematical models and modelling in the new curriculum. To be able to see the underlying ideas, it will be necessary to give a short presentation of the new curriculum and to show how the Gymnasium is incorporated into the Danish school system.

2. THE GYMNASIUM : MATHEMATICS INSTRUCTION IN THE GYMNASIUM

The Danish school system is divided into two separate systems. The *Folkeskole* provides the primary and lower secondary education of pupils aged 7 - 16. The upper secondary non-vocational education is provided mainly by the three-year system called the *Gymnasium*, admitting just under 30% of an age class (about 20 000 students).

Instruction in the Gymnasium aims at preparing the students for further education and at providing general education. The teachers hold university degrees, normally in two subjects, the levels of which correspond to the level of a Master's degree.

The Gymnasium is divided into two streams, a language stream and a mathematics stream. On the language stream mathematics is an optional subject, with four lessons (each of 45 minutes) per week during the third year. On the mathematics stream mathematics is a compulsory subject, with five lessons per week over the first two years. In the following this is referred to as B-level. A-level is obtained by choosing mathematics as an optional subject (five lessons per week) during the third year. On average about 70-80% of the students will continue to A-level.

The final examination for both levels consists of both an oral examination and a written examination paper, prepared by the central authorities.

3. MATHEMATICS STREAM : MATHEMATICS CURRICULUM AND SYLLABUS

Before 1988 the mathematics curriculum consisted only of main topics. The new curriculum consists not only of main topics but also of so-called aspects.

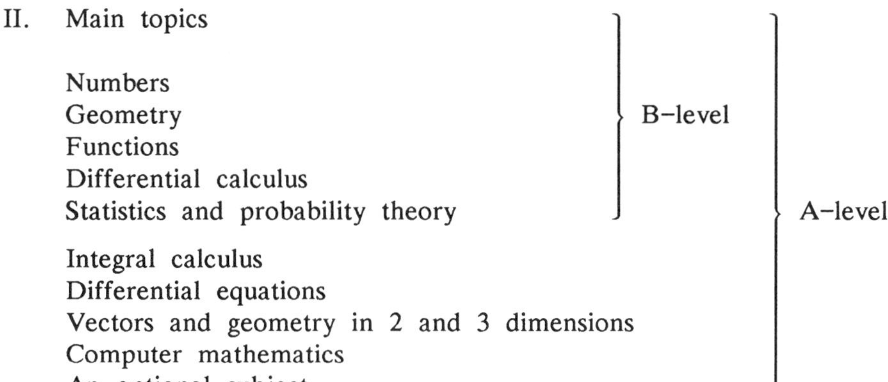

II. Main topics

Topics	B-level	A-level
Numbers Geometry Functions Differential calculus Statistics and probability theory	B-level	A-level
Integral calculus Differential equations Vectors and geometry in 2 and 3 dimensions Computer mathematics An optional subject		A-level

II. Aspects (both B- and A-level)

1. The historical aspect
2. The aspect of models and modelling
3. The internal structure of mathematics.

3.1 The aspects of models and modelling
The description of the aspect of models and modelling is the following.

"The aspect of models and modelling aims at making the students familiar with the building of mathematical models as representations of reality. They should be given an idea of the potential and limitations in the application of mathematical models. In addition, the instruction given should enable them to carry out a not too complex modelling process."

In the Direction for Teachers this description is elaborated.

"In not too complex situations the students should be able to carry out a modelling process by themselves. By way of examples, the following contexts can be recommended: simple optimisation problems, simple geometry problem situations where the geometric representation is not given in advance, and not too complex stochastic experiments. Elements of model building and problems related to the building and application of mathematical models should be discussed. Recommended are, for example, the purpose of modelling, selection of the segments of reality under consideration and idealisations thereof, mathematical representation including possible subsequent simplification and loss of information, verification problems. In connection with the treatment of the main topics, examples of models should be introduced, for example linear and exponential growth models. In addition, difference and/or differential equation models, linear programming or stochastic models can be included. General, dynamic or stochastic simulation programs can be studied. The given examples can also constitute self-contained instructional sequences. Likewise a major authentic model (or part of it) can be studied, for example physical, economic or ecological model. The social impacts of such models should be discussed."

The other two aspects are described in a similar way. The historical aspect aims at familiarising the students with elements of the history of mathematics and mathematics in cultural and social contexts. The internal structure of mathematics aims at providing the students with an understanding of the modes of thought and methods characteristic of mathematics, and their contribution to the development and structuring of mathematical topic areas.

3.2 Why compulsory aspects in the curriculum?

As already mentioned, the instruction in the Gymnasium has a double purpose. It qualifies the students for further education but it also qualifies them for society.

The same goes for mathematics. Most students leaving the Gymnasium do not later receive coherent mathematics instruction and, therefore, it is important to try to give them a balanced picture of mathematics. Hopefully, stressing the three aspects above is a way to achieve that.

The incorporation of the three aspects in the curriculum is intended to help the students understand better the role mathematics plays in society (in the past as well as today) and how mathematics can contribute to the understanding and solution of problems met out of school. Mathematical models, in particular, provide the basis for many decisions in society and play an increasing role in scientific disciplines.

In a technologically oriented society, people rely on things working and often do not expect to understand why they work. In applying mathematics, we considered it important and necessary also to know how and why mathematics works. That was the main reason for including the third aspect.

3.3 Consequences

The incorporation of these aspects in the curriculum requires that the teachers find new ways to organise and carry out their instruction. It also calls for teacher in-service courses and for new teaching materials.

If we look at the instruction in mathematical models and modelling as it is carried out, we will see that it is treated in the following ways.

- In connection with the main topics –
  for example growth models in connection with the treatment of linear, exponential and power functions

- In specially planned sequences –
for example models and model building, a study of a major authentic model (or part of it) and discussions of the social impacts and computer modelling (building new models or modifying known models, investigating their mathematical properties and discussions of model versus reality)

- In solving exercises –
for example applying known models to situations which are new to the students and building new models.

The written examination papers comprise applicational problems, but, in this situation, the students are not asked to formulate the problem and set up a mathematical model for its solution – this will be done in class.

Below we will give two examples meant as a guide for the written examination paper (B–level). In the first example the mathematical model is specified and in the second example it is partially specified but later the model has to be modified.

TOVVÆRK
3-slået polyester tovværk multifilament. Kvalitet 10452 højeste brudstyrke og speciel god slidstyrke, UV-stabiliseret.

diameter i mm	4	5	6	7	8	10	12	14	16	18	20	22	24	26
brudstyrke i kg	250	400	600	750	1000	1550	2250	3200	4000	4700	6000	7100	8600	10000

Figure 1

The table is from an advertisement for polyester ropes.

Show that the tensile strength as a function of the diameter can be described by

$$f(x) = b.a^x.$$

Determine a and b.

By how much is the tensile strength multiplied if the diameter is doubled?

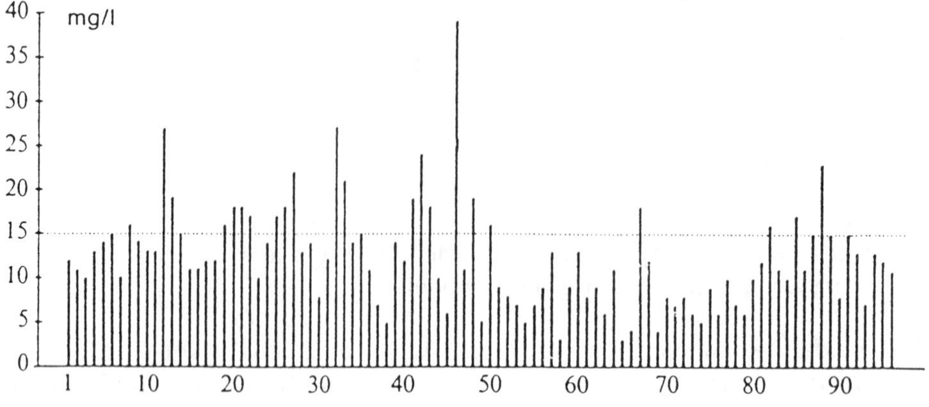

Figure 2

The figure shows the results from the analysis of 96 samples from a water treatment plant. The results show the so-called five-day biochemical consumption of oxygen (mg/ℓ). In 22 out of 96 samples the result exceeds 15 mg/ℓ. In the following it is assumed that the probability for a single sample to exceed 15 mg/ℓ is 22/96.

A control measure says

> No one of 12 samples chosen at random should exceed 15 mg/ℓ

Calculate the probability that the plant fulfils this measure.

If the control measure is altered into

> No more than two out of 12 samples chosen at random should exceed 15 mg/ℓ

what is then the probability that the plant fulfils this measure?

If the control measure is altered into

> No one of 12 samples chosen at random should exceed 20 mg/ℓ

what is then the probability that the plant fulfils this measure?

The introduction of the aspect of models and modelling and the two other aspects to the curriculum has also resulted in the publishing of textbooks about mathematical models, and it has also changed the content of other mathematical textbooks. As far as mathematical models are concerned, we see that the treatment of topics also comprises models and modelling and that special sections about model building and applications are added. Also, computer programs illustrating specific models and reference to general computer programs have been introduced in mathematical textbooks.

4. THE IMPACT OF COMPUTERS

All students in the Danish Gymnasium are now made acquainted with the use of computers and programs in connection with the instruction in a number of subjects and in mathematics in particular.

In the B-level mathematics curriculum, algorithms and numerical (iterative) methods are incorporated in the treatment of functions and differential calculus, and in the A-level mathematics curriculum this is continued in the treatment of integral calculus and differential equations. At A-level a new topic called computer mathematics is included. This topic is defined as a mathematical topic where algorithmic modes of thought and use of computers play an essential role. Examples are linear programming, chaos and fractals, three-dimensional pictures on a computer screen (Computer Aided Design). Some of these topics in computer mathematics can also be considered as important in relation to the aspect of models and modelling.

Appropriate instruction materials and programs have already been developed, and groups of teachers work seriously with questions related to the influence of computers in the process of teaching and learning mathematics. Some of these questions concern the use of computers in the learning of mathematical concepts and in the interaction between intuitive understanding and exact manipulation of mathematical concepts. Also, the use of computers calls for new ways of planning and carrying out mathematics instruction.

The Association of Mathematics Teachers plays a central role in this work with respect to the development of instruction materials and programs and to making the teachers acquainted with these materials and programs.

5. FINAL REMARK

As mentioned at the beginning, the new curriculum came into force in August 1988, so what actually happens now is that our model for incorporating mathematical models and modelling in a national curriculum is confronted with reality - students, teachers and time. So we are anxious to see how many times we have to run through the modelling cycle before the model satisfies the intentions.

REFERENCES

Hermann K. (1989). A Food Producing Factory. In W Blum *et al* (eds), *Applications and Modelling in Learning and Teaching Mathematics*. Ellis Horwood, Chichester.

Niss M. (1989). Aims and Scope of Applications and Modelling in Mathematics Curricula. In W Blum *et al* (eds), *Applications and Modelling in Learning and Teaching Mathematics*. Ellis Horwood, Chichester.

Hermann K and Hirsberg B. (1989). Recent Trends and Experiences in Applications and Modelling as part of Upper Secondary Mathematics in Denmark. In Blum W, Niss M, Huntley I (eds), *Modelling, Applications and Applied Problem Solving*. Ellis Horwood, Chichester.

CHAPTER 19

Incorporating the Aspect of Mathematical Modelling in the Danish Gymnasium Curriculum : Problems and Perspectives

M Blomhøj
Royal Danish School of Educational Studies, Copenhagen, Denmark

SUMMARY

Starting from an analysis of a new mathematical curriculum of the Danish gymnasium, two different arguments for including mathematical modelling are discussed. One is based on society's demand for adequately educated manpower, and one on the need for a democratic competence to the application of mathematical models in society. The author supports the last argument, and outlines some didactical positions which can be used as organising principles for an instruction in mathematical models which observes the given ends. Problems of implementation are discussed.

1. INTRODUCTION

In 1983 a new mathematics curriculum for the science side in the Danish gymnasium (upper secondary level, aged 16–19 years) was proposed by the Ministry of Education. At the same time it was made optional for the individual gymnasium to carry out pilot teaching along the lines of the proposed new curriculum. On the basis of this pilot teaching, the new curriculum has, with a few changes, come into force as from the academic year 1988/89 (Hermann and Hirsberg, 1989).

What is decisively new about the curriculum is that the teaching is now explicitly required to include three meta-mathematical aspects.

1. The aspect of history.
2. The aspect of modelling.
3. The internal structure of mathematics.

(The Ministry of Education, 1988)

In the new curriculum, it is required that the teaching must include these aspects in the treatment of mathematical topics, and that particular courses of lessons must be organised with one or more of the aspects in mind. Thus, the aspects are to be considered both in the compulsory mathematics of the first and second grade, as well as in the optional mathematics of the third grade of the gymnasium (see Note 1).

The new curriculum gave rise to widespread meeting and debate. As an element of my mathematical education at Roskilde University (Note 2), I participated in the debate with two of my fellow students, and in this connection we wrote two textbooks to be used for teaching the aspect of modelling (Note 3). On the basis of the experience from our work, I will discuss below some perspectives in, and problems with, incorporating the aspect of modelling in the teaching of mathematics on gymnasium level.

2. JUSTIFICATIONS FOR THE ASPECT OF MODELLING

As in most other West European countries, the gymnasium in Denmark has a dual purpose. The gymnasium should provide a general, all-round education, as well as prepare students for further education.

The construction of a new mathematics curriculum was a recognition of the fact that the existing mathematics teaching no longer managed sufficiently to advance the given ends. I think that the demand for changes in mathematics education derives particularly from two circumstances. Firstly, the base of admission to the science side of the gymnasium has widened considerably and admission has trebled since the mid-sixties. As a consequence, the secondary school leavers' choice of education is far more varied today. This means that appropriate mathematics teaching must prepare students for a wide range of short, medium and long term studies for which mathematical competence is required. Secondly, the notion of general education – understood as education enabling active and critical participation in a democratic society – has developed concurrently with the progress towards a society of high technology. Getting on in modern society increasingly necessitates the ability to integrate knowledge and experience from different disciplines. In this connection, teaching mathematics in isolation is futile. On the contrary, instruction must illuminate mathematics from different angles, so that the acquired knowledge becomes an integral part of the student's personal experience and skill.

In the light of this I find that two types of argument can be advanced which, from an extra-mathematical point of view, justify the introduction of modelling in the gymnasium mathematics curriculum.

(a) Technological knowledge about how to build and how to use mathematical models

Mathematics finds increasing application within other fields. The application takes place mainly by setting up and employing mathematical models. Therefore gymnasium mathematics teaching must prepare the students so that, possibly through their further education, they are enabled to set up and employ mathematical models.

The rationale behind this argument is that the demand for adequately educated manpower in the high-technology society includes the need of a large group, which has a technological knowledge about how to build and use mathematical models.

(b) Democratic competence in relation to mathematical models

The foundation for a wide range of important decisions concerning the individual citizen's employment and participation in democratic society is increasingly provided by the use of mathematical models (for example, production planning, economics, and pollution control). Therefore, mathematics teaching at gymnasium level must aim at enabling the students to take a qualified and critical position on the application of mathematical models in society (Notes 5 and 6).

By and large, higher educations in which mathematics is used neglect the critical perspectives in the application of mathematical models in their mathematics instruction. As long as this is the case, it is crucial that the mathematics teaching in the gymnasium contributes to the formation of a professionally critical judgement in relation to setting up and applying mathematical models. This is important for future experts who are to construct and employ mathematical models, as well as for future laymen.

These two arguments of justification for including the aspect of mathematical modelling at gymnasium level are, however, based on ends and purposes in teaching mathematical modelling so different in character that their realisation may often impose incompatible requirements on the concrete organisation of instruction.

If gymnasium mathematics teaching shall observe the end given in argument (a) most of the time available for teaching modelling aspects has to be dedicated to the students own work with setting up models and solving them analytically, or numerically on computers. Due to the limited time and to the students mathematical qualifications the examples have to be limited to well known problem areas in physics, chemistry and biology. In my opinion, such instruction will give the students an unbalanced picture of the modelling process, the validity of mathematical models and the role of mathematical models in society. Instead of facilitating this, such instruction will rather impede the formation of critical potential in students toward the use (and misuse) of mathematical models in society.

To me, the main reason for having mathematical models in the mathematics teaching of the gymnasium is clearly to be found in argument (b), but I find it important to add that the formation of a professionally critical judgement is closely bound up with the student's personal experience with setting up and applying mathematical models. This addition partly answers argument (a). Instruction in pursuit of the aims of argument (b) necessarily implies that the students are, to some extent, working with mathematical models themselves. However, this work will take place within the general aim that the students should gain an insight in the many different functions (whether intentional or not) which a mathematical model may have in society.

3. THE ORGANISATION OF INSTRUCTION

Of course, the more detailed organisation of instruction cannot directly be inferred from the general aims and purposes or any reason for pursuing these.

Below, I present some *principles* which can assist in organising instruction which meets the objective of this aspect of modelling. Some of the principles express my own didactic attitudes, while others are based on actual experience from the preparation and testing of the textbooks mentioned earlier.

(1) Instruction in mathematical models must, as a prevailing principle, exploit the complementarity between mathematical knowledge and skill on the one hand and knowledge about the process of modelling and the function of mathematical models on the other hand. If this principle is observed, I see a possibility that instruction in mathematical models may motivate the students both in the perspective of adult-life relevance and on the immediate level of experience and its epistemological implications.

(2) The planning of the instruction must be flexible enough to make use of the students' being motivated through their work with a model to acquire new subject matter.

(3) The teacher must communicate the subject matter interactively with the students during their work with the models to as large an extent as possible.

(4) Instruction must employ methods of work offering the students the opportunity to discuss mathematical as well as meta-mathematical aspects of their work among themselves.

(5) The process of setting up models must be central to the instruction, and the distinction between reality and model must be given special attention.

(6) Instruction in mathematical models at gymnasium level necessarily involves setting up, applying, and criticising pseudo-realistic models. However, it is crucial that such models are selected so that aspects which in one way or another are typical for the general problems of setting up, applying or criticising mathematical models are demonstrated.

(7) Instruction must comprise several courses in which the students are working through an entire process on their own (with proper support from the teacher) - from formulating the problem, setting up the model, manipulating the model, to clarifying the problem and criticising the whole process. Such intense treatment of separate problems opens up the prospect of attaching a range of personal, social and political aspects to the discussion about the application of mathematical models.

(8) Instruction should incorporate some authentic models with different functions and statuses. From their familiarity with the process of modelling, the students must discover that it is possible to criticise mathematical models and their application on many levels, and that the relevance of this criticism depends on the circumstances of the actual application.

The following section singles out some of the factors which, in my opinion, especially hamper the realisation of mathematics teaching which attempts to comply with the above principles.

4. THE PROBLEM OF IMPLEMENTATION

The biggest impediment to the realisation of any new curriculum is the inertia with which educational systems carry on the prevailing practice of instruction. With regard to mathematics instruction in the gymnasium, tradition prescribes a topically orientated and textbook-guided instruction which primarily aims at training students in skills required to solve the established exercises or the written exams. This tradition certainly encourages tool-orientated learning. Owing to the crammed list of topics and the students' ambitions of high marks, the teacher is impelled to focus on tool-orientated skills. Such instruction will, of course, affect the students' perception of mathematics and their attitudes to the process of learning, and also the teachers' attitude towards their role.

To be fair, in the new curriculum the list of topics has been somewhat reduced to provide time for the treatment of the three aspects, but the written examinations still comprise nothing but the traditional topics. Consequently, I see a substantial danger that the traditional modes of instruction will be carried on under the new curriculum and that an attempt will be made to cover the aspect of modelling by treating a number of trivial models which primarily lend themselves to the training of traditional mathematical skills.

Such instruction will in no way facilitate a critical and independent attitude towards mathematical modelling, and is therefore unable to satisfy the intentions behind the incorporation of the aspect of modelling. In an evaluation of the possibilities of changing instruction towards observing a higher degree of consideration for the intentions behind the introduction of the aspect of modelling, two concrete points are crucial: in-service training and the supply of teaching materials.

For in-service training it is important to note that the education of mathematics teachers for the gymnasium is traditionally a Masters level university education in mathematics. Only the new universities in Roskilde (RUC) and Aalborg (AUC) have integrated pedagogy and didactics in their mathematics - and only RUC and AUC explicitly require of their mathematics students that they study construction, application and criticism of mathematical models. Consequently, there is a massive call for in-service training among teachers - a call which is far from being answered in the present state of affairs.

Obviously, teaching material - however outstanding it may be - will never be a sufficient prerequisite for appropriate instruction in mathematical models but, on a short view, it is the scarcity of relevant teaching material which puts a limit on how fast instruction can change as intended. A large amount of the existing textbook material links the treatment of the aspects to the well-known presentation of the subject matter. I find that this is an extremely inexpedient situation, which only encourages continuation of the prevailing tradition of instruction.

An important reason for the want of alternative presentations is that the gymnasiums find it difficult to allocate resources to textbooks of mathematics which cover only a small number of the compulsory topics. Hence it is not economic to produce supplementary textbooks for mathematics teaching in the gymnasium.

Two types of teaching material are badly needed. Firstly, material in the form of exemplifying courses during which treatment of the subject matter has the intention that the students should acquire an insight in the problems concerning setting up, applying and criticising mathematical models. Such material will stimulate teachers and open up discussions about mathematical pedagogy. Secondly, presentations are needed which make a broad and representative selection of mathematical models available to the teachers.

Changing the prevailing practice of instruction is always a long term process and, in spite of the problems of implementation mentioned above, I find that the new curriculum gives a good basis for changing mathematics instruction towards observing formation of a professionally critical judgement in relation to setting up and applying mathematical models.

Notes

(1) With the new reform having come into force, mathematics teaching on the science side has remained compulsory in first and second grade while it has become optional in third grade. A more detailed description of the new curriculum can be found in the paper of Hirsberg and Hermann in this volume, or in Hermann and Hirsberg (1989).

(2) The teaching of mathematics at IMFUFA (Department of Mathematics and Physics at the University of Roskilde) has always been heavily influenced by the department's research interests which cover all aspects of setting up and applying mathematical models. Thus IMFUFA has many years of experience with students' project work covering both setting up models from scratch, critical analysis of existing models, and examination of the societal impact of mathematical models, along with consideration of didactic and cognitive problems concerning the incorporation of mathematical models in the gymnasium curriculum. On the basis of this, IMFUFA has, through its output of candidates and through participation in debate concerning the gymnasium instruction, played a leading role in the incorporation of the aspect of modelling in the mathematics teaching at the gymnasium.

(3) Blomhøj et al (1984) and Blomhøj and Frisdahl (1985) are the two books in question.

(4) My analysis of the new curriculum is based on the general analysis given by M Niss at ICTMA-3 (Niss 1989).

(5) Both arguments of justification have a premise side and a purpose side. It is sometimes useful to distinguish between these sides of

the arguments.

(6) The two given arguments for including the aspect of modelling in the mathematics curriculum at gymnasium level are closely related to the epistemological analysis of mathematical modelling given by Skovsmose (1989 p15).
"We have to deal with three different types of knowledge related to a process of mathematical modelling:
1. Mathematical knowledge itself.
2. Technological knowledge, which in this context is knowledge about how to build and how to use a mathematical model. Also, we would call it pragmatic knowledge.
3. Reflective knowledge, to be interpreted as a more general conceptual framework, or metaknowledge, for discussing the nature of models and the criteria used in their constructions, applications and evaluations."

REFERENCES

Blomhøj M, *et al.* (1984). *Linear Programmering.* Forlaget FAG, Frederikssund.

Blomhøj M and Frisdahl K. (1985). *Modelsnak – differentiallignings-modeller* (Model Talking – Differential Equation Models), Forlaget FAG, Frederikssund.

Blum W, *et al* (eds). (1989). *Applications and Modelling in Learning and Teaching Mathematics.* Ellis Horwood, Chichester.

Blum W, Niss M, Huntley I (eds). (1989). *Modelling, Applications and Applied Problem Solving.* Ellis Horwood, Chichester.

Hermann K and Hirsberg B. (1989). In Blum et al, *Modelling, Applications and Applied Problem Solving*, Ellis Horwood, Chichester.

Niss M. (1989). Aims and Scope of Applications and Modelling in Mathematics Curricula. In Blum *et al* (1989), *Modelling, Applications and Applied Problem Solving*, Ellis Horwood, Chichester.

Skovsmose, O. (1989). Mathematical education and democracy. Department of Mathematics and Computer Science, Aalborg University Centre, R 89–40.

Undervisningsministeriet (the Ministry of Education): 1988, Matematik. Bekendtgørelse og vejledende retningslinier. (Mathematics Curriculum and Guidelines), Haefte 19.

CHAPTER 20

Project Work on Models in Upper Secondary School

E von Essen
Himmelev Gymnasium, Denmark

SUMMARY

I will present two cases of curriculum projects on mathematical models with project work as an important teaching method. In both cases the students (aged 16–19 years) were on A–level courses in mathematics at the upper secondary school in Denmark.

1. FRAMEWORK OF THE MATHEMATICS TEACHING

The upper secondary school in Denmark (called the *Gymnasium*) is a three–year education. It is divided into a modern language side and a science side. On each side the students have to choose streams after the first year (at least, this was so until a change in the structure of the upper secondary school in 1988). The students are in the age range 16–19 years. The students involved in these projects were science students with A–level in mathematics and physics in the second and third form. The so–called A–level was the highest level of these subjects in the Danish Gymnasium. The students had five lessons per week in mathematics in the first form, five in the second form, and six lessons in the third form.

The syllabus contained (1) analytical geometry and vector theory, (2) functions (polynomials, trigonometric functions, logarithms, exponential functions), (3) differential and integral calculus, (4) probability theory and elementary statistics, (5) a topic chosen by the students, and also (6) a little algebra, number theory, and affine geometry. In 1984 a 'standard experimental mathematics curriculum' was published by the Ministry of

Education, and teachers were allowed to choose this curriculum if they wanted. In this publication the algebra, number theory and affine geometry was replaced by (6) algorithms and numerical mathematics involving the use of computers, and (7) some examples of mathematical models and some examples of historical aspects.

The course is completed by two examinations, one in written form with ordinary problems and one oral.

2. BACKGROUND

For many years the mathematics instruction at the upper secondary school in Denmark has concentrated on pure mathematics. However, many teachers at all levels have emphasised recently the need for including applied mathematics in the teaching. In a report from a national congress on mathematics in Denmark in 1981, the committee concerning upper secondary school found that all mathematics instruction ought to "give the student a knowledge of some authentic real-life applications of mathematics of essential social significance".

Probably, this national congress encouraged a number of teachers to include authentic applications in their mathematics teaching. Any teacher of the upper secondary school in Denmark may apply to the Ministry of Education for permission to change the syllabus (for example, include a new topic and leave out another) for his instruction of a class of students. Such teaching experiments are of great importance to the development of mathematics instruction. The two curriculum projects which are presented here are cases of such experimental instruction. So, rather than being an academic treaty, this paper reports on a piece of developmental work carried out in actual school practice.

In these two curriculum projects I wanted to give authentic applications of mathematics an important position in mathematics instruction, and to offer the students an opportunity to work on their own with mathematical topics – in this case with group projects – without continual control from the teacher. I also wanted to investigate the advantages, disadvantages, and difficulties of such a working method in mathematics teaching.

In both cases I obtained permission from the Ministry of Education to devote 40 lessons to such work, and obtained an equivalent reduction of the examination requirements.

3. THE FIRST PROJECT : 1983–1985

The first project involved a class of students in the second/third form in 1983–85. The main points of this teaching experiment were mathematical models and applications of mathematics at the point of production or in the service sector. The latter was the central point, and the choice of business necessarily influenced the organisation of the mathematics instruction. Consequently, in the first months of the second form, the students were informed of various possible choices. After some discussion in class, most of the students preferred to investigate models of fish populations and fisheries, especially in The North Sea.

I contacted the Danish Institute for Fisheries and Marine Research. A fishery biologist helped me obtain some material on their work – elementary lecture notes from the education of fishery biologists, scientific papers, and an annual report from the Institute. In addition I contacted the newly established North Sea Centre based in a small town called Hirtshals in Northern Jutland. A visit to the centre with the class was arranged in August 1984.

In the second form we dealt with differential and integral calculus, differential equations, and so on. The students were then given notes on mathematical models. Most effort was put into topics like different types of models, and the phases and difficulties in modelling. The theory was illustrated by a few elementary examples taken from biology, such as exponential and logistic growth of populations. Finally, computer methods for the solution of differential equations (such as Euler integration) were discussed.

During the first weeks of the third form, the class worked on fishery models. The 14 students split into three groups, each group working on an individual topic. One group chose to focus on simple models of recruitment, growth, mortality, and so on (models with just one differential equation). Another group concentrated on how a multi-species model was constructed from the simple models. Thc third group studied the role of the different models in international fishing politics, prognoses for the fish stocks depending on the fishing efforts and suchlike.

The students decided to work entirely on their own. My role was just that of an adviser, who was consulted when the group lacked relevant literature or had problems of one nature or another.

As an introduction to the study of the fishery models, all the students read extracts from an annual report of the Institute for Fisheries and Marine Research, a non-mathematical article on the so-called North Sea Model, a book on the economic perspectives of the Danish fishery trade, and a scientific paper (written in English) for fishery biologists, with an extensive use of rather complicated mathematics.

In August 1984 the students and I spent a week in Hirtshals, studying fishery and fishery models. Of the greatest importance was the visit to the North Sea Centre, where we spent a whole day. We attended a lecture on the subject, and were shown around the centre. The students worked at the computers there, with a simplified fishery model. We also paid a visit to the North Sea Museum, which is a part of the North Sea Centre.

In addition we visited some local fishing industries (producing frozen and tinned fish) where we had the opportunity of discussing fishing quotas and the condition of the fishing industry in Denmark. We also attended a fish auction.

During the excursion we lived in a scout cottage in the neighbourhood, where the students undertook the cooking, cleaning, and so on. Part of the time the groups read relevant literature on the subject, and began work on their reports.

In the weeks that followed, the groups continued to work on their projects during part of the mathematics lessons, and their usual homework in mathematics was suspended. The students studied articles or chapters from books on fishery models and their significance, and some of them worked with simple computer models. In October each group made an outline of their report, and in November their work was completed and the group reports were collected into one big report, a copy of which was given to each student (with my corrections and other comments). Finally each group presented the main points of their work to the other groups.

4. THE SECOND PROJECT : 1987–89

The second project involved a class of students in the second/third form in 1987–89. As with the first project, the second project concerned mathematical models, and project work played a major part. Again the students were introduced to the concept of a model by reading notes. Instead of one long period of project work, which involved studying just one model or application, the remainder of the 40 lessons was divided into three periods – two short periods dealing with two quite different models, partly chosen by the students, and one period with project work in groups. The subjects of the group projects were chosen by the students, and this time the groups studied models which had no relation to one another.

For the first period the students decided to work with models of the structure of stars. The instruction was closely coordinated with the physics teaching of the same topic. A very concise text in English was used, the students worked in small groups, and were guided to several questions in the text. As in the previous project, an excursion for a week was arranged. Part of the time was devoted to the continued study of models of stellar structure. We attended a lecture on this subject at the Astronomical Institute of the University of Aarhus, and we visited an observatory. On the basis of these experiences, and some literature on the subject, the students divided into groups and wrote a short report.

It was stated in the description of the curriculum project (approved by the Ministry of Education) that at least one of the models treated should be of real social significance. As we had not much time left the selection was limited, and the students opted to deal with fishery models. Owing to trouble with the difficult text on models of stellar structure, the students decided that this topic should be taught in the ordinary way, by instruction in class. We just went through a chapter from a textbook, and worked with a simple model on the computers at the school.

During the last period the students worked in small groups on one or more specific models of their own choice. The chosen themes were national economy, business economy, population dynamics, fisheries and chaos. This time the project period was shorter but comprised the same stages, and there was no attempt to relate the different projects. The students recieved a copy of just their own group's report, but no opportunity was given to present their reports to the other groups.

5. THE EXAMINATIONS

At the written examination, with ordinary mathematical problems set by the Minstry of Education, it seems there was no difference in the standard of achievement of these students compared with that of normal classes. In one instance, however, one problem had to be replaced by another, because the original problem concerned a topic omitted from the syllabus.

At the oral examinations, only half of the texts used in the instruction have to be offered, and those offered are selected by the students and the teacher in common. The students felt rather uneasy about offering their reports for the examination, so (in both cases) we offered only the notes with the introduction to models and a little on fishery models. Consequently there was nothing extraordinary about the examinations.

6. EVALUATION OF THE PROJECTS

In both cases, after termination of the project work, the students answered a questionnaire on their opinion of the instruction in mathematical models. Almost all expressed a high degree of satisfaction with the working method, and stated that they had benefitted from the instruction. All were pleased to have had the opportunity to work on their own. Some of the students in the first class, however, regretted having chosen so much independence for so long a period. This is one of the reasons for choosing a shorter period of project work in the second case.

Students in the second class were happy with the working methods, but some found that it was not so very different from the ordinary mathematics lessons. I think they are right. The project work was too small a part of the instruction in models.

As a negative aspect of the working in groups, some mentioned that the cooperation in their group was poor. In particular they had some difficulties in structuring their work. Not until shortly before the deadline for the reports did they begin to work seriously on the subject.

Motivation has varied a great deal. Some of the students have had resources for and interest in putting an extraordinary effort into the project work – and with a great return for their effort. Others have welcomed the possibility of relaxing a bit.

Regarding the skill and insight achieved, the students are quite satisfied, and they find the study of the models relevant and interesting. In their comments on the questionnaire, many of them mention the importance of experiencing the application of mathematics in practice on location. There is no doubt that the students in the first class got a greater and more powerful experience in this respect than those in the other class. Many of the students in the second class have regretted their choice of studying models of stellar structure, not only because the text was difficult, but also because this application of mathematics was quite as abstract to the students as the pure mathematical theories they were usually taught.

Another point that many of the students (in both classes) mentioned is the excursion. The concentrated work for one week, with only a few topics concerning their school subjects, the reading of relevant literature, visits to places where they can see applied mathematics in practice, writing the report, were all found to be most valuable by the students. This fact is especially thought-provoking, because of the contrast with the usual work at school.

7. FINAL DISCUSSION

Among the objections I have met to project work and teaching of mathematical models, are the following.

(a) The students learn too little compared with the time invested.
(b) It takes too much time from the teaching of (pure) mathematics.
(c) It cannot be tested in a reliable way.
(d) The teachers' knowledge of the different applications is inadequate.

Comments

(a) Usually our students do not have to work on their own in the mathematics instruction. I am convinced that students who have carried out project work like this are better prepared for further studies involving mathematics or applied mathematics, and for any kind of employment requiring independent work.

In addition, the students' view of, and attitude to, mathematics have been strongly affected by the experimental teaching. They have left school with an impression of mathematics being more in touch with real life and of greater importance than most students. You could not obtain this just by *telling* the students that mathematics is of great importance and has many real life applications – they have to experience it in practice.

Furthermore, it seems that the teaching of simplified, adapted examples from the real world is not sufficient. It is remarkable that the experience of the importance of fishery models to the fishing industry is convincing to the students – a lecture on models of stellar structure at the university is not. The models have to deal with the *real* real world.

(b) Perhaps there is a tendency to incorporate more and more topics within the syllabuses. Consequently the students' knowledge of the different topics tends to be more and more superficial. A syllabus with emphasis on models and project work could give the students an opportunity to concentrate upon fewer topics, which might raise the standard of mathematics instruction.

(c) At the oral examination, it turned out to be quite possible to test the students' knowledge and understanding of the models and applications they had dealt with. However, the students felt rather uneasy about offering their reports or other texts from the experimental teaching for the examination, probably because most of their textbooks were quite different, with just a few simplified examples of applications. After the revision of the curriculum in 1988, a number of textbooks with more emphasis on applications and models have been published. I think that the use of such books in the instruction will reduce this problem.

Probably, an ordinary written examination is less suitable for testing applications and modelling achievements.

(d) No teacher can be familiar with all details in every branch of applied mathematics, but that will not be necessary. On the contrary, it is my experience that it may improve the relations between the teacher and the students to be in a situation where the teacher does not know all the answers, or where the questions have more than one answer. I am sure that it was inspiring and stimulating for my students in their project work to have the opportunity to obtain knowledge and an understanding of the topic which exceeded mine.

Consequently I agree with the students, that it is profitable to include examples of authentic applications in mathematics instruction, and that it is profitable to include a period of project work or other kinds of independent exploratory work in groups.

REFERENCES

Hermann K and Hirsberg B. (1989). Recent trends and experiences in applications and modelling as part of upper secondary mathematics instruction in Denmark. In Blum W, Niss M and Huntley I (eds). *Modelling, Applications and Applied Problem Solving.* Ellis Horwood, Chichester.

CHAPTER 21

Teaching Linear Regression Models at Secondary School Level

RM Bottino, P Forcheri, MT Molfino
Instituto per la Matematica Applicata del CNR, Genoa, Italy

SUMMARY

An educational methodology is analysed, which allows integrated mathematical and computer science concepts and notions to be applied and/or introduced through the teaching of linear regression models.

In particular, two different approaches to the methodology itself are discussed. Furthermore, the experiments we carried out with upper secondary school students are described briefly, together with the results obtained.

1. INTRODUCTION

Mathematical modelling is not present in the Italian secondary school curriculum. However, the educational importance of this topic has now been recognised. As is well known, mathematical models constitute an important class of tools which allow real situations to be described and interpreted. By investigating these situations, students can be guided to examine them quantitatively, to recognise the common aspects of various phenomena, make assumptions from examination of the data available, and make predictions and verify the consistency of an assumption by analysing critically the result obtained.

In recent years, these considerations have led to the development and testing of teaching sequences aimed at including mathematical modelling in schools.

The work we have carried out falls within this sphere and concerns, in particular, the introduction of linear regression models into mathematics courses during the last three years of upper secondary school (age 16–19).

Several motivations justify this choice, and numerous phenomena can be analysed through this type of model with satisfactory approximation. It is also possible to provide both graphical and geometrical examples, and theoretical explanations, using notions and concepts applicable to the age range under consideration. Furthermore, the topic allows integrated mathematical and computer science notions and concepts to be applied and/or introduced to advantage. The linear regression models provide a means of revising, from an operative point of view, concepts and mathematical techniques included in the school curriculum which the student frequently finds difficult to understand. The manipulation of algebraic expressions, analytical geometry, approximations and errors, and the use of linear and nonlinear graph papers, can all be mentioned as examples. With regard to computer science, the linear regression models are reasonably simple and permit concepts that have already been introduced – such as algorithm and program – to be applied and revised, and new concepts – such as elements of computer graphics and file handling – to be introduced. All this is in order to facilitate the formation and investigation of assumptions regarding the general trend of a phenomenon.

Finally, the linear regression models allow activities to be performed involving many subject areas. In fact, the data to be analysed may originate from real or experimental situations appertaining to many different subject areas, such as physics, chemistry, biological sciences, and economics. The application of the same mathematical model to different situations confirms the importance of the introduced topic, and requires that the detailed analysis of the results obtained accounts for the constraints of the environment in which the data were chosen.

These considerations have led to the development of a teaching package based on regression models, limited to cases which are either linear or that can be linearised. Two different approaches to the package are discussed below.

2. LINEAR REGRESSION MODELS IN THE SECONDARY SCHOOL

The aim of our work is the practical introduction to the concept of quantitative models, by reference to examples taken from different situations, with a brief account of the distinction between reproducible and non-reproducible phenomena. The introduction of these concepts is limited to linear regression models (or those that can be linearised), which are studied in detail.

The methodology that we propose is the result of pedagogical considerations regarding the problem of the search for a model. From this point of view, three particular aspects arise: the process of abstraction, which forms the basis of the development of an assumption; the way in which it is formalised, which requires the ability to handle mathematical objects; and its validation (or rejection), which requires an ability for critical analysis.

As we have observed in numerous cases, students are assisted in the development of mental processes of this type if they are made to work on concrete situations which can be analysed intuitively in graphical form. The survey of cases from different domains naturally leads the students, through generalisation, to understand the concepts of modelling. The students are introduced, through the use of examples, to the idea that there are other types of models, and that the ability to make assumptions is the root of the modelling process.

Subsequently, examples are analysed to induce students to employ formal methods by which the fit of a model can be evaluated quantitatively.

The first, intuitive, phase allows the practical significance of the work to be clarified, and eliminates the risk that the successive phase, in which concepts are formalised and quantified through a detailed study of the linear regression model, will be considered as a mere set of notions to apply in a mechanical manner. Consequently, the work is developed according to the following methodology.

(a) Analysis of real phenomena
(b) Identification of a model to apply to the phenomena under consideration by using qualitative instruments (graphics, data evaluations, and so on)
(c) Determination of the formulae which express the coefficients of the regression line and the coefficient of the correlation
(d) Identification of a model to apply to the phenomena under consideration, by using quantitative instruments (using a computer program to calculate the regression line coefficients)
(e) Testing the assumed model for validity

3. DEVELOPMENT PROPOSALS

The different points in the methodology can be developed and studied to different levels of detail, according to the teaching context in which they are performed. We can illustrate two different proposed teaching schemes: the first develops mathematical and computer science aspects, while the second develops application concepts.

3.1 Mathematical and Computer Science Aspects

The aim of the teaching scheme, described in the following paragraphs, is to explain mathematically the regression line coefficients (point (c) of the sequence) and, in part, simultaneously to introduce and apply computer science knowledge.

In fact this proposed scheme has been developed for incorporation into the mathematics course of the last three years of the upper secondary school. The scheme also takes into account discussions currently in progress in Italy regarding the use of computer science notions and concepts within the mathematics curriculum.

With regard to the concept of the model, our proposal provides for the development of the didactic methodology according to a deterministic approach. This choice is motivated by the fact that the Italian upper secondary school provides for different course options. The number of hours and the level of learning in the mathematics course vary according to the option taken. We believe that the topic 'models of linear regression' should become part of the common subject matter in all course options. This objective, however, can only be realised if the topic itself can be integrated with those already in existence, and if it requires only a limited number of teaching hours. Consequently, we have chosen an approach requiring the minimum amount of background, building on the mastering of topics normally covered in the various school course options. The choice of a stochastic approach (Capelo *et al*, 1987) would have needed notions on probability and statistics as prerequisites which are not included in the current Italian secondary school curriculum.

In terms of the mathematical content, in particular the regression line concept and the determination of the line coefficients by the least squares method, it was decided to introduce these topics employing a method that would be in line with the minimum requirement principle thus avoiding reference to the concepts of calculus. The method employed (Bottino *et al* 1988, Dacunha-Castelle and Duflo 1982) allows the coefficients of the regression line to be obtained through the computation of the minimum value of a quadratic function of only one variable. The student is then able to solve this problem geometrically by determining the vertex of a parabola.

With regard to computer-science concepts, the proposed scheme provides for the revision and application of basic knowledge, for example, the use of a computer, the concept of an algorithm, notions on programming in a given language, and so on, and the introduction of advanced concepts such as those associated with data storage and computer graphics.

The necessary requirements for the development of the proposed scheme in the classroom are shown in figure 1; the mathematical and computer science contents are shown in figure 2. The points relating to the didactic methodology in which they are developed have been indicated alongside.

REQUIREMENTS	
MATHEMATICS	COMPUTER SCIENCE
Symbolic expression manipulation	Basic concepts of algorithms
Analytical geometry (cartesian plane, straight line, parabola)	Practical knowledge of a programming language: writing programs involving the concepts of sequence, decision, repetition and non-structured data
Intuitive concept of a function	Use of a personal computer

Figure 1 : Requirements

CONTENT	
MATHEMATICS	COMPUTER SCIENCE
Approximation and practical mastering of real numbers Representation of real numbers on the linear paper (a)	Representation of real numbers on a computer (a,b)
Arithmetic mean Standard Deviation (b)	Description and implementation of numerical algorithms Program design and debugging (b,d,e)
Centroid and its properties (b,c)	
Method of least squares Minimum of a function Regression line (c)	Data files Reading and writing of sequential files (d)
Correlation coefficient (e)	

Figure 2 : Content

This proposal is the result of experience over a period of several years, through experiments conducted initially with university students and subsequently reviewed for secondary school application (Forcheri *et al*, 1975). The first test in the classroom was performed during the 1981/82 school year, with students of a scientific upper secondary school (age 15–16). In this case, the teaching subject matter to be tested was prepared by two final year university students, who also followed the work in the classroom and collaborated with the mathematics teacher. Low-level-language programmable pocket calculators were used.

The positive results of this work led us to propose, in conjunction with other researchers, a didactic unit on linear regression models, which has been implemented within the framework of the IRIS Project (Computer Science Initiative and Research in Schools) promoted by CEDE (European Centre for Education) (Bottino *et al*, 1988).

In the realisation of this didactical unit, an effort has been made to separate the difficulties, and graduate the introduction of new concepts. The unit is organised into three different phases: realisation of an experiment and analysis of the resulting data by means of graphical and numerical considerations; computation of the regression line coefficients and the correlation coefficient, with application of the results to the analysis of the previously obtained data; implementation of the program to compute the coefficients and an introduction to data file handling and related procedures (in Pascal or BASIC).

Each phase is subdivided into classwork, laboratory work, homework, and evaluation tests. These activities in turn are supported by worksheets in which new concepts are explained, exercises and questions are proposed. The teachers are provided with a guide which gives them detailed suggestions for the development of each phase. The didactic unit covers approximately 15 hours of work. We have found this working method, which gives both the teacher and the student very detailed directions to follow, to be useful for the complexity of the topic faced. In fact, the proposed methodology cannot be easily managed by a teacher with little experience in this field. In the first experiment undertaken this need was less obvious, as the presence of the university students allowed us constantly to monitor the classroom activity in a more flexible manner.

3.2 Application Aspects

The topic linear regression models can also be developed through the use of didactic software. In this case the pedagogical goal is to provide the student with the ability to form and make assumptions by studying real situations (points (d) and (e) of the methodology). It is particularly suited to students who have not yet developed abstract reasoning ability, and can either be used by the mathematics teacher to introduce the topic or by the teacher of another subject requiring a means to analyse data collected in the laboratory. In this case, the importance of employing software to focus the attention of the student on the data and the results of the problem, rather than on the computation process itself, becomes clear. These considerations have induced us to study the problems concerned with the design and implementation of eductional software suitable for introducing and applying the concept of models, and to develop and test systems of this type.

In this case, the use of application software for solving problems of best fit in general, assuming the concept of models as known, is not suitable for meeting the objective.

The needs of a student differ considerably from those of an everyday computer user, who employs the machine to obtain quick and correct results to a calculation which, otherwise, would be too long or tedious to resolve longhand. In general, students are not sufficiently familiar with advanced computing machines, and needs them for different purposes; they are, or should be, interested in the methods used to obtain the result, rather than the result itself. They should be persuaded to try different methods to reach the same conclusion and, furthermore, should be capable of introducing variations into a method in order to compare these differences. Both deductive analysis and intuitive understanding should be used to examine the results.

A systematic method for the development of software to satisfy this need seems difficult to formulate. The tests carried out, however, have provided us with general directions. In particular, we have verified that students are greatly assisted if they are provided with a software environment which deals with a limited number of situations, leaving them free to perform the desired tests according to the assumptions made, and to reject them quickly should they be found to be incorrect (Antoy *et al* 1982, Forcheri *et al* 1983 and 1989). This type of environment is provided by interactive software that gives the student a certain number of commands which, by changing the parameters, automatically allow the following.

- Pairs of data points, entered via the keyboard, to be stored on a file
- Pairs of data points, obtained as the result of previous operations, to be stored on a file
- The main statistical parameters of a set of stored data points to be calculated
- A set of previously stored data point pairs (choosing the appropriate units) to be shown graphically
- Pairs of data points to be eliminated
- A data file to be shown in numerical form
- Different substitutions on a data file to be chosen, in order to verify whether the relationship between the data points obtained can be considered as being linear
- The line coefficients that best fit a set of stored data points to be calculated

- The data, and the corresponding regression line that has been calculated, to be graphically represented
- The correlation coefficient to be computed

We have implemented and tested this type of software in different educational situations. We have utilised the PAST system (Antoy, 1982), written in Pascal on a PDP11/40, and the LREG system, which is essentially a limited version of the PAST system operating on an Olivetti M24. Both packages have been tested with students of upper secondary schools.

An example exercise concerns the use of the LREG system in the teaching of fluid mechanics to students of 17 to 18 years old. This exercise focusses on the application of the regression concept to the calibration of a manometer. The calibration of a manometer is carried out by creating several pressures and, for each pressure, taking two measurements, p_1 and p_2: the first with a master-gauge, and the second with the given manometer. From the two sequences of measurements, students discover whether the given manometer can be calibrated by analysing the sequence of the corresponding absolute and/or relative errors, or by computing the regression line of p_2 on p_1. Usually, the error-based method is used, as it does not require too many computations and, in addition, the mathematical concepts on which this method is based are known by the students. However, the comparison of the regression line with the equation $p_1 = p_2$ would give more information from a quantitative point of view.

The teacher of fluid mechanics in the technical institute where the exercise was carried out decided, in collaboration with the mathematics teacher, to introduce the regression model as a tool for analysing laboratory data. Initially, the students were introduced to the concept of mathematical models from a theoretical point of view. The LREG package was used to analyse various types of data, to make students form their own opinion (from a graphical point of view) as to whether a linear model could be reasonably hypothesised or not, to obtain (when appropriate) the regression line coefficients and the correlation coefficient, and then to discuss the model with respect to the graphic representation and the quantitative result obtained.

When this work was completed, students were introduced to the problem of calibrating a manometer and were given instructions on conducting the laboratory experiments. The laboratory work was carried out by the students themselves, divided into groups. Finally, the collected data were analysed using the LREG system. A total time of approximately 15 hours was required to complete this work.

The first exercise performed during the 1985-86 school year produced good results, and is now used each year.

4. CONCLUSIONS

The two different lines of development of the methodology proposed have, until now, been examined separately. This choice was made since the current situation in Italian schools dictates that only limited innovative teaching sequences can be tried. It must also be remembered that no experiments are possible in the final year of upper schools - the final year is normally used by the teachers to prepare students for the school leaving examination, based on an established syllabus. In addition, even if today the school world is considerably aware and interested in testing schemes which provide for the introduction of both computer science and mathematical concepts, it is not ready to use didactic software.

Teachers are neither experienced in, nor aware of, the effective potential of these types of tools. Nevertheless, the situation is changing, and this leads us to believe that it is worthwhile working towards the integration of the two directions indicated. We propose to carry out tests spread over a number of years, in which the software can be used to allow the students to work on the concepts from an intuitively graphical point of view; the concepts are then formalised and, finally, the software is employed again for the applications. This will be the direction that our future work will take

REFERENCES

Antoy S, Forcheri P, Margiocco M, Molfino MT, Pedemonte O. (1982). Software Didattico per la Statistica. Considerazioni sull' uso e la produzione. *Proceedings of AICA Annual Conference,* Padova, 99-101.

Bottino RM. (1982). An introduction of mathematical concepts with the aid of computers for studying an economic topic. *Proceedings of 34 CIEAEM Conference,* Orleans, 219-225.

Bilghese M. (1982). Tesi di laurea in matematica. Facoltà di Scienze, Università di Genova.

Bottino RM, Forcheri P, Gazzaniga G, Ironi L, Lemut E, Molfino MT. (1988). Modelli di regressione lineare. Progetto IRIS, SEI, Torino.

Capelo AC, Dos Santos C, Gazzaniga G, Ironi L. (1987). Crescita Esponenziale e Logistica. *L'insegnamento della matematica e delle scienze integrate,* Vol 10, **8**, 794-838.

Dacunha-Castelle D, Duflo M. (1982). *Probabilités et Statistiques*. Masson, Paris.

Forcheri P, Lemut E, Molfino MT, Pedemonte P. (1975). Progetto CAI. Technical report, **4**.

Forcheri P, Lemut E, Molfino MT, Pedemonte O. (1980). Mathematical teaching experiences by using computers. In Lewis and Tagg (eds), *Computer Assisted Learning*. North-Holland, 113-121.

Forcheri P, Molfino MT. (1982). Using Computers as Didactic Aids in Secondary School Science Courses. *Proceedings of 34 CIEAEM Conference,* Orleans, 268-272.

Forcheri P, Molfino MT, Pedemonte O. (1983). CAL in Teaching Statistics in High Schools. *Proceedings of the 4th Canadian Symposium on Instructional Technology,* Winnipeg (Canada), 336-370.

Forcheri P, Molfino MT. (1989). Settore Matematica. In Olimpo and Ott (eds), *Analisi del Software Didattico una raccolta di esempi*. Instituto Geografico De Agostini, Novara.

Mazzarello T. (1982). Tesi di laurea in matematica. Facoltà di Scienze, Università di Genova.

CHAPTER 22

System Dynamics Modelling in Secondary Schools

G Ossimitz
Universität Bildungswissenschaften Klagenfurt, Austria

SUMMARY

A recent curriculum reform introduced system dynamics as a new subject of mathematics education in a branch of Austrian upper secondary schools. This paper gives an overview of the basic ideas of that new curriculum area, describes four levels of handling dynamical systems at school, and closes with some remarks about the problems of teaching system dynamics.

1. SYSTEM DYNAMICS IN THE AUSTRIAN MATHEMATICS CURRICULUM

A recent reform of the mathematics curriculum of Austrian upper secondary level *Gymnasiums* (Grade 9–12) introduces, in grade 11 of the *Realgymnasium* (a branch of Gymnasium with some impact upon mathematics and natural sciences), a new curriculum area *Untersuchung vernetzter Systeme* (investigation of interrelated systems). The new curriculum stresses the following ideas.

- **Systemic thinking** Pupils should be equipped to grasp some of the interrelations in dynamical systems and be aware of the fact that even quite simple systems might behave in a strange manner.

- **Application orientation** The curriculum expects that practical examples are used for studying dynamical systems. Economics, ecology, biology and physics are explicitly stated in the curriculum.

- **Descriptive mathematics** In the context of system dynamics, mathematics is basically used as a means of describing and communicating system models in various ways.

- **Numerical simulation** The computer should be used by the pupils as a tool for simulating and displaying the development of systems.

- **Experimentation and reflection** Pupils should change model-parameters, vary models and discuss the results of numerical simulation, as well as consider the assumptions underlying the process of modelling the system.

- **Project-orientation** Both simple and more complex modelling tasks are within the scope of the curriculum.

Together with a colleague, I am engaged in a project to develop practical materials for teaching system dynamics at school according to the new curriculum. In this paper I would like to give an overview of what could be taught at school, and a short discussion of some important questions about teaching system dynamics.

2. WAYS TO DESCRIBE A SYSTEM

The problems and questions arising when teaching system dynamics depend to a large extent on the way system models are denoted. So let me outline four important ways to describe system models – verbal descriptions, causal loop diagrams, flow diagrams, and systems of equations.

2.1 Verbal Descriptions

The least sophisticated way to describe a system is merely to use ordinary text, without any special notation. To describe something verbally is a rather unfamiliar task in mathematics courses, yet I believe it is a very important aspect of learning to deal with systems.

2.2 Causal Loop Diagrams

Causal loop diagrams are graphical displays, with arrows to denote causal relations between system elements. Causes and effects are interrelated in a kind of network, which usually contains closed loops of arrows. Such *feedback loops* indicate that an effect directly or indirectly reacts upon the cause. The feedback might be an escalating (positive) or stabilising (negative) one. General characteristics of causal loop diagrams are as follows.

- Loop diagrams show the relevant elements and relations of a system. Sketching a diagram focuses our attention on certain aspects of the system, while other aspects are neglected.

- Causal loop diagrams are a method of qualitative modelling. No quantification has to be undertaken for analysing loop diagrams. Causal diagrams are also suitable for describing non–quantifiable models.

- Escalating and stabilising feedback loops can easily be identified in causal diagrams. They support a qualitative feedback analysis.

- There is no single *true* causal diagram for any modelling situation. Often the same situation can be depicted in various, even contradictory, ways.

2.3 Flow Diagrams

About 1960, J Forrester at MIT introduced another kind of diagram to denote systems, which he called *flow diagrams.* In flow diagrams, different types of system elements (levels, rates, auxilliary variables) are distinguished. Thus, in flow diagrams, both qualitative and quantitative aspects of system modelling are combined. On the one hand, the graphical display gives a qualitative impression of the model; on the other hand the clear distinction between levels and rates incorporates important aspects of quantitative modelling into flow diagrams. There even exist dynamical simulation programs, such as STELLA and DYNAMO, which allow the modelling of executable system models at the flow diagram level.

2.4 Equations

The most formalised way to denote dynamical models is to use the mathematical language of equations. These might be essentially either difference equations or differential equations. The equations of dynamical models are usually structured in a discrete, computational style. They state explicitly how a new value is calculated from known values.

$$\text{new value} = f(\text{known values}) \tag{1}$$

So–called *level equations* have a specific structure

$$\text{new\_level} = \text{old\_level} + dt*(\text{inflows} - \text{outflows}) \tag{2}$$

Here dt denotes the duration of a finite timestep (which need not be one time unit), old_level is the value of the level variable one timestep before the new evaluation, and inflows and outflows are averages per unit time in the interval between the old and the new evaluation of the level variable.

3. PROBLEMS OF TEACHING SYSTEM DYNAMICS

3.1 Questions of nomenclature and notation

One of the most difficult tasks when introducing system dynamics at school level is the development of a proper nomenclature and notation. Let me sketch some important problems.

Positive and negative feedback loops This nomenclature might suggest that positive feedback loops are good and negative feedback loops are bad in an empirical sense. Actually often the opposite is true: many positive loops tend to escalate until the system completely breaks down – consider hyperinflation, banking crises, the arms race, overpopulation and starvation. On the other hand, negative feedback loops stabilise and regulate systems.

The term *rate* for flows in dynamical models might also be misleading. In system dynamics a rate is the change of a level variable per time unit, observed in a time interval of length dt. In ordinary English the term rate is used in a broader sense (consider for example the birth rate, defined as the number of births per 1000 people per year). This problem might be avoided by using the term flow instead of rate in the system dynamics context.

Time specifications in system equations The variables could be denoted with or without any explicit specification of the time point to which the variable refers. Most modern simulation programs do not require time specifications.

For practical teaching I suggest that, for the first few models, the time specification should be included in order to stress the difference between levels and rates with respect to old and new values in the dynamical simulation. Later, the time suffix might be dropped.

Should the finite timestep of discrete simulation be denoted by the symbol dt or by the symbol Δt? The symbol dt is obviously more convenient for computer simulation than the mathematically more correct Δt, since no special character is needed.

For practical teaching I would suggest choosing one of the following possibilities: either the symbol dt is used all the time in the system dynamics context (as it is done in this paper) or it is strictly limited to denote (infinitesimal) small timesteps in continuous models.

3.2 Mathematical versus Applied Approach

When teaching system dynamics in school, the teacher can choose among a broad variety of system models. Some of these models are already part of classical mathematics or physics, like the harmonic oscillator, gravitational models, models of heat flow, the classical predator-prey system and so on. They are originally stated, and often solved, as differential equations.

On the other hand, there are many simple system dynamics models which do not fit the classical theory of differential equations: a simple inventory system with some delivery delay, a population model with some delay between birth and fertility. Models containing empirical, nonlinear functions are usually beyond the scope of treatment with differential equations, but they are easily accessible to numerical simulation.

For practical teaching models could be chosen from the mathematical or the applied side. The former approach would support the teachers' prior knowledge about classical mathematical systems, whereas the applied approach would yield a more lively picture of system dynamics. In our project we would rather support the second aspect.

3.3 Qualitative versus Quantitative Modelling

How important should qualitative reasoning about models be when teaching system dynamics? One can do a lot of work at the level of causal loop diagrams, discussing various ways of describing system behaviour almost entirely without any numerical simulation. (In the system dynamics textbook of Roberts (1983) the first equation appears on page 231!) On the other hand, it is also possible to go straight to computer models without much thought about loop diagrams and suchlike. The Austrian curriculum allows for both approaches. I personally think that a qualitative approach is not inferior to a strict computer-oriented one, since qualitative reasoning supports a deeper reflection about systems and their behaviour.

3.4 How to Simulate?

Simulation of system models on computers can be done either with ready-to-use models or with models being constructed by the pupils themselves. In the former case, the pupils' activity is restricted to varying parameters and to interpreting output. The model structure itself is given, and cannot be altered by the pupils. In the latter case, students have to do their own constructive modelling work before the system can be simulated. This approach is more tedious, but gives the pupils more insight into the modelling process and the internal model structure, whereas ready-made models can be used merely as black boxes. On the other hand, ready-made models can be constructed with more sophistication than those of a student.

When pupils create their own simulation models, an appropriate computer simulation environment must be available. For the easiest models, any modern spreadsheet program (with graphics included) would suffice. The columns of the spreadsheet refer to the system variables; any line of the spreadsheet describes the state of the whole system at some given instant. The increase of one timestep dt is denoted by the transition from one line in the spreadsheet to the next.

Most dynamic simulation environments are far more powerful than spreadsheets. An increasing number of software products, more or less suitable for teaching system dynamics, is offered. These differ greatly in their features. We have evaluated a number of simulation programs, and found that most of them have serious weaknesses for teaching system dynamics. Some programs use a very specialised and unusual syntax for describing the models – we recommend a notation which is based upon the quasi-standard being introduced by Forrester. Many programs offer only inferior model-editing features. Some simulation programs – like the original DYNAMO – are actually programming languages, with strict and complicated syntactical conventions, rather than convenient modelling environments such as STELLA (Richmond et al, 1987). Other simulation environments are actually programs (or a number of subroutines) written in BASIC or Pascal. Here, modelling actually means programming in that language – which seems to me to be an unacceptable overhead if pupils are actually to do system modelling. Thus, let me conclude with a short checklist which might help in choosing the right product.

- How simple is it to develop a model? How much syntactical and technical knowledge is required for mastering the simulation environment?
- How are systems input? Is the syntax of the system equations (or diagrams in the case of iconic input) convenient?
- Which output features are supported by the system (tables, graphs and so on)? Can output parameters be altered easily?
- How easy is it to alter model parameters and/or initial values?
- Does the system support higher-order integration techniques for better simulation of continuous models?
- Which special system dynamics functions (delays, pulse-functions and so on) are supported?
- Which ready-to-use models are delivered together with the product? How well are they documented?

3.5 Project-oriented Teaching

To teach system dynamics without referring to a real-world context is a mere waste of time, since dynamical modelling makes sense only when practical applications are involved. Thus, a close interrelation with other subjects is necessary. This implies a project-oriented style of teaching. A more flexible structure in the timetable of lessons, certainly compared with the strict time-schedule of Austrian schools today, would be very helpful.

REFERENCES

Forrester J. (1968). *Principles of Systems.* MIT Press, Cambridge, Mass.

Richmond B *et al.* (1987). *An Academic User Guide to STELLA Software.* High Performance Systems Inc, Lyme, N. H. 03768.

Roberts N *et al.* (1983). *Introduction to Computer Simulation - a system dynamics Modelling Approach.* Addison-Wesley, Reading, Mass.

CHAPTER 23

Applications and Geometry – The example Conic Sections

H Schupp
University of the Saarland, Saarbrücken, FR Germany

SUMMARY

Various aspects of geometry are studied at school – modelling applications form only one among many principles in this area. Geometric instruction must try to present this richness of aspects. The subject matter should be selected and instruction planned according to the extent they allow pursuit of different goals and their mutual furtherance. We use the example Conic Sections to demonstrate how closely an application-oriented school geometry can be linked with other objectives of that discipline.

1. THESES

We present four theses. The first three are, we hope, self-evident and lead consequentially to the fourth, which is the central point of this contribution.

First Thesis
Application-oriented school geometry must not be reduced to geometric measuring.

Such a reduction is customary in my country – not in the official curricula, but in practised instruction – nor should geometry be reduced to the mere use of formulas.

We have to ensure that our students observe and cope with geometric qualities, relations and processes (especially spatial ones) everywhere in their environment and within other disciplines (see Graumann 1989).

Second Thesis
School geometry must not be reduced to the single aspect application and modelling.

There are further aspects which should be worked out with comparable intensity.

- Geometry as an essential discipline in the history of mathematics and as an important part of culture
- Geometry as an (at least locally) understandable mathematical theory
- Geometry as an heuristic source (with fascinating problems at all levels)
- Geometry as a domain, with a plethora of content and methods
- Geometry as an arsenal of algorithms, especially constructions
- Geometry as an aesthetic and creative activity.

Third Thesis
Geometric content should be selected - at school level - with regard to how far it represents the great variety of aspects of geometric method.

As a good example of such a rich and meaningful geometry in higher grades (FRG: grades 12, 13) we cite Conic Sections (Schupp 1988). As a counterexample (at school level) think of the Linear Geometry proposed and introduced during the New Math phase (for a profound critique see Seyfferth, 1981).

Fourth Thesis
Geometric instruction should be organised and performed in a way which points out the different aspects of geometry, and leads to their mutual stimulation.

2. EXAMPLE OF A TEACHING UNIT

We shall try to illustrate the last thesis by sketching the teaching unit Shadow of a Sphere.

2.1 We study the shadow of a sphere in central illumination - realised, for example, by a ball and a torch (figure 1 gives a front view). In opposition to the shade, which is always a sphere segment, the shadow of the sphere cast on a plane base has a hitherto unknown and steadily-changing shape, except in position L which, of course, gives a circle.

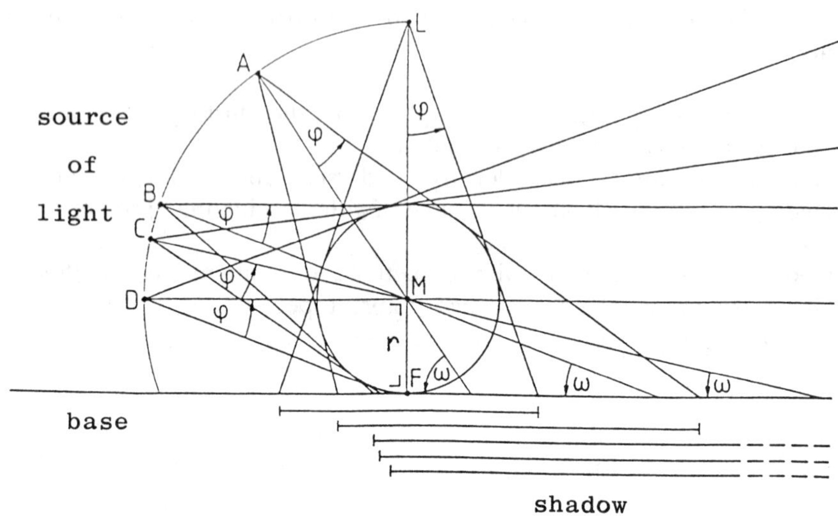

Figure 1

Figure 1, when interpreted three-dimensionally, shows us that the original physical situation leads to a mathematical problem: What do the plane sections of a cone look like?

Also, the figure sheds light on the fact that we have to contrast several cases characterised by the relation between cone angle φ and section angle ω.

Let us start by considering position A. In this case ($\omega > \varphi$) we call the section an ellipse. We add the two Dandelin spheres (which do not appear out of the blue, but suggest themselves as the illuminated sphere and its equivalent at the opposite side of the section) and identify the ellipse as a geometric locus - a set of all points which have the same sum of distances from two points (the tangent points of the spheres). This allows us to construct such a figure (pointwise), to work out its symmetries (one of which is not at all trivial - Albrecht Dürer didn't recognise it and spoke of the "egg-line"), its vertical and focal points, and to interpret it as a generalisation of the circle.

By comparable reflections we get the corresponding qualities of the parabola (position B, $\omega = \varphi$) and of the hyperbola (position C, $\omega < \varphi$). As in the physical situation, there appears only one branch of the hyperbola, and these two section types may not at first be distinguished by the students.

2.2 This approach is a product of the 19th century. Two hundred years ago mathematicians succeeded in solving geometric problems by transforming them into algebraic ones – and analytic geometry was born.

The cone can be characterised by either of the equations

(i) $((\mathbf{x} - \mathbf{s})\mathbf{e})^2 = (\mathbf{x} - \mathbf{s})^2.\cos^2\varphi$

(ii) $(\mathbf{xe})^2 + 2\,|\mathbf{s}|\,\cos\varphi\,(\mathbf{xe}) = \cos^2\varphi\,\mathbf{x}^2 - 2\cos^2\varphi\,(\mathbf{xs})$

where **x** is the vector of a conic-section point, **s** is the vector of the cone vertex, and **e** a unit vector in the direction of the cone axis (see Schupp, 1988).

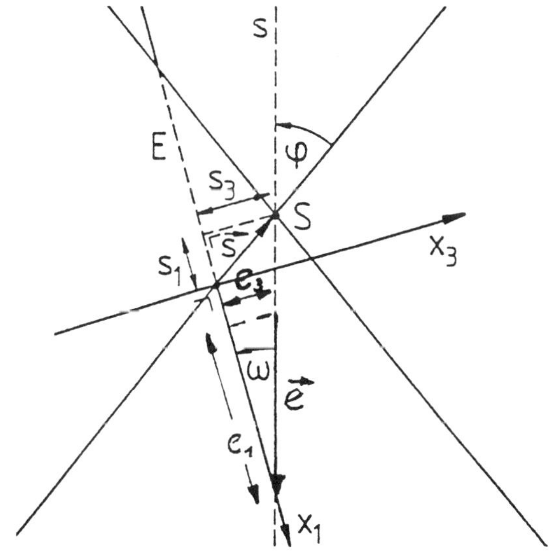

Figure 2

We now introduce a carefully chosen three-dimensional coordinate system (figure 2) in which

$$\mathbf{s} = \begin{bmatrix} s_1 \\ 0 \\ s_3 \end{bmatrix}, \quad \mathbf{e} = \begin{bmatrix} e_1 \\ 0 \\ e_3 \end{bmatrix} \text{ (with } e_3 = \cos\omega\text{) and } \mathbf{x} = \begin{bmatrix} x_1 \\ x_2 \\ 0 \end{bmatrix}.$$

When we perform the required substitutions in (ii) and simplify, we finally get

$$\text{(iii)} \quad x_2^2 = 2 \left[\frac{|\mathbf{s}| \cos\omega}{\cos\varphi} + s_1\right] x_1 - \left[1 - \left[\frac{\cos\omega}{\cos\varphi}\right]^2\right] x_1^2.$$

This suggests the definitions

$$\varepsilon = \frac{\cos\omega}{\cos\varphi} \quad (> 0)$$

which is obviously an indicator of the section type, so that $\omega > \varphi \Rightarrow \varepsilon < 1$ (ellipse), $\omega = \varphi \Rightarrow \varepsilon = 1$ (parabola), $\omega < \varphi \Rightarrow \varepsilon > 1$ (hyperbola), and

$$p = |\mathbf{s}| \; \varepsilon + s_1 \quad (> 0).$$

Each conic section can now be characterised by the equation

$$\text{(iv)} \quad x_2^2 = 2p \, x_1 + (1 - \varepsilon^2) \, x_1^2.$$

We may use (iv) to draw the different section types, for example with constant p and ε = 0, 0.5, 1, 1.5 respectively, which nowadays may be drawn by computer (figure 3).

Another advantage of (iv) is that we can derive from this equation several known and some unknown qualities of the sections.

However, (iv) doesn't fit the original situation – in which p depends on ε (see the definition) as well as p and ε on ω, φ and r (the radius of the sphere, see figure 1).

Therefore we calculate ε as shown above, but p as follows.

$$p = r \, [\varepsilon - \cos(\varphi+\omega)] \left[\cot\varphi + \tan\frac{\varphi+\omega}{2}\right] \quad (>0)$$

(The technical details are suppressed but are given in Schupp, 1988 and 1990.)

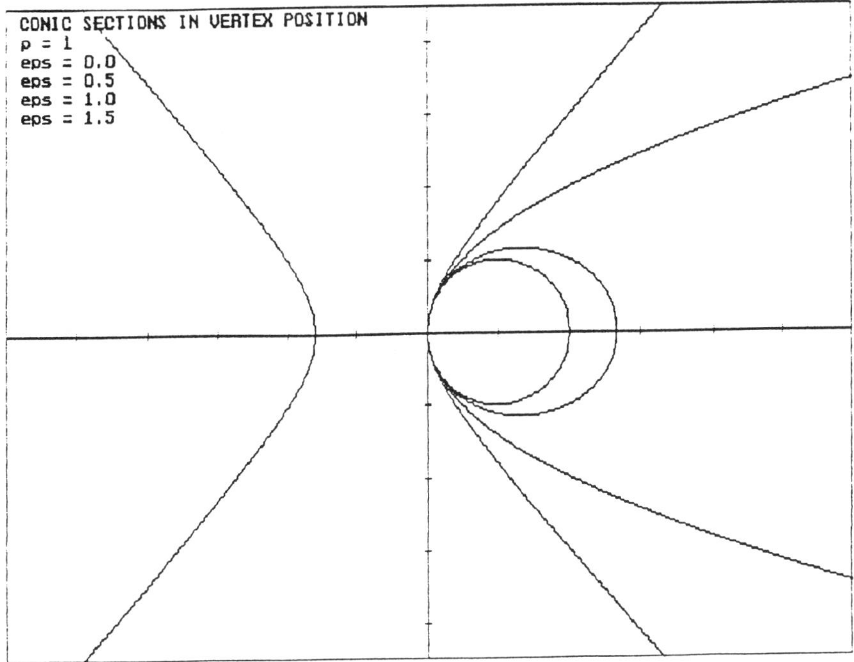

Figure 3

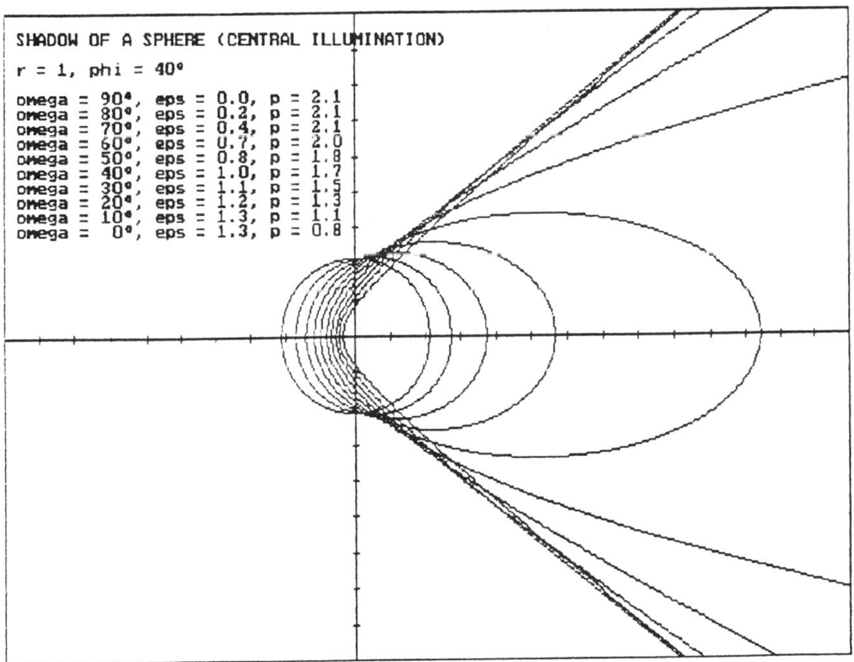

Figure 4

Another problem is that the shadow lines of the sphere, caused by the moving source of light, don't have a common vertex (as in figure 3) but a common focus. To place this point at the origin (instead of the vertex) we firstly have to calculate the distance between vertex and focus (which is $p/1+\varepsilon$) and then to substitute $x_1 + p/1+\varepsilon$ for x_1 in (iv).

We are now able to simulate the original experiment (demonstrated with $r = 1$, $\varphi = 40^\circ$ and $\omega = 90^\circ, 80^\circ, \ldots, 0^\circ$ (figure 4)). In comparison with the original we now get a complete overview and a more precise, as well as more dynamic, picture.

Thirdly, as figure 1 shows, we can move the source of illumination over D ($\varphi = 0^\circ$) until it reaches the base ($\varphi = -\omega$). Then we can complete our analysis and continue our computer simulation. What does the shadow look like in this case? Where does it tend?

2.3 How did the Greeks, 2000 years ago – without computer and without analytic geometry – manage to study the true shape of a conic section which they needed, amongst other things, to solve problems like the doubling of a cube or the trisection of an angle? Archimedes turned the intersecting plane, together with the section, around its axis (figure 5) and constructed the two intersection points P_1 and P_2 by means of the front view point P' (via P). He was able to derive from that construction the fundamental qualities of the sections.

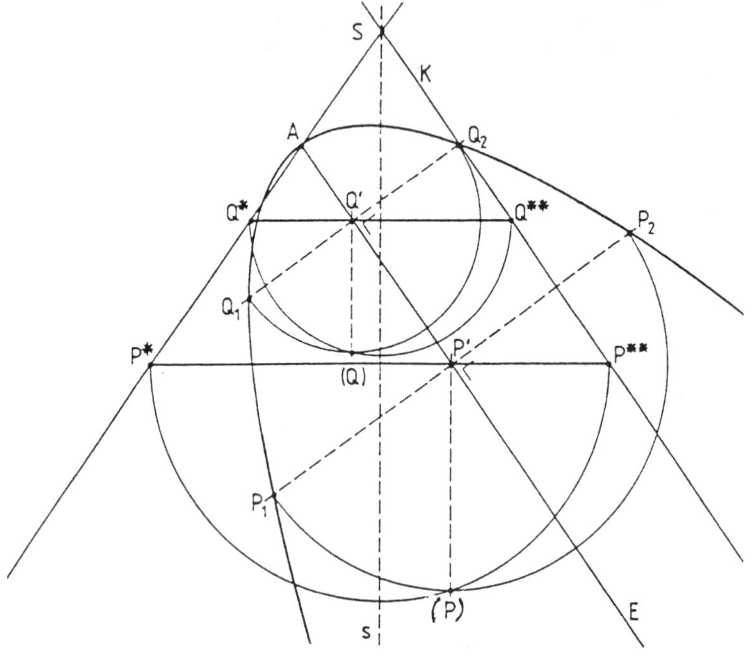

Figure 5

2.4 Let us come back once more to our original situation. We are familiar with the types and basic qualities of the shadow figures, and with the special role of the tangent point of the sphere. However, we know even more, since by our way of modelling, our results hold whenever a cone is cut by a plane – for example in a sharpened pencil, in a conic glass partially filled with water, in a conic box filled with sugar, in a sundial (with the top of the pointer as the cone vertex and the daily course of the sun as the cone producing circle), and so on. Also we are able to discriminate the respective type of the section – for example a hyperbola in the case of the pencil and of the sundial. We have enlarged our ability to analyse the geometry of our environment.

It is possible to widen our knowledge; when we increase the distance between ball and torch, we come more and more to a parallel illumination (as by the sun) – the angle φ gets smaller and smaller, and we shouldn't be surprised when finally there is only one type of section – the ellipse. The cast shadow of a sphere in parallel illumination is always an ellipse (the shade always a half sphere), likewise the sloping section of a sausage, the surface of a liquid in a cylindrical glass, and so on.

3. POSSIBLE CONTINUATIONS

Our example contains the seeds of further good examples in support of our main thesis.

3.1 The focal qualities of the conic sections discovered by reflecting on the consequences of the cone model motivates the following questions. Where do we have such loci in reality? Where does the name focus come from? These lead us from theoretical results to their applications. For example, in the case of the parabola, they lead to the bundling or producing of parallel rays (parabolic reflectors in astronomy, car head lights); in the case of the ellipse, they lead to the qualities of a whispering gallery (unrealistic, but helpful) and from there to the kidney-stone shatter, maybe even to the orbits of planets and satellites (Kepler theorem), in the case of the hyperbola, they lead to a method of finding out the centre of an earthquake.

3.2 When we try to get the true shape of an ellipse, as a cylinder section, from its top and front view (figure 6) we discover the affine relation between circle and ellipse; we recognise the ellipse as the perpendicular-axis affine image of a circle and can quote the following equations.

$$x' = x; \qquad y' = k.y \qquad \left[\text{with} \quad k = \frac{1}{\cos\alpha}\right]$$

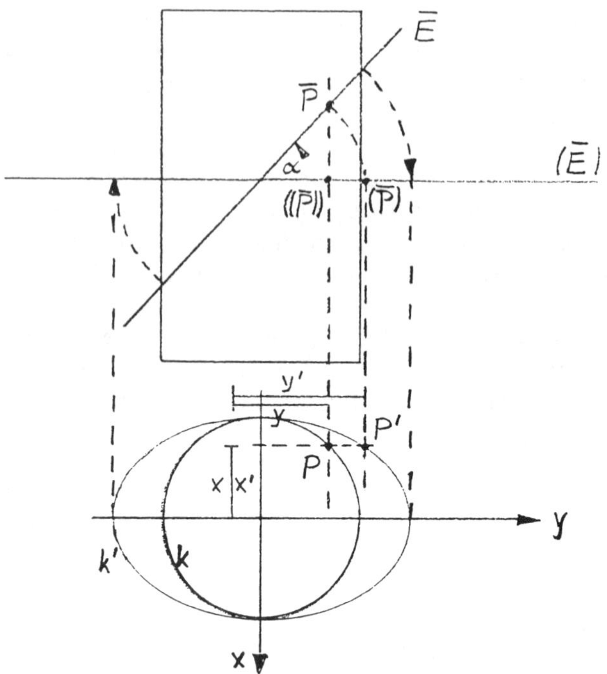

Figure 6

This relation gives us a lot of additional mathematical results (for example construction of points or tangents of an ellipse, conjugated diameters) and has, in addition, important applications (for example area and approximate circumference of an ellipse, parallel perspective images of spheres, cylinder, cones).

4. FINAL DIDACTICAL REMARKS

I would like to emphasise the importance, not of the sequence of the single aspects within a geometric unit – this may be influenced by the preknowledge of the students, by the strategies of the teacher, perhaps even by the discoveries and interests of the students, and therefore will vary from unit to unit – but of their coming together as intensively and as early as possible, for their mutual furtherance.

It is encouraging that such a broad and multi-dimensional approach is supported and even demanded by the psychology of learning.

It permits a wealth of student activities at different levels and in different fields, with stimulating variations:

- to carry out physical experiments,
- to search for relevant phenomena in our environment and to clarify them,
- to construct and to analyse top and front views,
- to characterise lines by equations (first under the teacher's guidance, but later independently)
- to translate equations back into lines and geometric qualities,
- to build, to test and to evaluate computer programs (in a student-oriented programming language),
- to find new problems by re-solving old ones and reflecting on the solutions.

REFERENCES

Graumann G. (1989). Geometry in everyday life. In Blum W *et al*, *Applications and Modelling in Learning and Teaching Mathematics*, Ellis Horwood, Chichester.

Schupp H. (1988). *Kegelschnitte. Mannheim*, BI – Wissenschaftsverlag.

Schupp H. (1990). Der Schatten einer Kugel – oder: Geometrieunterricht bei Licht betrachtet. To appear in *Praxis der Mathematik*, **32**.

Seyfferth S. (1981). Zur Beziehungshaltigkeit der linearen Algebra in Hochschule und gymnasialer Oberstufe. In *Journal für Mathematikdidaktik*, **2**, 3, 195–224.

Wittman EC. (1987). *Elementargeometrie und Wirklichkeit* – Braunschweig/Wiesbaden, Vieweg.

CHAPTER 24

Automatic Discrimination of Written Languages from Random Strings, with an Application to Error Detection

T Nemetz
Mathematical Institute of HAS, Budapest, Hungary

SUMMARY

This paper provides an application-oriented modelling of written texts. Four experiments are presented with didactical comments, as preparation for introducing the first-order Markovian approximation. A simple version of this approximation is used to assign language-dependent integer check values to strings of letters. These check values are typically small, if the string is a small part of a meaningful text, and they are rather large if the string looks like a random sequence. This explains why one looks for codes which turn the decoded version of the erroneously received message into a random-looking sequence. Such a code is presented here, and involves matrix operations in modular arithmetic. The check value of the decoded text is used to decide if the transmission was error-free or not. Numerical illustrations are also provided.

1. INTRODUCTION : GENERALITIES

The selection of appropriate topics for teaching modelling and applications is a delicate task. It is generally agreed that the topics to be discussed must be real-life problems, but the handling of such problems is usually very time consuming. This is due to the many facets of the applications, such as

- knowledge required of the specific background of the subject,
- difficulties in collecting data and its organisation,
- the 'why' and 'how' in selecting models,
- possible debates due to the subjectivity in interpreting data in the light of the model chosen.

Two educational requirements of the selection are

- the work to be done must be motivating for the students - it should be clear that the problem is 'really real',
- the problem should be 'evergreen' - keeping its basic characteristics from year to year but, at the same time, the concrete realisations naturally changing from occasion to occasion.

In this paper, we put forward an example which illustrates that coding theory provides an inexhaustible source for such problems. The knowledge required in our concrete example is knowledge of the mother-tongue of the students. We deal with written, living languages as the information source, but this can be readily applied to others, for example to programming languages.

2. PRELIMINARIES

Most misspellings in written texts can easily be corrected by school children. All the same, we can give two simple instances where automatic, machine-operated error correction, or at least error detection, is needed.

- The speed of transmission in modern telecommunication is very high. Telefax systems, for example, transmit one full page of text in less than one minute, a time during which no human could even dream of reading it.
- There are errors which simply cannot be corrected (or even detected) without substantial precautions. The following sentence illustrates such a case:
  WE ARE NOΘ ABLE TO COVER THE COSTS OF YOUR PARTICIPATION AT ICTMA-4.

This example should suffice to show that human error correction is unacceptable for modern telecommunication. Therefore, messages have to be carefully preprocessed to allow automatic detection. Such a preprocessing is effective if a small error in the transmission results in a scrambled mess at the destination. We will show how this goal may be attained by utilising the redundancy of natural languages, and this will be achieved by a code that does not increase the length of the text. In order to reach our goal, the children must become acquainted with some basic statistical properties of the language. This is achieved by the following three experiments, which were originally designed to introduce basic notions in stochastics.

3. INVESTIGATING CERTAIN RANDOM PROPERTIES OF A WRITTEN LANGUAGE

Before attempting to build a model for any phenomenon, one must try to become acquainted with its properties. This is the aim of the following set of experiments, which allow meaningful conclusions in a very short time. A double lesson (2×45 minutes) was enough to run the experiments with varying age groups.

The experiments are formulated for written English using 26 capital letters and the space (which is denoted by –, for clarity). Punctuation marks are omitted, and the texts chosen do not contain digits. This should be viewed as a necessary simplification of the phenomenon under investigation. It can also be considered as a preprocessing of the text which could almost be uniquely decoded.

3.1 Experiment: Shannon's original guessing game

The name refers to the basic work of Shannon (1954), where this experiment was used to investigate the redundancy of printed English.

The teacher chooses an arbitrary text and gives the first 20–30 letters to the students. The students are asked to guess the subsequent letter by hit-or-miss type questions. (Question allowed: "Is the following letter ..."; Forbidden: "Does the following letter belong to the set ...".) The lecturer responds – Yes or No. Guesses are repeated until the letter is known, then the next letter is targetted.

Illustration:
Communicated part of the text:
"MOST–OF–THE–".
Questions and answers for the following letter:
A? No, M? No, T? No, E? Yes.
With an item of extra information one can envisage the following continuation:
X? Yes, A? Yes, M? Yes, P? Yes, L? Yes ...

A suggested size for the new letters is 30–50 while keeping a record of the number of questions needed. Records look like the following.

5,3,1,1,1,1,1,1,1,17,11,3,2,1,1,1,8,4,1,1,1,...

3.2 Experiment : Every Tenth Letter

We suggest working with another part of the same text as before. The teacher prepares a subsequence of every tenth letter of the text. Now,

Shannon's guessing game is repeated with this specific sequence, giving the first few characters and leaving the others to be guessed.

The goal of this experiment is to arrive at a conclusion as follows.

In the case of 3.1, we can be very clever, much cleverer than any machine which could be invented. At the same time, the best guessing strategy in case 3.2 needs no human activity. No information on the past history of the sequence can be utilised. The optimal strategy is to guess the most frequent letter first, then the second most frequent second (if necessary), and so on. The frequency order of the letters can be established by the students themselves, by inspecting longer texts (for example a text of one page - around 2000 letters - will do).

The experiment should go on until the students arrive at a similar conclusion. Besides keeping a record, the number of necessary questions should be averaged.

The frequency order of the English language was estimated, from the text in the proceedings of a recent conference as follows.

ETIANOSRCHLDMUGFPWYKXBVZQJ

Guessing in this order will yield an average number of questions around seven.

Extension:
Having established the frequency distribution of the language, a useful extension is to pose a simple substitutional crypotext for the students to break. Pupils aged 12–14 are usually enthusiastic to attack this problem, and many upper secondary students will also do so.

3.3 Experiment : Random text
Shannon's guessing game, just as before, is performed on a random text. Here 'random' means that the subsequent letters are chosen independently with equal probabilities. An effective way of demonstrating this is to sample with replacement in front of the class, and let the class guess the letter just drawn. Such a sampling technique is a *must* when one aims at certain stochastic notions. For our present goals, however, a kind of pseudo-random text is preferable.

In preparing this pseudo-random text the teacher should apply the *randomisation procedure* of point-five-below to the same text as was used in experiment 3.1. The students can even be told how the text was obtained, keeping secret, of course, the actual parameters of the procedure. We assert that the number of necessary questions will behave in the same way as in the case of random text.

Conclusions:
There is no clever, best–guessing strategy. All strategies will provide the same expected number of guesses, and this expected value is 14. Accordingly, the average counted from the records of the experiment should be around this number, which is considerably larger than in the two previous cases.

Suggestion:
In a live demonstration of this experiment, one should consider 20 or so letters to be guessed. It is recommended, however, that microcomputers be used to aid the experiment. The randomisation procedure is easy to program, and the students can begin with their own meaningful text. Such a version is the most convincing.

3.4 Experiment : Guessing based on the preceding letter only
In experiment 3.2 the teacher had chosen every tenth letter of a meaningful text, and the children have determined these letters. Their task is now to find out the letters immediately following these letters. Thus, the children have partial information on the text, and can utilise it if they wish. It becomes obvious quite quickly that this is useful information. Seeing a Q, for example, the only reasonable first guess is U, while it can hardly be the best after an A.

Conclusion:
The frequency order of the letters of the alphabet depends mostly on the preceding letter of a text. Guesses are to be made according to this conditional frequency order. The order itself could be reliably estimated from a text of about 20 000 characters which, in turn, could be provided by text editors. Writing a program for this purpose is a straightforward exercise at upper secondary level.

It is important to note that the best–guessing strategy could be conducted by machine. The experiment may conclude by demonstrating that the average number of necessary questions is less than in the previous two cases.

Time can be saved by using ready–made, conditional frequency order tables. Here we present one, which has been made using the first 35 000 preprocessed character of reference. The first column lists the alphabet, and these are followed by the letters in their conditional frequency order, which appeared as follow–up letters. The sign – denotes a space.

-	TACIOSMEWPBFDRLHNKGUVQYJXZ
A	TNLRC-BDGIMYPKSVUWFJ
B	ELYASUORIJ-BT
C	HOATEPISR-LCKYU
D	-EIRUGOSAHYVDWC
E	-RNDSMALCXIVTEPFWYGOHQBUK
F	-OIEUFRALMYSTMV
G	-EHIAROUSNLGYTF
H	EIAO-TURMYNLS
I	NCOTSLEVRAFDGMBZPK-U
J	EOT
K	N-IEKGS
L	-EDAIYLOTUSVMFGPRW
M	AE-IPUOMSCYB
N	-TGDSIOCEAVMNFKLUYHJQR
O	NF-RUMWTLCODPSBGIVAEHKY
P	EROALT-PIUMHNDFS
Q	U
R	E-AOISNTURMLDKPVYGBCF
S	-TEISOCUHPAFKMRLYDW
T	HI-EOSUARYTWLCBMP
U	LRCNASTMDEGPIB-
V	EIOA
W	H-EILAONSTD
X	PTA-CIS
Y	-IESOLCMPTADRZN
Z	EIAO

Table : Conditional frequency order of English letters

4. THE MARKOV MODEL OF A WRITTEN LANGUAGE

Many people share the opinion that one should try to construct as perfect a model as possible. In contrast to this, we will follow the pragmatic way: we will be satisfied by a rough model, provided it offers a reasonably good solution to our error detection problem. Therefore, however surprising it may seem, we do not utilise any of the semantics involved.

As a starting point we postulate that the following.

A: A written language is a sequence of symbols belonging to a given finite set, called the alphabet. The symbols are called letters.

Our experiments tell us that there is considerable uncertainty in predicting subsequent letters, therefore we claim the following.

B: A written language is a random sequence.

When analysing experiment 3.2, we have concluded that the relative frequencies of different letters are quite different. We formulate this observation at two different levels, C1 for school children and C2 for upper secondary (possibly university) students.

C1: In a written language, there are letters which are typically very frequent (like the space, A, E), and others, which are rare (like J, Q, Z). Of course, these letters depend on the language in question. The frequency order of the letters in longer texts is almost the same.

C2: The frequency distribution of the letters in a given language is a stable, non-uniform distribution.

Remark:
Postulate C rules out the random sequences of experiment 3.3 from the set of meaningful written texts. This is not strong enough, however, for our purposes. We need to go a little bit further and utilise the correlation between subsequent letters. This step of abstraction was prepared by experiment 3.4, and its conclusion is formulated again in two ways.

D1: The frequency order of the letters of the alphabet following a given letter in a written text depends strongly on the given letter. These frequency orders are essentially the same in all long texts of the given language.

D2: The first-order transition probabilities of the language are given by a fixed set

$$\Pr\{\text{subsequent letter} = y \mid \text{present letter} = x\}$$

of probability distributions.

This characterisation is attributed to Markov. The phrase Markov chain originates from here.

Surprisingly enough, this rather crude model provides good potential for solving our error detection problem, as will be demonstrated in the last section of this paper.

5. A RANDOMISATION PROCEDURE

Written text is scrambled by multiplying its blocks by an invertible matrix over a finite field. This sentence is, of course, not meant for the children, although this is the whole procedure. It could be presented to the students in a number of ways, which we will not comment on.

The procedure has three parameters

P : a prime number, not less than the alphabet size
N : length of the blocks on which the procedure operates
A : an invertible N×N matrix with modulo-P considered entries

There are two variables

X : N-length text block forming the input of the procedure
Y : N-length scrambled block, the output

The input and the output are connected by the matrix multiplication

$$Y = AX \quad , \quad X = A^{-1}Y$$

The multiplication between numbers and letters has a simple explanation: all letters are substituted by the (integer − 1) which shows their order in the alphabet which, in turn, is supplemented by some fictitious elements in order to ensure that it has exactly P letters.

As an illustration, we consider the parameters

P = 31, N = 10, and a self-inverse matrix A

$$A = \begin{bmatrix} 10 & 2 & 3 & 21 & 14 & 2 & 17 & 0 & 14 & 3 \\ 0 & 24 & 5 & 12 & 30 & 18 & 3 & 29 & 3 & 6 \\ 11 & 27 & 6 & 0 & 9 & 8 & 20 & 3 & 21 & 12 \\ 11 & 6 & 9 & 4 & 3 & 17 & 23 & 21 & 26 & 13 \\ 4 & 14 & 30 & 24 & 20 & 14 & 10 & 7 & 16 & 30 \\ 29 & 19 & 26 & 23 & 1 & 20 & 12 & 13 & 28 & 10 \\ 12 & 11 & 17 & 23 & 6 & 27 & 30 & 14 & 27 & 17 \\ 3 & 11 & 14 & 24 & 30 & 27 & 27 & 11 & 25 & 10 \\ 27 & 9 & 25 & 17 & 21 & 1 & 29 & 16 & 20 & 27 \\ 19 & 5 & 2 & 22 & 8 & 21 & 10 & 28 & 8 & 10 \end{bmatrix}$$

The alphabet is supplemented and ordered as follows

ABC$ = "-abcdefghijklmnopqrstuvwzyz.,?*"

The conversion between letters and mod-P numbers is a single command in GW BASIC. The substitute for the k-th letter of the input X$ is given by

```
INSTR(ABC$,MID$(X$,K,1))
```

We wanted to use the randomisation for error detection. For this purpose we interpret the scrambled version as a code for the original text. This is to be sent over the communication channel, and the receiver may get this error free. In this case, when the inverse matrix-multiplication is performed, one gets the original text back. If, however, there was an error in transmission, the decoded version would again look like a scrambled, random sequence. This can easily be demonstrated by letting the children introduce artificial errors in the output sequence, then perform the inverse transformation, and compare the decoded version with the original meaningful text. This assertion can be put more formally.

TEXT-INPUT:X(1),X(2),...,X(N)
TEXT-OUTPUT:Y(1),Y(2),...,Y(N)
RECEIVED:Y(1)*,Y(2)*,...,Y(N)*
ERROR:E(1),E(2),...,E(N) where E(I) = Y(I)*-Y(I),I=1,2,...,N
DECODED:X(1)*,X(2)*,...,X(N)*

Assumptions:
- A is randomly chosen among all invertible mod-P matrices
- The error does not depend on the actual text
- There is an error, so that E≠0.

Claim:
The decoded vector X* is a random sequence, which is statistically independent of the original text X.

Remark:
The exact mathematical formulation is given in Nemetz (1981). Here we would like to underline that a demonstration of this phenomenon is more instructive for the students than a mathematical proof, even at university level.

The formal presentation of the procedure may be omitted in the classroom. It suffices to present it as a computer program, which transforms strings into strings, and to show that its repeated application reconstructs the original one. It is also relevant to demonstrate that, after carrying out any change in the transformed string, the repeated application results in a completely different string.

The following five lines illustrate the effect of randomisation, and the consequence of very minor errors in the received sequence. The first and last numbers will be explained in the next point.

CV(X)	X:input	Y:code	Y*:received	X*:decoded	CV(X*)
16	MATHEMATIC	KIWSRU*AXD	KIWSRU*AXD	DWZADDMAWK	95
30	AL-PREDICT	S?INNHXILZ	S?INAHXILZ	EYGHFWSVII	116
31	ION-AS-A-S	OTKRW--K-H	OTKRW-AK-H	MWMHE?HZE.	116
43	OCIAL-INST	FSUQEJJLKG	FSUQEJJKLG	?H.FUOVIW-	94
42	RUMENT-END	.MUH*P?CLT	*MUH*P?CLT	QUOGZNEHB*	100

6. AUTOMATIC ERROR DETECTION : CHECK VALUE

We have demonstrated that an error in the received signal turns the decoded message into a random mess, while no error results in the original text. Therefore, if we wish to decide if there is an error in the transmission, we have to be able to distinguish between written text and random sequences. For this purpose we assign a *check value* to blocks of the letters. Markov's approximation of the language is relevant. Formally, we introduce the notation

Ord (y/x) = Frequency order of the letter y among letters subsequent to x

These conditional orders are given in the table above.

Check value of the block X(1),X(2),...,X(N) is defined as

$$CV(X(1),X(2),\ldots,X(N)) = \sum_{I=1}^{N+1} Ord(X(I+1)/X(I))$$

Example

```
M  A  T  H  E  M  A  T  I   C
   1+ 1+ 1+ 1+ 6+ 1+ 1+ 2 + 2  = 16
```

as opposed to the decoded erroneous message received

```
D  W  Z  A  D   D  M  A   W  K
  14+ ?+ 3+ 8+ 13+ ?+ 1+ 18+ ?   =  ? > 95
```

These numbers are shown in the closing example of the previous section. The other four lines support the conjecture that the check value of a block is much smaller in the case of a written text than in the case of a random sequence. Now the teacher should utilise the observations on the average number of the necessary questions in experiment 3 and 4. A subsequent discussion could conclude by setting a threshold for accepting the message to be error free.

Remark:
Using the conditional (empirical) distributions, one can calculate the expected value in the first case. In the second case we do not even need a calculation – all strategies require an average (1 + alphabet size)/2 questions.

7. CONCLUDING REMARKS

We have performed the four experiments in a number of classes and the students unanimously enjoyed them. It is recommended that 60–90 minutes be devoted to them. The youngest age group with which we have successfully performed them was 11 years, but the age group 13–16 seems to be a more appropriate target group.

The concept of the check value and its calculation proved to be simple for the suggested age group. Children understand that these values are dependent on the language. Different texts in the same language result in similar small check values, while a different language increases these values. Our last table was prepared on the students' initiative.

English				*Non-English*		
1	2	3	CV	H	G	F
20	33	42	..-35	3	5	3
23	36	23	36-45	0	14	12
12	6	10	46-50	1	15	21
10	11	8	51-55	13	11	19
11	4	8	56-60	6	10	12
13	6	5	61-70	12	9	15
11	4	4	71-..	23	15	9
0	0	0	NO CV	42	21	9

Table: English check values for consecutive blocks of length ten of different texts

```
                English                                Non English

EEEEEEEEEEEEEEEEEEEEEEEEEEEEEEEE ..  - 35 NNNN
     EEEEEEEEEEEEEEEEEEEEEEEEEEE 36  - 45 NNNNNNNNN
                       EEEEEEEEE 46  - 50 NNNNNNNNNNNN
                      EEEEEEEEEE 51  - 55 NNNNNNNNNNNNNN
                        EEEEEEEE 56  - 60 NNNNNNNNN
                        EEEEEEEE 61  - 70 NNNNNNNNNNNN
                          EEEEEE 71  - .. NNNNNNNNNNNNNNNN
                                 NO    CV NNNNNNNNNNNNNNNNNNNNNNNN
```

Graph of the cumulated data

The randomisation procedure is not elaborated didactically here. Although it offers many opportunities to deal with mathematical questions outside of the usual school curriculum, we have included it mainly for completeness. As we have remarked at the end of section 5, it could be introduced and treated as a computer program alone.

REFERENCES

Nemetz T. (1981). Run-test discrimination between written Hungarian and random sequences. In Révész *et al* (eds), *Lecture Notes in Statistics: The First Pannonian Symposium on Math Stat P.* Springer Verlag 182–194.

Shannon C. (1954). Prediction and entropy of printed English. *Bell System Technical J*, **30**, 50–64.

CHAPTER 25

Quality Control - Bathtub Curve - Weibull Distribution

P Bungartz
University of Bonn, FR Germany

SUMMARY

The goal of this proposal for teaching stochastics is the demonstration of the usefulness of mathematical methods in real applications. Pupils, and teacher, discover that knowledge of probability theory and statistics enables them to understand real applications in technical or industrial processes. If young students evaluate some data sets using mathematical methods at least once, they get an insight into the usefulness of the theorems, algorithms, and formulae that they learn. They will therefore be more motivated to study mathematics, including genuine scientific matters such as proof. Quoting from Niss (1983, p248), "They should acquire understanding of those factors within mathematics (such as ideas, concepts, edifices of theory, methods etc), as well as those outside mathematics, which are important to the applicability of mathematics, its potentials and limitations."

1. METHODOLOGICAL REMARKS

Let me explain a project which I introduced to one class of a German Gymnasium. The question posed was, "How long do you think you will enjoy your motor car?" Students, who may only recently have obtained their driving licences, will discuss the benefits and drawbacks of different types of cars. Generally speaking, a car is as good as its components, the clutch, gears, indicators, lights, tyres, brakes, and so on. Normally a breakdown is caused by the failure of one of these components. A good car, in our sense, is one with components of about the same efficiency and a similar length of working life. In order to evaluate the efficiency of a car, a method of testing components should be developed. It is therefore advisable that the students visit a motor manufacturer or supplier, to ascertain what procedures are adopted to test the components. Sometimes a mathematician will be available to explain the Weibull test, but often the explanation given is poor - a computer gives an interpretation of the data. In the class we studied various publications to get an insight into this method of quality control. However, to teach the Weibull distribution, you will need to have studied normally and exponentially distributed random variables, because the Weibull distribution is a special type of exponential distribution, and you need the knowledge of evaluating data in a probability paper (see Bungartz 1985).

2. HISTORICAL REMARKS

The testing method we used was developed by the Swedish engineer Weibull in 1949, and was originally constructed to gauge the data of testing apparatus. Between 1950 and 1960 Johnson, in the USA adapted it to estimate the loss of motor cars. The Chevrolet division of General Motors carried out tests which were successful, and in 1968 the Weibull test was officially introduced into quality control. It became the basis for predetermining and estimating the number of failures of motor cars. Since that time the majority of producers of machines and components, for example Volkswagen, Daimler-Benz, Ford and Opel, have adopted this method, and many publications have described its operation.

3. MATHEMATISATION

The bathtub curve

The failure rate $\lambda(t)$ of some component is defined by the quotient of the number of failures and the working hours/years. In general one would expect an initial decrease in λ. However, after sale the first failures of a machine, the guarantee loss, must be borne by the producer, and this is why λ should decrease quickly. In the main span of the machine's working life you expect only a few failures, which are equally distributed over a long time. At the end of the normal working life the failure rate $\lambda(t)$ can increase exponentially, and at this stage repairs must be paid for by the owner. Describing this graphically you find the 'bathtub curve' as shown in figure 1.

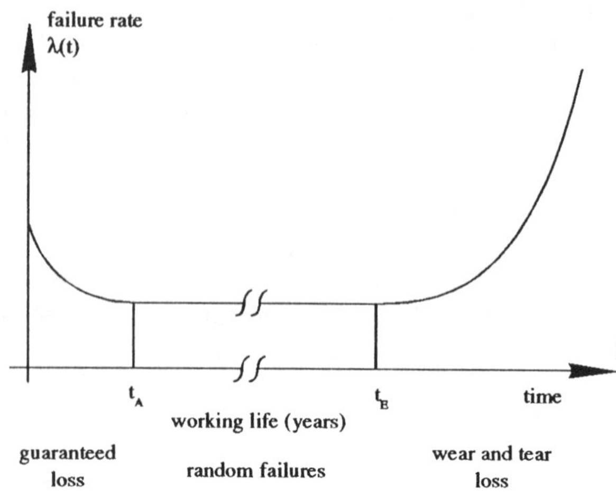

Figure 1 : bathtub curve

The Weibull distribution

We look for a function $\lambda(t)$ with the above graph. The rapid decrease at the beginning, the steep increase at the end and the constant behaviour in the middle lead us to $\lambda(t) = b.t^{b-1} : b \epsilon R,\ b>0$. If $0 < b < 1$ the graph decreases steeply (guarantee loss); $b = 1$ means the constant failure rate during the normal working life and $b \gg 1$ shows the steeply increasing rate at the end of the service life (wear and tear loss). The bathtub curve can be composed partwise by means of the following graphs of $\lambda(t)$, depending on b (see figure 2). Mathematically, the failure rate is a rate of change: $\lambda(t) = -(dN/dt)(1/N)$, $N=N(t)$ is the number of components in work at time t. For the effective working life, $t \epsilon (t_A, t_E)$, $\lambda = \lambda_0$ = constant. Integrating this we obtain $N = N_0 e^{-\lambda_0(t-t_0)}$, where N_0 is the value of N at time $t = t_0$. $N(t)/N_0 = R(t)$ is called the *reliability* of a component. You may interpret it as the probability that a component is still working at time t. The unreliability, often called the failure probability, is $F(t) = 1-R(t) = 1-\exp(-\lambda_0 t)$ (see Messerschmidt-Bölkow-Blohm, p106 and Pfanzagl 1974, p32).

For the guarantee loss and the wear and tear loss you get, by integration of $b.t^{b-1} = -(dN/dt)(1/N)$, $N(t) = N_0\exp(-t^b)$, $R(t) = \exp(-t^b)$ and $F(t) = 1 - \exp(-t^b)$. The parameter b is called the failure steepness. You can interpret F(t) as the probability that a component you look at will break down at time t, or as an estimate of the share of all components which will have failed up to time t. F(t) is a cumulative probability function called the life distribution. The probability density of this distribution is $f(t) = b.t^{b-1}\exp(-t^b)$ which is obtained by differentiating F. The following figures give some examples of the different functions depending on the parameter b, which should be discussed in class.

Weibull introduced a second parameter T > 0 and called it the characteristic working time, because F(T) = 63.2 means that 63.2% of all components have failed up to time T. The correct Weibull distribution therefore is $F(t) = 1 - \exp(-(t/T)^b)$.

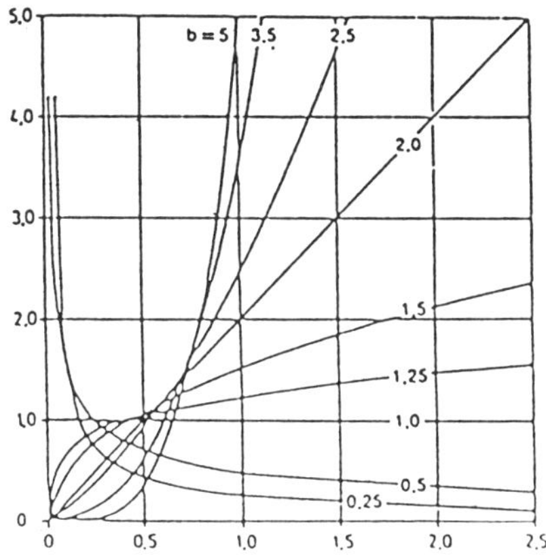

Figure 2 : failure rate $\lambda_b(t)$

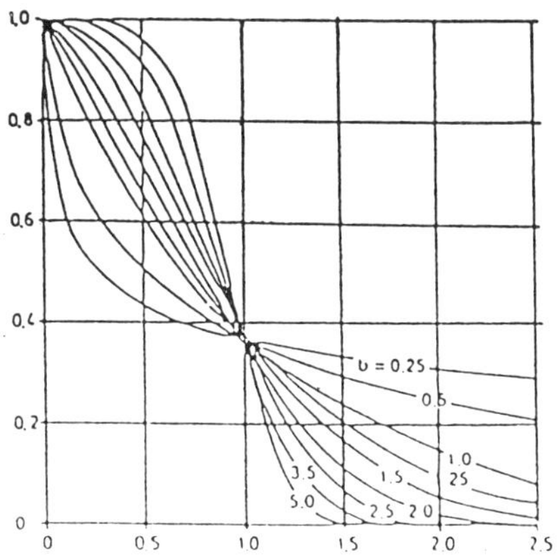

Figure 3 : reliability $R_b(t)$

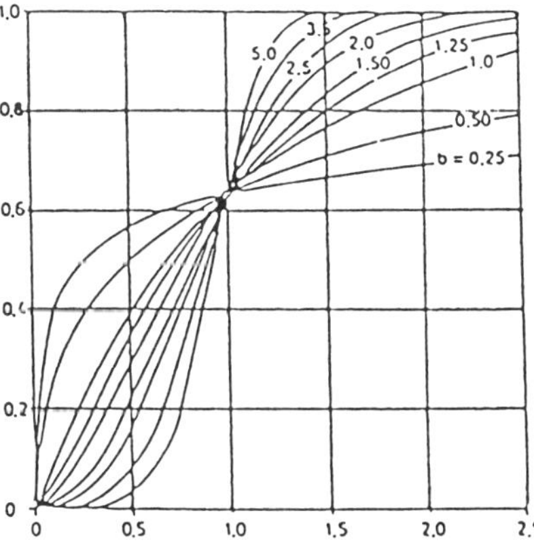

Figure 4 : distribution $F_b(t)$

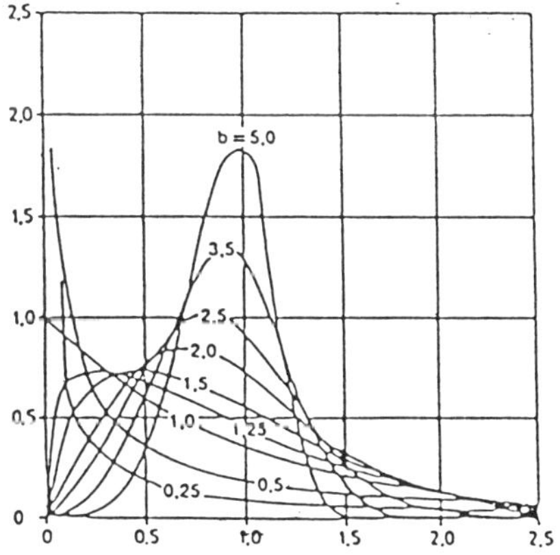

Figure 5 : density $f_b(t)$

4. APPLICATIONS

Real data sets can be taken out of the above-mentioned publications. As an example, let us use this method for headlight switches on a motor car (table I). The technical data are ordered in classes, and absolute, relative and sum frequencies are computed. In lessons at school, raw data should be taken and made into such a table.

Table I : Failures of headlight switches, n=200

Class $(\times 10^3)$ (km)	*absolute*	*relative*	*sum*
	frequencies		
0 - 10	1	0.5	0.5
10 - 20	5	2.5	3.0
20 - 30	21	10.5	13.5
30 - 40	14	7.0	20.5
40 - 50	28	14.0	34.5
50 - 60	40	20.0	54.5
60 - 70	28	14.0	68.5
70 - 80	32	16.0	84.5
80 - 90	14	7.0	91.5
90 - 100	8	4.0	95.5
100 - 110	7	3.5	99.0
110 - 120	1	0.5	99.5
120 - 130	1	0.5	100.0

As with the treatment of normally distributed data sets, we here use a probability paper of the Weibull distribution (which can be purchased at most stationers). The points $(C_{j,u}\ ,\ S_j)$ = (upper class border , sum frequency) taken from table I are marked on the paper. If they lie approximately on a straight line, then the data are Weibull distributed. Following the instructions included with the paper, which gives details for the use of the Weibull method, one finds b = 2.75, and T = 0.68. An interpretation is that the data belong to a wear and tear loss, since $b > 2$, and that about 63.2% of all switches are broken after approximately 68 000 km. This information may well alarm the producer. The working life of the switches, about three years, is too short. The producer should check with the production manager in order to improve it, otherwise no customers will purchase this type of switch.

5. MATHEMATICAL DEVELOPMENTS IN SCHOOL LESSONS

In order to understand what we did using the instruction paper on this method, we constructed our own probability paper for the Weibull distribution. If the pupils are well acquainted with the use of normal distribution paper (see Bungartz 1988 and Pfanzagl 1974) it is easy to construct a Weibull paper. A double logarithmic transformation of F(t) gives us $\ell n\ \ell n(1/(1-F)) = b(\ell n\ t - \ell n\ T)$ the equation of the Weibull line $y = bx + c$ in a coordinate system with double logarithmic y-axis and logarithmic x-axis. The parameter b is the increase of the data line, and T is found by looking at the intersection of the 63.2% line and the data line. If one is interested in knowing the time t (or the kilometers) when 10% of the components are broken, look at the intersection point of the 10%-line of the Weibull paper and the data line. If one wants to understand the scales on the paper more thoroughly, further mathematics - the Gamma function - is needed. Using the scales, the empirical standard deviation s can be computed by looking at the relative deviation d = sT. For the switches we get s = 24 000 km, which agrees quite well with the data, and the mean time $\bar{t}$ = 60 000 km. Between 40 000 km and 60 000 km about 64% failures are to be expected.

6. FURTHER WORK

We learned to use the Weibull method evaluating one special data set, including a verbal recommendation to the producer of these components. We had to study additional mathematics in order to be able to fulfil the engineers' requirements. However, sometimes the data sets are more complicated, and different questions arise. Some examples of these questions are the following.

Mixed distributions
When evaluating data one may discover that the points $(C_{j,u}, S_j)$ lie not on one straight line but on two well-balanced lines - one for the upper points and one for the lower points. Obviously you have to separate the data and make two separate evaluations.

Estimation of the real working time
Some components of a car must be replaced several times during the total life, for example, lights, tyres, exhausts, brakes. The producer needs a statistical estimation of the efficiency of his production processes, so data are collected from repair shops concerning broken components only, and again the Weibull method can be used.

Minimum working time
Some components have, by means of their construction, a minimum working life, for example, tyres, brake linings, exhausts. Failures usually occur after a period of working time t_0. If such data are analysed on Weibull paper, the points lie on a curve and not on a straight line. The data have to be amended and the Weibull distribution transformed (see Verband der Automobilindustrie 1976 and Reichelt 1978).

Confidence regions
If you evaluate data you have to compute confidence intervals for every parameter; for the data line in the Weibull paper, you get two confidence curves which bracket the data line. However, the formulae for the confidence computation cannot be deduced at school level because the chi-square distribution, t-distribution and F-distribution are needed.

7. CONCLUDING REMARKS

Pupils benefitted from gaining an insight into the applicability of mathematical theorems, methods and ideas they learned at school. They understood the Weibull method reasonably well, without becoming experts in quality control, and some computer fans attempted to develop software for this method. They all learned, however, that you can only apply such a method if you understand it. You often have to learn more mathematics!

REFERENCES

Bungartz P. (1985). Toleranzen im Getriebe eines Autos und die Normalverteilung. Journal für Didaktik der Mathematik.

Bungartz P. (1988). *Risiko Kernkraftwerke,* Reprint. Math Inst, Bonn.

Deutsche Gesellschaft für Qualität. (1975). Lebensdauernetz für Weibull-Verteilung, Erläuterung und Handhabung, DGQ 25. Beuth-Verlag, Postfach 1145, Berlin 30.

Ohl HL. (1977). Weibull-Analyse, *Schriftenreihe der A. Opel AG 9.*

Messerschmidt-Bölkow-Blohm GmbH. *Technische Zuverlässigkeit.* Springer-Verlag, Berlin.

Niss M. (1983). *Proceedings of the Fourth Int Congress on Math Ed.* Boston, 247-249.

Pfanzagl J. (1974). *Allgemeine Methodenlehre der Statistik* 1, 2. Berlin.

Reichelt C. (1978). Rechnerische Ermittlung der Kenngrößen der Weibull-Verteilung, *VDI-Zeitschriften, Reihe 1, 56.*

Verband der Automobilindustrie. (1976). Zuverlässigkeitssicherung bei Automobilherstellern und Lieferanten, Frankfurt/M.

CHAPTER 26

Models of a Real-life Situation

J Fishman
The College of Staten Island, New York, USA

SUMMARY

The use, or misuse, of nonrenewable resources is one of the crucial problems facing humanity today. The process of analysing data and developing mathematical models of trends in the consumption of a nonrenewable energy resource, such as natural gas, leads to interesting insights into the characteristics of exponential growth, and raises questions about the future availability of this resource. The models which are adequate for an earlier period require modification to take recent data into account.

1. INTRODUCTION

What would be the advantages of applying secondary school mathematics to problems of air and water pollution, waste disposal, housing, global warming, and so on? We may find it possible to integrate and motivate the learning of mathematical concepts and skills in the process of analysing a well selected real-problem situation. Moreover, since other subject areas are already involved, to some degree, shouldn't we feel compelled to contribute our expertise toward an interdisciplinary approach, enhancing the understanding of these problems whenever possible?

I would like to discuss such a problem situation and present some of its potential for mathematical model building. The concept of finite, nonrenewable resources is relatively new, and still not widely understood. Although there now seems to be a present abundance of oil and natural gas, there remains concern about the future availability of these nonrenewable fossil fuels. The consumption of natural gas in the United States, throughout the first three–quarters of the 20th century, has grown exponentially, under the universal assumption that supplies would always be available. We have come to regard continual growth and increase as being the normal order of things.

A study of trends in natural gas consumption provides material for applications and model building for secondary school mathematics. The changes in trends in the last two decades is an interesting challenge to the earlier exponential growth models. The exponential function as taught in the secondary schools is narrowly construed; students never sense its descriptive power in relation to several aspects of modern life.

The following five problems suggest ways in which mathematical models can be presented and developed in the classroom, in order to provoke discussion, analysis and further understanding of an important issue. In each of these problems there is potential for teacher questioning and student exploration beyond that which is developed here.

2. PROBLEM 1 : PRELIMINARY MODELS

Let us assume that the school is encouraging an interdisciplinary study of energy sources. The social studies department has acquired much information regarding trends and patterns in energy consumption. The data have been systematised, in tabular and graphic form, and the mathematics department has been asked to involve its students in developing further insights.

Figure 1 is simple to interpret mathematically, but it is also implicitly open to historical, economic, scientific and even literary references. The graph describes trends in the relative importance of each of the energy sources in the United States. The following are some questions which may be proposed in the mathematics (and social studies) classes.

The use of wood and coal seems to have grown exponentially and then declined exponentially. Investigate the earlier growth in the use of wood as fuel. What can we reasonably predict about oil and gas? Notice the slow beginning in significance, then sharp rises – possibly exponential – but then a levelling off. Will these fossil fuels also follow the same path of growth and decline? Can the United States (and other industrial nations) realistically believe that the uses of energy sources can be looked at as overlapping exponential growth curves – that is, when one source is no longer adequate, can the nation optimistically assume that another source will become available? How do we account for the levelling of consumption of oil and gas in recent years? Absolute values of magnitudes of consumption are required for further analysis.

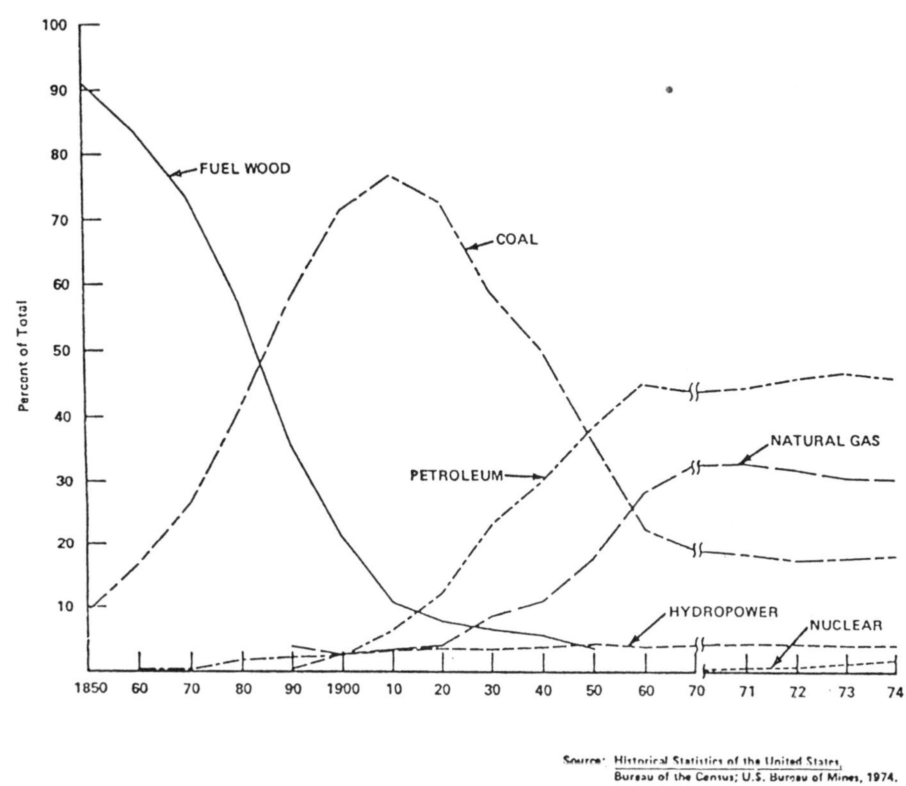

Figure 1 : US Encrgy Consumption Patterns, 1850–1974

Table 1 shows the consumption of natural gas in the developed and developing regions of the world: these data also have much of interdisciplinary interest. Mathematics students can find relationships among regions, regarding trends in the consumption of natural gas. Notice the slight decline in consumption in North America and the relative stability of Western Europe. However, the total world consumption, after a temporary slow down, continues its exponential growth – at an average annual rate of approximately 2.8 percent. If this rate continues, in what interval of time will world consumption double?

The developed nations continue to consume a large part of the world's consumption of natural gas but this proportion is declining: compare the proportion in 1973 and 1985.

REGION	1973	1974	1975	1976	1977	1978	1979	1980	1985
North America	24.66	23.63	22.20	22.07	22.28	22.63	23.08	23.22	20.78
Central and South America	0.95	0.99	0.98	1.03	1.12	1.13	1.30	1.38	1.82
Western Europe	5.15	5.84	6.07	6.40	6.50	6.48	7.17	6.64	6.74
Eastern Europe and USSR	9.95	10.86	11.95	13.27	14.25	15.10	16.47	17.32	23.36
Middle East	1.18	1.34	1.42	1.50	1.72	2.00	1.44	1.17	1.84
Africa	0.32	0.34	0.37	0.54	0.50	0.74	0.78	0.78	1.96
Far East and Oceania	0.94	1.12	1.28	1.48	1.76	1.86	2.14	2.26	3.79
WORLD TOTAL	43.15	44.12	44.26	46.28	48.13	49.94	52.36	52.77	60.28

US Department of Energy. 1987 **International Energy Annual.** Washington DC : 1988, 18-19

Table 1 : World Natural Gas Production/Consumption 1973–1985 (10^9 cubic feet)

3. PROBLEM 2 : FINDING A MODEL WHICH DESCRIBES TRENDS IN NATURAL GAS CONSUMPTION IN THE UNITED STATES

An analysis of gas consumption data provides an opportunity to find the function which best fits the data – to develop a model (an equation, graph or further tables) which clarifies past activities, and justifies making conjectures about the future.

(a) It can be observed in table 2 that magnitudes of consumption have grown from 1918 to 1970, and then remained relatively stable. Is it possible to develop an equation which describes the trends in consumption for these two periods of time, or one for the entire period? For now, let us consider the longer period. The annual increments in consumption seem to be increasing. A polynomial function could generate a sequence of terms with increasing differences. The exponential function, of course, also generates sequences with increasing differences. If we graphed the data, there still remain different possibilities for describing the graph. It is worthwhile for students to explore these possibilities.

Year	Marketed Production (10^9cf)	Year	Marketed Production (10^9cf)	Year	Marketed Production (10^9cf)
1918	0.72	1941	2.81	1966	17.21
1919	0.75	1942	3.05	1967	18.17
1920	0.80	1943	3.41	1968	19.32
		1944	3.71	1969	20.70
1921	0.66	1945	3.92	1970	21.92
1922	0.76				
1923	1.01	1946	4.03	1971	22.49
1924	1.14	1947	4.58	1972	22.53
1925	1.19	1948	5.15	1973	22.65
		1949	5.42	1974	21.60
1926	1.31	1950	6.28	1975	20.11
1927	1.45				
1928	1.57	1951	7.46	1976	19.95
1929	1.92	1952	8.01	1977	20.03
1930	1.94	1953	8.40	1978	19.97
		1954	8.74	1979	20.47
1931	1.69	1955	9.41	1980	20.18
1932	1.56				
1933	1.56	1956	10.08	1981	19.96
1934	1.77	1957	10.68	1982	18.52
1935	1.92	1958	11.03	1983	16.82
		1959	12.05	1984	18.23
1936	2.17	1960	12.77	1985	17.20
1937	2.41				
1938	2.30	1961	13.25	1986	16.79
1939	2.48	1962	13.88	1987	17.16
1940	2.66	1963	14.75		
		1964	15.46		
		1965	16.04		

Table 2 : United States Marketed Production (Consumption) of Natural Gas (Reports of the American Gas Association)

(b) Analysis of the data is facilitated by looking at annual consumption at the end of each decade – see table 3 (consumption for 1890, 1900 and 1910 has been added). Considering only these quantities as a sequence of terms, we can use the notion regarding polynomials: when a given sequence eventually produces a set of finite differences that is constant, one can find the generating expression. If the polynomial is of the form $C = C_0 + a_1t + a_2t^2 + \ldots + a_nt^n$, where C_0 is the gas consumed at the end of the first decade, and t is time, with each 10 year interval taken as one unit, the method of differences could be used to find the values for the constants, a_1, a_2, a_3, ...

1890	1900	1910	1920	1930	1940	1950	1960	1970	1980	1987
0.3	0.3	0.5	0.8	1.9	2.7	6.3	12.8	21.9	20.2	17.2

Table 3 : Natural Gas Production/Consumption (10^9cf)

Taking into consideration that we are working with approximations, it is found that the sequence of terms does not produce a set of differences that are constant, and therefore it is concluded that the relationship between time and gas consumption is not given by any polynomial function.

(c) Consider if there is a constant multiple between the terms and that the data, at intervals of 10 years, are described best by the geometric sequence of the form: C_0, C_0R, C_0R^2, ..., C_0R^n, where C_0 is the initial amount in the sequence and R is the fixed ratio of any two successive terms, C_{n+1}/C_n. Even though these ratios are not all the same, as we would expect from real data, they lie within a narrow range. The geometric mean of these ratios (for each of the seven decades) is 1.84, and the equation which generates the data in the sequence of 10 year intervals is $C = 0.3(1.84)^t$.

(d) Another important observation is that there is an approximate doubling of annual consumption every 10 years (except for the depression years 1930 to 1940). These two observations (the approximate constant multiple between terms and the doubling of data in fixed periods) should suggest to the student, who has some familiarity with the exponential function, that the *annual* growth of consumption up to 1970 is best described by the exponential function $C = C_0(1+r)^n$, who should then find r, the average annual rate of growth, to be approximately 0.063. Thus, it is concluded that the equation best describing yearly consumption of natural gas, from 1890 to 1970 is $C = 0.3(1.063)^n$, where n is the number of years.

(e) How do we account for the absence of growth in natural gas consumption after 1970, and the decline into the 1980s? Is this only a temporary slowdown, or does it presage new developments in the consumption of natural gas in the United States? In what way do this recent data conflict with the model? What modifications in the model will account for all the data? Before we pursue this further, let us explore a graphical model.

4. PROBLEM 3 : THE GRAPHICAL MODEL

(a) Graphs of the data provide further insights into consumption trends and the characteristics of exponential growth. The graph of the exponential equation $C = 0.3(1.063)^n$, plotting magnitudes of consumption against time on regular graph paper, shows how consumption increases slowly at first then begins to rise sharply once the magnitude becomes sufficiently large (see figure 2). In a real-life situation, one is not usually aware of exponential growth until this later stage.

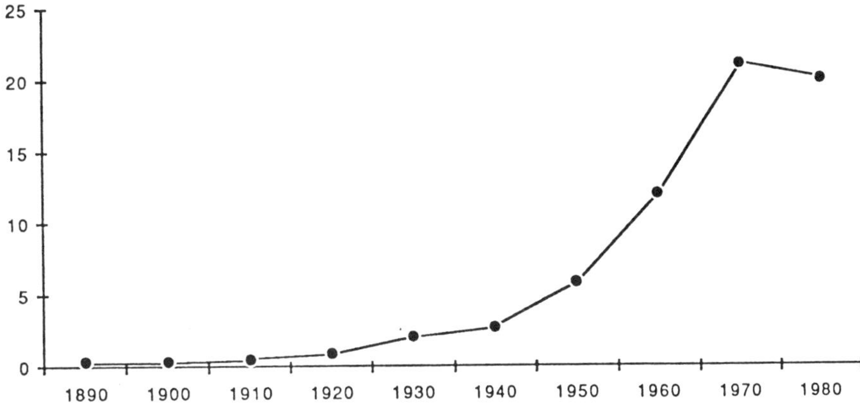

Figure 2 : Gas Consumption (10^9 cubic feet)

(b) Transforming the exponential equation into a logarithmic equation, log C = log 0.3 + x log 1.063, which is linear in form and, plotting the consumption data on semi-logarithmic paper (see figure 3), there is a very close approximation to a straight line, indicating the constant rate of growth in consumption (from 1900 to 1970), with log 1.063 as its slope. The linear graph also shows the rapid rise in magnitudes after a slow beginning (Kastner, 1978).

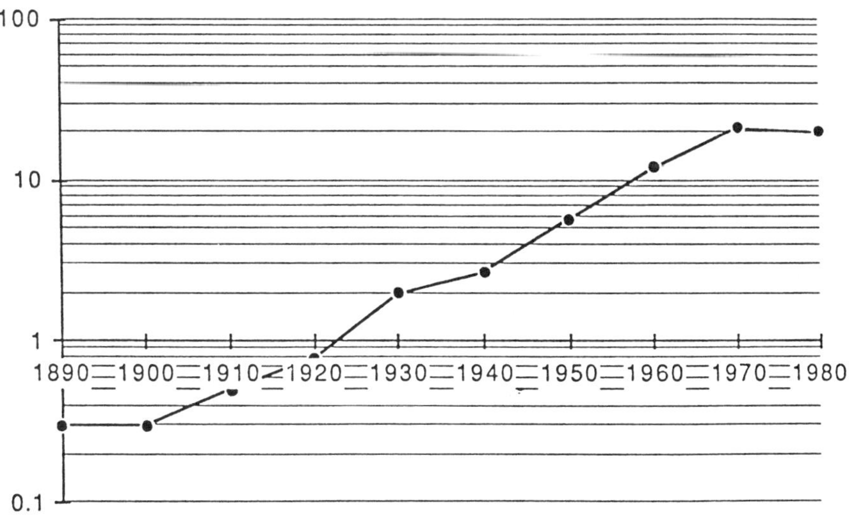

Figure 3 : Gas consumption (10^9 cubic feet)

5. PROBLEM 4 : ENRICHING THE MODELS WITH FURTHER ANALYSIS OF THE CHARACTERISTICS OF EXPONENTIAL GROWTH

The mathematics students have made good progress in the process of analysing the data, and it is worthwhile staying with the topic because there are other aspects of the model which should be observed.

(a) An important characteristic is that the approximate annual increments of consumption, with an average fixed rate of growth, lead to a doubling of consumption in fixed periods of time – a remarkable mathematical, as well as ecological, phenomenon of exponential growth. This has been observed in table 3. If the US had continued consuming natural gas at the same rate as it did from 1900 to 1970, annual consumption would have reached approximately 40 units (10^9 cubic feet) by 1980 and 80 units by 1990 – a dangerous situation which, so far, has not come to be.

Find the doubling period of natural gas consumption in the United States for the years 1900 to 1970. The doubling interval is actually a little more than 11 years, but we work with an interval of 10 years to simplify the model. What assurance is there that the present slowdown in consumption is not temporary, and that given changed conditions this nation will once again be on the exponential growth track? It has been estimated that somewhere between 740 and 1200 units of discovered and undiscovered natural gas are all that will be available to the US in the future. How many more doublings could theoretically take place before all the supplies were depleted? Even with continued consumption at 17 to 20 units a year, how long could natural gas last?

(b) The last term in the sum of the doubling geometric sequence, $S = a + 2a + 2^2a + \ldots + 2^{n-1}a$, is always greater than the sum total of the previous terms. We find, $S + a = 2^n a$. This is approximately true for the sequence of data for natural gas consumption in table 3.

(c) Cumulative consumption is doubling as well. This should be a dramatic observation for anyone concerned about the supply of resources available for future generations. The cumulative consumption for each decade can be easily computed or, by using integration (considering consumption as continuous), can be approximated. For example, for the years 1960 to 1970, with consumption growing at the rate (r) of 5.6 percent for this decade, the total gas consumed for the 10 year period, beginning with the annual consumption of 12.7 units, can be found by integrating

$$12.7 \int_0^{10} (1+r)^t \, dt$$

The cumulative consumption of natural gas during this period is 164 units (see table 4). In less than three doublings, the US would have been completely out of natural gas if the same rate had continued.

The Cumulative quantities for each decade		Total Cumulative Consumption
1900–1910	3.89	3.89
1910–1920	6.43	10.83
1920–1930	12.82	23.14
1930–1940	22.13	45.27
1940–1950	40.5	85.77
1950–1960	87.14	172.91
1960–1970	164.2	337.11

Table 4 : Cumulative Consumption

6. PROBLEM 5 : THE MODEL MODIFIED – THE COMPLETE CYCLE OF US GAS CONSUMPTION

Finally, we must consider how the model can be modified to deal with the data of approximately the last two decades, showing a slight decline in natural gas consumption. Students should think about this situation before going on; the previous work should lead to the conclusion that continued exponential growth of a finite resource can only result in a catastrophe of sudden depletion of resources. As with other aspects of growth in nature, this generally does not occur; other factors begin to inhibit continued growth. The following modifications of the model provide many opportunities for student explorations.

King Hubbert (1969), who pioneered the application of mathematical physics to geologic problems, estimated, in the 1960s, the ultimate quantity of natural gas to be 1290 units, and predicted the decline of production/consumption as an inevitable trend in the availability and usage of the fossil fuels.

According to his interpretation of the complete cycle of production, a peak production of about 25 units per year would occur about the year 1980 – see figure 4. Hubbert describes the curve of the cycle as bell–shaped, where the early phase is one of a positive exponential rate of increase and the declining phase an exponential rate of decrease. The amount of natural gas ultimately available for consumption is given by the total area under the curve. Of course, the bell curve is the general form; the actual results would never be so smooth.

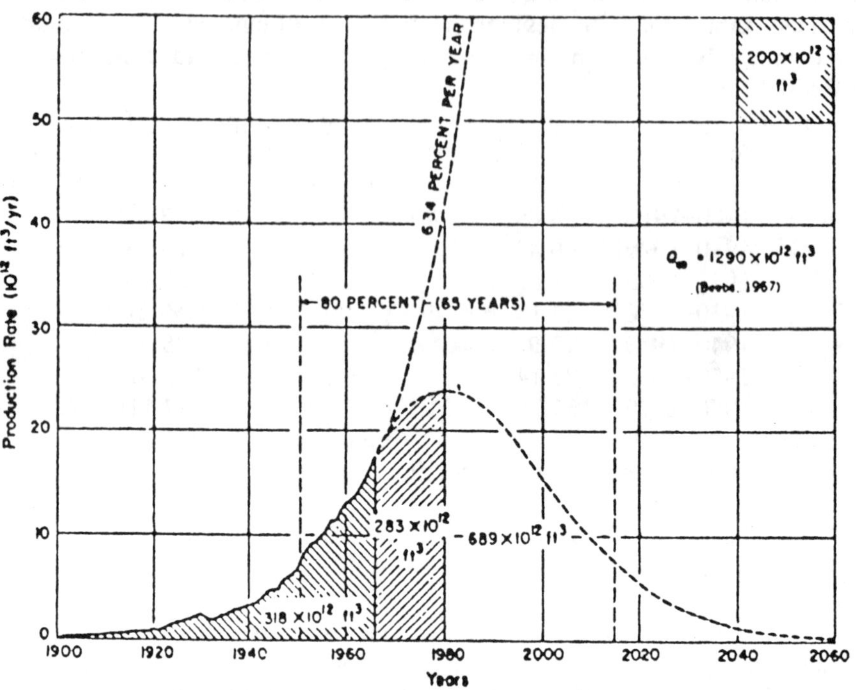

Figure 4 : The Complete Cycle of Natural Gas Production

Also shown in figure 4 is the curve of what the production would be until about 1985 if it were to continue the growth rate of 6.3 percent per year, which had prevailed during the 20th century up to 1970.

7. CONCLUSION

The fossil fuels took hundreds of millions of years to be formed and deposited in sedimentary sands, muds and limes. However, it has been only in the last 50 years that most of the consumption has taken place. The rate of exploitation of these resources will determine how long the petroleum era will last. If it is true that exponential growth does indeed describe several aspects of modern life, then we have reason to be pessimistic about the dire economic and ecological effects if nations go about their business ignoring these realities.

This perspective and concern does not receive unanimous support. Critics of this pessimistic view claim that technological development, alternative resources, and human ingenuity will provide solutions for any serious problems. What the reality of total resources of natural gas is, only future events will confirm. It is surely a debate worth considering and thinking about seriously; it should be one of many such topics in the secondary school curriculum, including mathematics.

REFERENCES

Blum W and Niss M. (1989). Mathematical Problem Solving, Modelling Applications, and Links to Other Subjects. In Blum W, Niss M, Huntley I (eds). *Modelling, Applications and Applied Problem Solving*, Ellis Horwood, Chichester.

Fisher WL. (1989). Assessing the Natural Gas Resources Base. *Gas Energy Review (AGA)*. June, 2-5.

Guilliote HP. (1986). The Methods of Finite Differences : Some Applications. *Mathematics Teacher*, September, 466-470.

Kastner B. (1978). *Applications of Secondary School Mathematics*. National Council of Teachers of Mathematics. Reston, Virginia.

Hubbert MK. (1969). *Energy Resources*. Resources and Man. WH Freeman, San Francisco.

Simon JL and Kahn H. (1984). *The Resourceful Earth : A Response to Global 2000*. Basil Blackwell, New York.

CHAPTER 27

Alton Towers : A Practical Approach to Modelling in Mechanics

A Kitchen
University of Manchester, UK

SUMMARY

Alton Towers is a large theme park in Staffordshire. It contains a large number of fairground rides of many designs, from simple helter-skelters to huge roller coasters. The following paper gives the history of a project using the park as a vehicle for teaching mechanics to students of between 16 and 18 years of age.

1. INTRODUCTION

Nine years ago I started to teach mechanics in school, after having worked as a computer consultant for many years. Coming into school I became aware that, for many people, an important part of mathematics was absent from the syllabus. Calculus, algebra, geometry and Newtonian mechanics were all being taught but they were being taught in isolation from the real world. This was especially true of mechanics - strings were light, friction had a limiting value which depended solely on the normal reaction, static and dynamic friction were considered to be interchangeable. All these simplifications are valid in some cases but we, the teachers, had made them for our students with very little explanation, and certainly without letting them have a chance of making their own assumptions. This meant that when they met a real-life situation with all its variables, they did not know how to handle its analysis. I, on the other hand, initially could not understand their difficulties. It was rather like teaching someone to ride a bicycle. Once one has learned how to do it and has been riding for many years, one forgets how impossible it seems to a beginner. All the decisions such as which way to turn the handlebars or how to lean when going round a corner have become completely instinctive. So much so that it is hard to explain what occurs. This is really what happens with problem solving and modelling. It is all too easy to look on them as instinctive, not needing to be taught, and therefore to concentrate on the mathematical theory of any situation. We were concentrating totally on the analysis in our teaching, by making the model for the students and by ignoring the need for model building, interpretation and validation. These are, however, just as important as the analysis, and need to be taught. This is especially true in mechanics.

2. MECHANICS TEACHING IN BRITAIN

In many countries, however, mechanics as we know it in Britain is not considered to be part of mathematics at all, but simply a branch of physics. Even in Britain, mechanics is taught both in mathematics, and as an option in physics. I, as a mathematician, tended to see mathematics as a theoretical subject, which could be taught totally divorced from any sort of practical work. By implication, it might appear therefore that practical work was unimportant. The physicists do the opposite and put great stress on the practicality of mechanics, bringing in mathematics where needed as a necessary evil. We both need the other, but the emphasis we put on the material is very different. One of the great benefits of applied mathematics, therefore, is that it can form a link between pure mathematics and science, and acknowledge the importance of both. Certainly, Isaac Newton did not make the distinction between the two. Part of the great power and genius of his work lies in the fact that his *Principia* were not bounded by any narrow subject boundaries, and we must follow his example in our teaching.

3. WHY ALTON TOWERS?

It was obvious, however, that I could not just bring in the sort of practical work already covered by physics. For one thing, my department was not equipped for this sort of work, and for another I felt that the formal practicals set up in the laboratory were too restrictive in their outlook. I decided to look for a real-life situation that could both be modelled by the students and would also enrich their experiences of things like force, acceleration, velocity and circular motion. These were concepts that they were all too ready to talk about in the abstract but for which they had no real intuitive feel. I decided to go to a large amusement park nearby. A few goes myself on rides like the Corkscrew, the Spider and the Pirate Ship convinced me that I was in the right place. Not only could the students have a lot of fun, experiencing forces and velocities in an exaggerated way, but they could also learn some of the stages of the modelling process in a context that was meaningful to them. My plan was to take the students to the park, let them choose a ride and then use modelling techniques to study it. I hoped that the students could use their mathematics to describe the reasons for the specific design choices and economics of the ride and, I hoped, see how a change in any component would affect the ride as a whole.

This was not as straightforward as it sounds, however. I soon discovered that, where I looked at a ride and instinctively simplified it into its component parts, ignoring all the superfluous details, the students looked at the same ride and saw a confusing mass of rods, bars, cars and suchlike that they had no idea how to simplify. I decided that, as they began to make a mathematical model, they should look at the reality from the viewpoint of making a physical model of the ride. This would, of necessity, involve them in making decisions as to the important parts of the structure as far as the mechanics of the situation were concerned and the physical model would, therefore, help them to formulate a theoretical model.

4. RUNNING THE PROJECT

The project itself takes about eight hours of school time and an equal amount of the students' own time. First, the students spend two hours in the classroom discussing how to model one particular ride. This is done with the help of videos and an actual model of the ride itself. I usually use The Spider, discussed in detail in the next section. The class work through the modelling stages together, and see what can be done.

The students then have a day visit to Alton Towers. In the morning they are expected to try every ride and, by lunch, they decide which they wish to study in detail. The afternoon is spent collecting the data needed to model their chosen ride. Back in school, a further two hours of class time is available for individual work with a teacher on hand to guide where necessary. The students then have a further eight hours of their private study time, spread over at least three weeks, to write up their project.

5. THE SPIDER

This has proved an ideal ride to use as an exemplar in the introductory phase of the project.

Photograph 1 : The Spider

This ride appears to be very complex at first glance. The cars seem to travel in a completely random way, first speeding up, then slowing down changing direction the whole time. The students' first reaction is that the whole ride is far too complex to model in any meaningful way. However, by encouraging them to look at how they could make a small model of the ride the important components become clear.

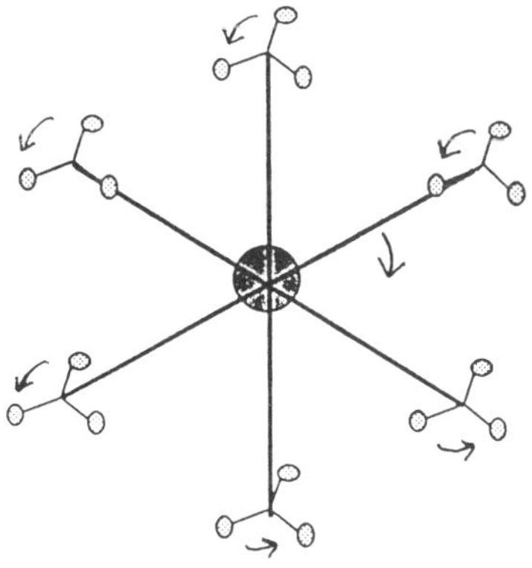

There are six main arms which rotate about a central hub at a constant speed. At the end of each arm, a series of smaller arms rotate. The riders sit in cars attached to the ends of these smaller arms.

Firstly, how can we model the large arms? We don't need all of them, just one will do. A pin through a plastic strip will give us an arm that will rotate. This gives us our first decision. How long should it be and in which direction should it rotate? Obviously in a model the length should be manageable and, in my case, is such that it will fit easily on an overhead projector screen. Next the direction. We look at the actual ride and discover that it rotates clockwise.

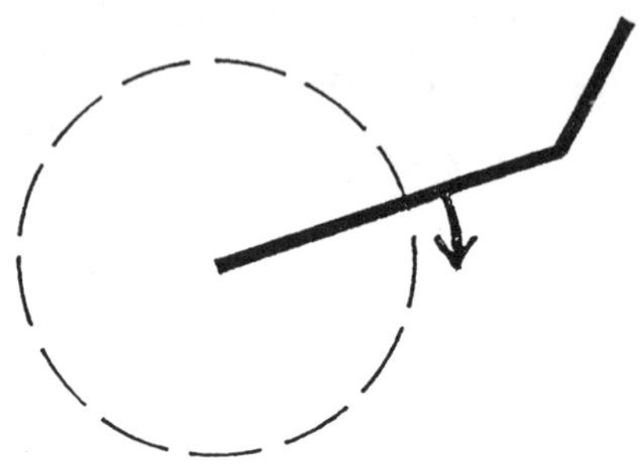

Now we need the smaller arm. Again just one will do. Pinning it to the end of the large arm will give us a very simplified model. However, when I now rotate the arm, the motion does not bear any resemblance to the ride. What is different? The rotation of the main arm and the small arm need to be linked. When the large arm has made one complete rotation, the small one has performed three rotations anticlockwise. A set of gears will solve this problem, and allow us to slow the whole ride down to a pace that we can handle.

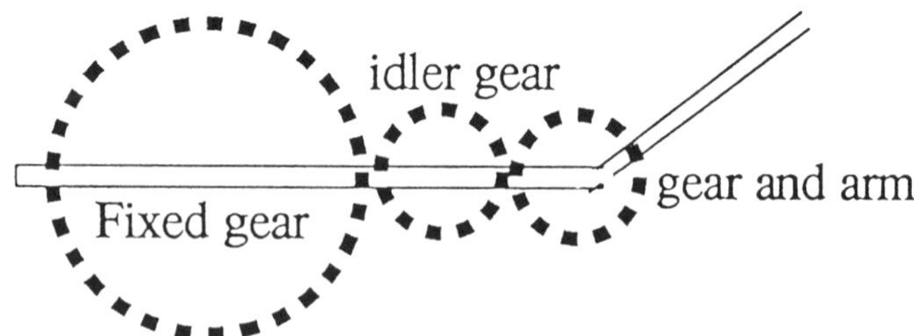

Turning the large arm through one complete revolution shows that the small arm rotates three times relative to the large arm, but only twice relative to the ground. The rider's perception of how many rotations undergone will differ according to the frame of reference chosen.

How can we study the actual path of the ride, the speeds attained and the forces on the riders? Again, a look at our model will help us to get an initial feel for the answers.

As I turn the large arm through 15°, the small arm rotates through 45° relative to it. I can mark the locus of the path with a pen as I turn the arm. The distance between each dot will represent a single unit of time, as the arm is rotating with constant speed.

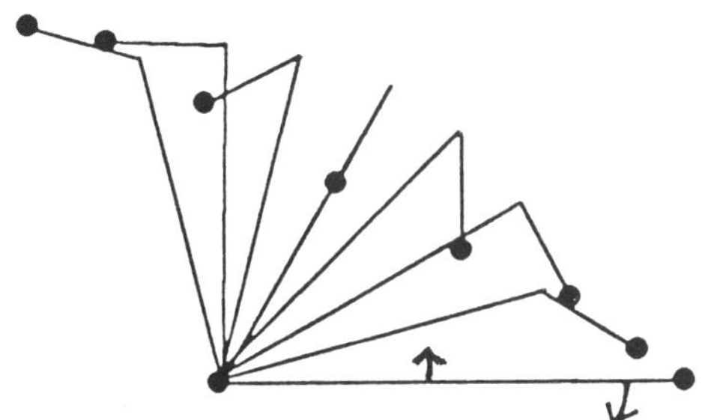

You can see that the path is a cycloid. The dots are close together at the outside, showing that the car is travelling slowly, and far apart as the car swings in toward the central hub, making the car speed up. However, the acceleration is a different matter. Remembering that acceleration is rate of change of velocity, we can see that the direction of the velocity changes rapidly at the outside of the path. We will need to look at the mathematical model to find out more about this. Obviously, things like radius of arm, speed and direction of rotation, have a great deal to do with the type of ride experienced. It must be exciting, but not too nauseating; safe, but not tame.

6. THE MODEL

Where do mathematics and real modelling come in? It is obvious that the ride needs to be studied before it is built, in order to decide the best measurements for the various components. One can't add on a couple of metres to an arm when the ride has been built.

Set up model:
Let the rider be a particle B on the end of a pair of arms given by OA and AB.
Let the length of the large arm be R and that of the small arm be r.

Let their angular velocities be W and w measured anticlockwise. Define axes OX and OY.

Analyse the problem:

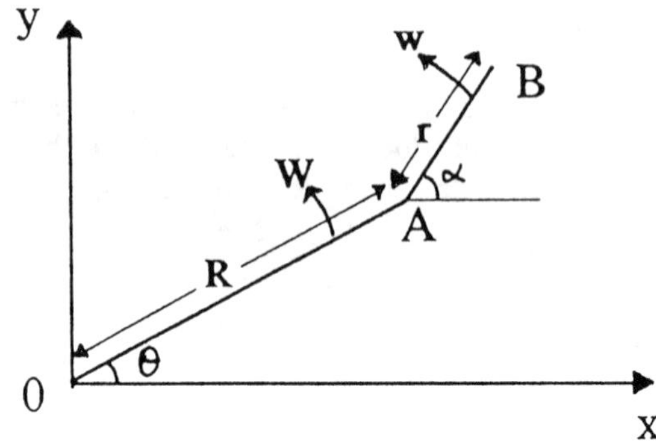

$$x = R\cos(Wt) + r\cos(wt) \quad \Rightarrow \quad \dot{x} = -RW\sin(Wt) - rw\sin(wt)$$

$$y = R\sin(Wt) + r\sin(wt) \quad \Rightarrow \quad \dot{y} = RW\cos(Wt) + rw\cos(wt)$$

$$\text{speed} = (x^2+y^2)^{\frac{1}{2}} = (R^2W^2+r^2w^2+2RrWw\cos(W-w)t)^{\frac{1}{2}} \quad \text{at time } t$$

$$\ddot{x} = -RW^2\cos(Wt) - rw^2\cos(wt)$$

$$\ddot{y} = -RW^2\sin(Wt) - rw^2\sin(wt)$$

the acceleration has magnitude

$$(R^2W^4 + r^2w^4 + 2rRW^2w^2\cos(W-w)t)^{\frac{1}{2}}$$

Interpret:

The maximum and minimum values of $\cos(W-w)t$ are +1 and −1. This gives maximum acceleration of $RW^2 + rw^2$ when $\alpha = \theta$, and a minimum acceleration of $RW^2 - rw^2$ when $\alpha = \theta \pm \pi$. At first glance it might appear that this also holds for the speed. However, the speed has the term $2WwRr\cos(W-w)t$, and this also depends on the signs of W and w. If they are of opposite sign, one going clockwise and the other anticlockwise, then the term itself is negative when $(W-w)t$ is $2n\pi$ (when the car is furthest away) and so the velocity is a minimum, and vice versa.

To summarise, the acceleration is always greatest when both arms are fully extended, but the velocity is only greatest at this time if the arms rotate in the same direction. As far as the quality of the ride is concerned, it is more exciting if the maximum velocity and acceleration do not coincide. This will become obvious if we look at a computer simulation of the path.

This is easy to program, and will show the many different loci that will occur as W, w, R and r are varied.

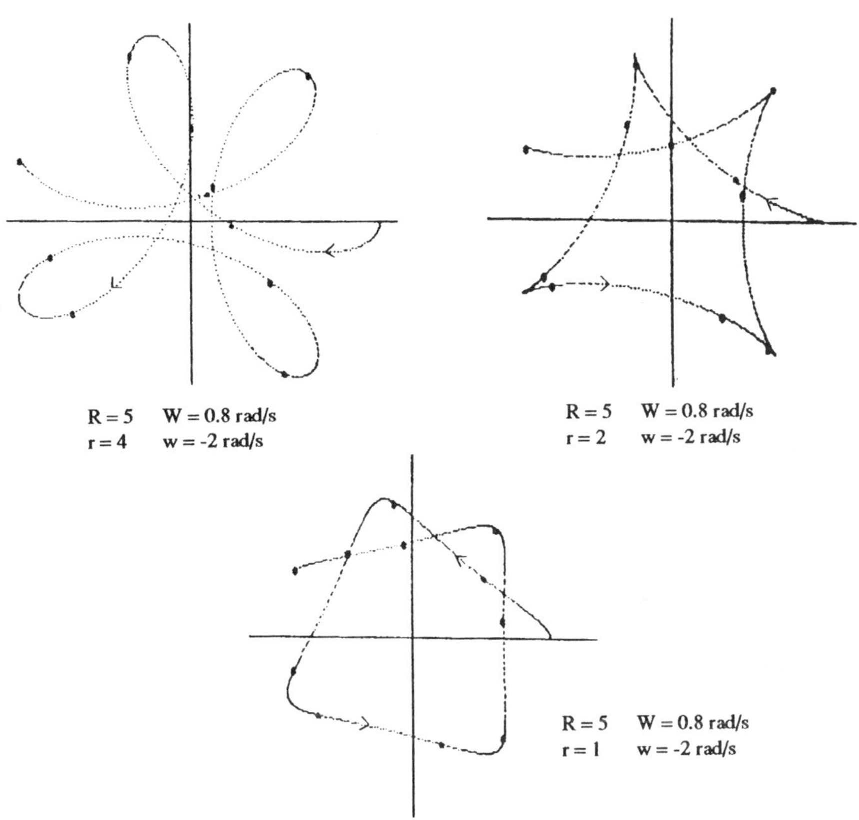

Initially, we will look at the effect of varying r, keeping the other variables constant. When R is 5 and r is 4, then the motion is a series of large loops. The rider is at times travelling clockwise relative to the ground and at other times travelling anticlockwise. The speed never drops to zero, although it is least when the rider is furthest from the central hub of the ride. As r decreases, so the loops become more and more pointed until, when RW = rw, they turn into cusps. At these points the speed is zero. As r continues to decrease, the path flattens out until the limiting value of r = 0, where the locus is a circle. Let us look further at the acceleration during the ride – this is of paramount importance to the designer. The riders must accelerate with the car if they are not to become a projectile. Some means of providing a force of a suitable magnitude and direction must therefore be found. The riders can either be firmly strapped into the car, or the back and side rests must provide all the forces needed, and there should be little or no backwards acceleration. The forces must also not be so variable that a whiplash effect is felt. Looking at the path of the spider again, we can see that the accelerations, and hence the forces required, do not vary greatly in magnitude and that their direction does not vary much from that of the small arm. The maximum acceleration is just over $11 m/s^2$, just over that experienced in free fall.

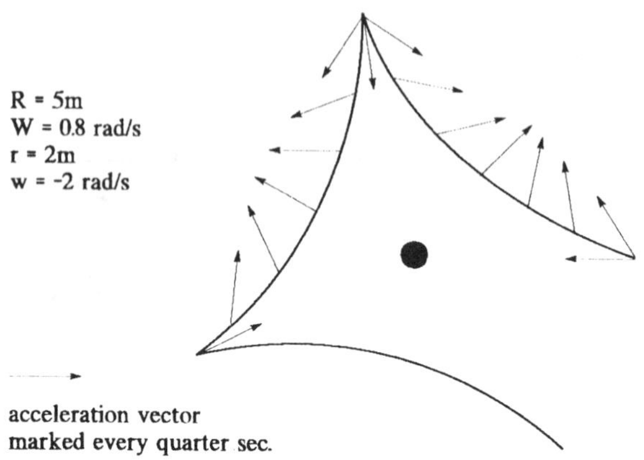

Looking at the path of a single car gives some idea of the sensations present, but we must remember that our model ignored the other cars, each of which is travelling round in space. Their paths are similar, but they all had a slightly different starting point. If we superimpose two such cars on adjacent long arms, we can see how they interelate to one another. At certain points, the two cars appear to be travelling towards each other at a high speed. Remember, the marks on the diagram show positions at one second intervals.

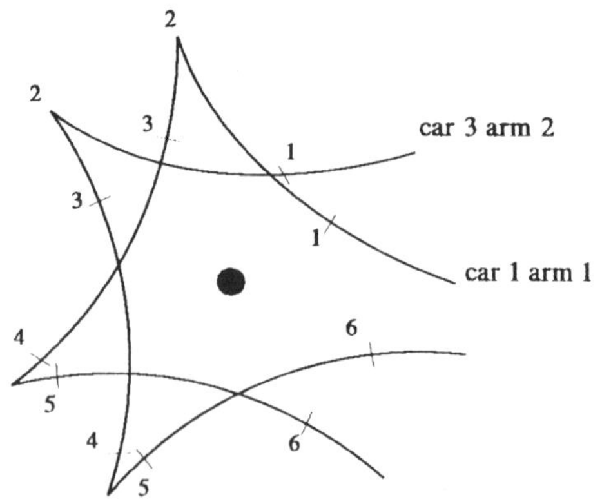

We mentioned briefly that the spider had only six arms. This is because the greater the number of arms the shorter the small arm must be. For safety's sake, it must be able to rotate in a circle without impinging on a neighbouring set of arms.

So far, we have looked at our model, analysed it mathematically and interpreted the mathematics. The actual ride can now be re-examined to see if the real values of the variables fit in with our interpretation. This is just a simple model of the motion, and takes no account of the up and down motion of the ride.

7. THE PIRATE SHIP

Photograph 2 : Pirate Ship

Another interesting ride is the Pirate Ship. Here the rider sits in a giant swing boat, that swings to and fro. It gradually increases the amplitude of the swing until, at the highest part of the ride, the rider is level with the centre of rotation of the ship. Watching the video, we can see that the simplest model is, obviously, that of a particle on the end of a string and, for a passenger in the very middle of the boat, it gives a reasonable model. However, it should be obvious that the passengers at the very front of the ship have the same speed at any instant as those in the middle and that their motion is not symmetrical about the vertical. As the riders are not restrained in any way, but just sit on a bench seat with a bar to hold on to, it produces a strange sensation in the riders who feel, not unnaturally, that they are being left behind the ship on the way up, and are falling forward on the way down.

I leave the mathematics of this to the reader. Suffice it to say, that it soon becomes apparent from this ride who understands conservation of energy and who has just learned a few formulae parrot fashion. There are many different rides, and all of them produce real situations which can be modelled.

8. WHAT HAS BEEN ACHIEVED?

So much for the theory. How has all this worked out in practice? I have taken between 30 and 40 seventeen-year-old students from school to Alton Towers each year for the past eight years. The standard of work produced by the students has improved dramatically over the years. It may be that my expectations of them have changed. It may be that the students have become aware, from those that went the year before, that a great deal of value can be gained from the project. Certainly, the class each year has tried to outdo the work of the year before. It has amazed me that each year brings new work on the same rides. The fact that many of the students have used their projects as discussion points when applying for industrial sponsorship has helped the motivation of the students. Initially, the first batch of students felt rather unsure of the whole purpose of the exercise. Now, however, each class looks forward to the experience. Many other schools are now working on the same lines and using this, or a similar trip, as part of an enrichment to a more conventional A-level course. I can definitely recommend it. Perhaps I should finish by pointing out that it is not just suitable for sixth formers. I have learned a great deal myself.

9. CONCLUSION

What I have tried to demonstrate in this paper is that an approach using both physical and theoretical modelling can be very powerful. A fairground ride is an excellent subject for modelling. It is rich in mathematics and also enjoyable. It enables the students to experience reality and to model from it. They can then extend the model to the limits of their own knowledge, and use it as a motivation to learn more. Why not visit an amusement park yourself? Try all the rides. Find some problems to solve. Have fun. That's what mathematics is all about.

CHAPTER 28

Modelling in Mechanics: a Cross-curricular and Problem-solving Approach to Learning Mathematics

JS Williams
University of Manchester, UK

1. INTRODUCTION

Previous work, reported at ICME-6 and in Williams (1989a), described the work of the Mechanics in Action Project (MAP) in terms of the following three problem-solving modes.

- Intuitive or informal problem solving
- *Ad hoc* empirical modelling
- Newtonian modelling

It was shown there, how practical work in classrooms provides opportunities for students to learn new mathematics and to apply mathematics in new ways in thc context of solving a real problem of scientific or technological interest. In particular, it was argued that this approach made empirical modelling and Newtonian modelling more meaningful, and so accessible, to a wide ability and age range of students. It was also indicated that the identification of the three distinct modes relates to students' development, and has methodological implications. The purpose of this paper is to develop this theory.

Our theory of pedagogical practice aims to do two things. Firstly, it should *describe* paradigms of good practice. (It is here presumed that good practice has indeed been identified by the author and project colleagues!) Secondly, it should provide a *prescription* for the design of future practice. This will involve prescribing types of activity as well as the teacher's interventional role.

For illustration, I will refer to two tasks as examplars of the type of work in which the project has been involved. These have been used with students throughout the secondary age-range, from about 11/12 to 17/18, though the details of presentation vary.

The first task (Mason, 1988) involves the modelling of a bouncing ball, in the context of answering such questions as the following.

- Which balls bounce best?
- How do you measure bounciness?
- How high will a ball rebound?
- Describe the distance between bounces.

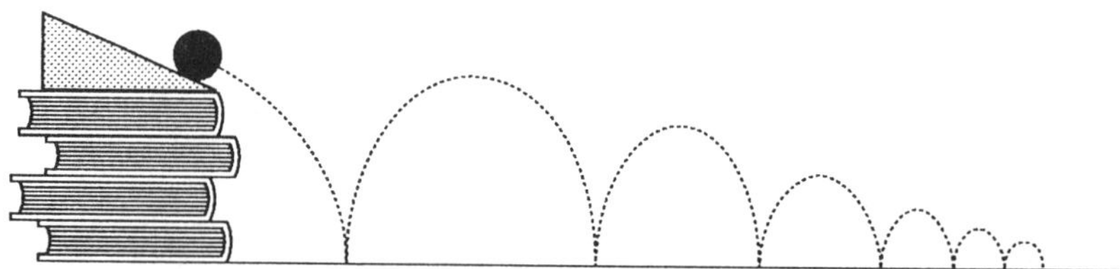

Figure 1

The second task concerns Galileo's experiment, in which we ask how long it takes a ball to roll down an inclined track.

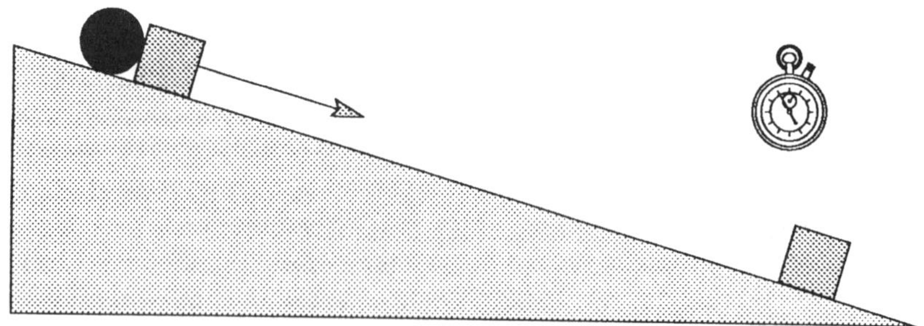

Figure 2

We have extensive experience of working with these practicals in different forms, with different teachers and pupils. The practicals usually involve many of the following features.

- Real problem solving in a scientific or technological context
- Hands-on practical work by small groups of pupils, connected with personal everyday physical experience outside the classroom
- Accessibility to simple informal and intuitive solutions, requiring only elementary mathematical skills
- The potential for empirical modelling involving increasingly formal and powerful mathematics
- The further potential for modelling with Newtonian mechanics

2. FEATURES OF PRACTICAL WORK

2.1 Real problem solving with mechanics in science or technology

The first principle of the theory is that the tasks should draw on, or contribute to, work with mathematics in a scientific or technological context. This can provide motivation for students; it also establishes a link between mathematics and other important experiences.

The second principle is that the tasks should provide a challenge for the students. This involves posing a real problem, by which I mean a problem of immediate interest with a tangible solution, which is understandable and verifiable *by the pupils themselves* in their classroom. Our experience is that using real apparatus and pupils' personal experience provides the key. Video is second best. Micro simulation robs pupils of the process of modelling and validation. Purely Realistic, rather than real problems, are still too abstract for most pupils. As Mason (1988) says, most approaches to modelling in fact fail to involve the pupils in the process, and reduce them to passive observers of modelling.

2.2 Hands-on work in small groups

Such a challenge inevitably involves *group work* in which pupils can draw on the strengths and skills of a team. Discussion and negotiation, division of labour and coordination of results and collaborative decision-making are seen to be essential elements in their success.

Many students need help with this. Practical mechanics provides a focus for the group. There are instruments to use, there is apparatus to set up and manipulate.

2.3 Accessibility to informal solutions

The tasks are designed to be approachable using only very elementary mathematics - use of measurement and simple instruments, drawing, tabulation and so on. Such skills are often, in real life, only incidental to a technological task. The major part of the work is done by play, and by trial and improvement. All pupils can engage in the task with some success at this level, without making formal mathematical demands - everybody succeeds.

3. THE POTENTIAL FOR MODELLING

Each task involves, within it, a concretisation of some significant mathematical model. The presentation of the task, and the intervention of the teacher, is then planned so as to reveal the pertinence of the mathematics. Some of the mathematics will be applied by the students to the situation, but there will also be potential for the teacher to help the student to see new mathematics in context. Consequently, the task provides a vehicle for the teacher to reveal new mathematics.

For example, the bounce marks of the ball along the paper reveal a pattern. The student perceives that the marks are getting closer together, so that the bounces are getting smaller. It is clear, also, that the differences in bounce distances are decreasing. Eventually the distances fall to zero as the ball starts rolling.

By cutting strips of paper laid along the path the student can make an 'instant graph', and explain how the above interpretations are revealed in the shape of the graph in figure 4.

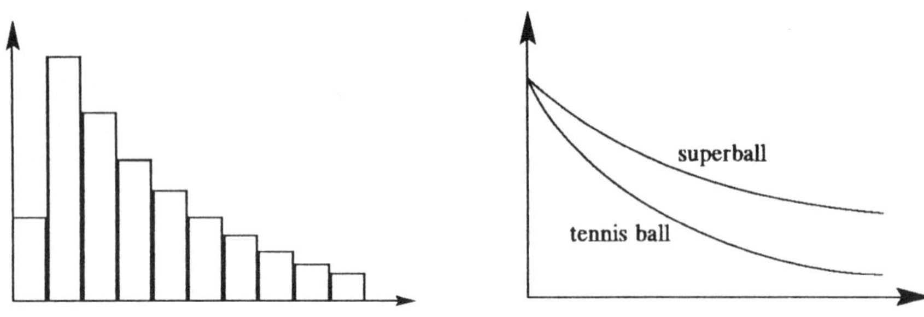

Figure 3 : concrete graph **Figure 4 : qualitative graph**

The appreciation of the meaning of graphs such as those in figure 4, both on a tennis ball and superball is an important step. The difficulty that children have in moving from quantitative graphs to such qualitative graphs is well known (Bell *et al*, 1987). Here, the richness of the children's experience of bouncing balls facilitates their grasp of the concepts. Rather than take the children's conceptual development in the hierarchy from quantitative to qualitative graphs, the *modelling* approach allows the children to develop the concepts directly, out of concrete experience. What is proposed here is an alternative learning path, which is only made acceptable within the context of modelling.

Eleven and 12 year old children can go a little further than making and interpreting the graph, perhaps to sketching a graph for a bouncier ball or a ball dropped from a higher point. More advanced pupils will look at the proportions of consecutive distances, and will treat the graph quantitatively.

Calculating with the model

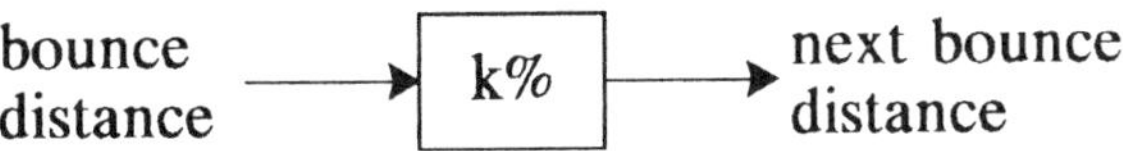

gives a powerful *predictive* model. Students can predict the distance between, or the successive heights of, any number of bounces.

At A-level, and sometimes prior to A-level, students will fit an algebraic formula to the data with the aid of a microcomputer (Williams, 1989b).

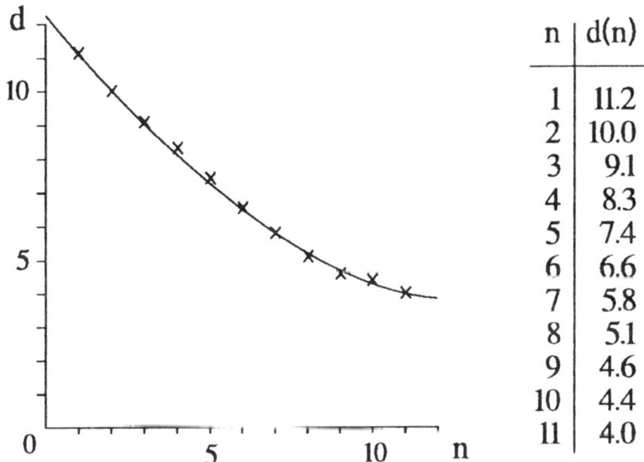

n	d(n)
1	11.2
2	10.0
3	9.1
4	8.3
5	7.4
6	6.6
7	5.8
8	5.1
9	4.6
10	4.4
11	4.0

Figure 5

Students will further generalise the geometric progression, and fit formulae such as $d(n) = d(0) \times k^n$ to the data. At each stage, the teacher must judge the state of readiness and motivation of the students, and decide to what extent to intervene with exposition of new mathematics. Although much will be discovered by the pupils, and much can be done in response to pupils questions, there is clearly an active role for the teacher here.

The context, therefore, suggests a series of learning stages in the development of concepts of graphs (and functions), from intuitive through to formal.

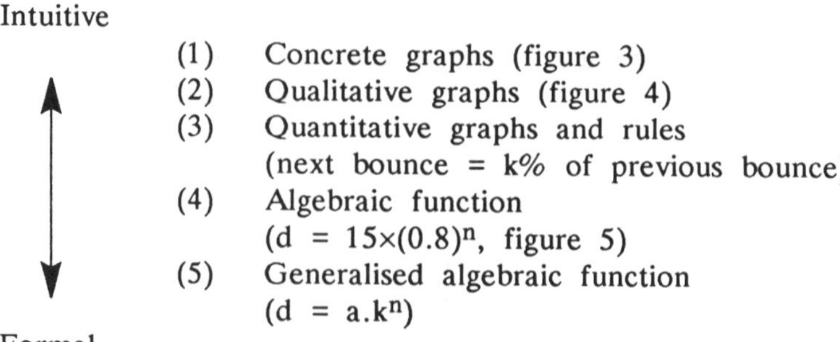

Notice that the more intuitive and concrete the models are, the more they *describe* rather than *predict*. The more formal models are more powerful in terms of their ability to predict a wider range of outcomes.

Notice that the students first experience the variables in the problem in an intuitive and natural context. An informal appreciation of the relationship between the variables involved is experienced *before* the mathematical relationships are extracted or formalised. The practical gives this experience, and provides a continuing reference point to which students attach formal mathematics. Discussion between pupils, and with the teacher, provides the language necessary to acquire the new ideas, but the practical experience provides the focus of this discussion, and the concrete link with reality which helps pupils to anchor their new mathematics.

It is not uncommon, of course, for mathematics educators to try to design concretisations for new abstract material structures to be taught, or discovered. This tradition goes back to Bruner (1966) and Dienes (1960). The difference, here, is that we design meaningful real–world tasks and practical problem–solving experiences in which the concretisation is embedded. This provides the experience with an added dimension of purposefulness.

4. MODELLING WITH NEWTONIAN MECHANICS

From about the age of 15, students are learning science and will, themselves, occasionally transfer their knowledge and skills to the tasks mentioned. Notice, in the illustrative tasks, how *mechanics* provides a link between the three subjects of mathematics, science and technology, and also a tangible, practical, physical environment to explore.

At a certain level, then, the students can be taught, or be asked to apply, Newtonian mechanics to the task in hand. As I reported at ICME–6, see Williams (1989a), this lends a new dimension to the power of the mathematical modelling involved.

In the second illustration, Galileo's experiment, one can see the bridge between empirical modelling and Newtonian modelling most clearly.

Young, secondary pupils will collect data and model the motion of the ball with a distance–time graph or function. They interpret their graph in terms of the speed of the ball, and may validate the model by interpolation and extrapolation.

Further work on the graph may yield the *kinematics* of the ball's motion. The gradient of the distance–time graph gives the velocity, and the velocity–time graph gives a straight line, implying constant acceleration.

The calculus of constant–acceleration motion can now be applied to a *class* of practical physical situations. The usual model

$$\frac{dv}{dt} = a \qquad v = u + at \qquad x = ut + \tfrac{1}{2}at^2$$

can now be interpreted and applied to a ball rolling along a horizontal table, up an incline, falling under gravity, and so on. This gives a whole class of problems which motivate the study of calculus. However, the rolling ball also gives an anchor for the whole model; students can, throughout their study of calculus, refer back to this experience in discussing gradient, integration, limit, second derivative, and so forth.

The further Newtonian study of dynamics introduces the model of *force* as the cause of changes in motion. In this case, applying Newton's Second Law to a particle model gives

$$\text{acceleration} = g \sin \alpha - \frac{\Gamma}{m}$$

Assuming zero friction, the ball has constant acceleration $g \sin \alpha$, independent of its mass.

Interpretation and validation of this model is now an interesting exercise! Sixth formers who don't experience the significance of energy loss, due to friction, will never understand the relevance or significance of a model in which bodies are assumed to be smooth. Students who do not experience significant errors due to rotational energy will never understand the significance of a model of a body as a particle.
The presentation of a *real situation* is vital to all this motivation. It is essential that students experience the process of *validation* if they are to be motivated to refine a model and so search for new mathematics. This is just as true at 6th form level as it is at high–school age.

5. CONCLUSIONS

The tasks we have designed (Savage and Williams, Cherouvim *et al*) involve challenging, real problems which students can explore and, to some extent, solve purely practically. Very simple, elementary mathematics may be used in the process. However, the tasks can involve the application of much formal mathematics in shedding further light on the problem. By varying the precise presentation of the task, the teacher can suggest lines of activity at a variety of levels. The teaching of functions and graphs was illustrated in this way. It is expected that students will apply well-known mathematics to the problem, as well as weakly-known mathematics, albeit with teacher support. It is also part of the design that students will be motivated to learn *new* mathematics, in the context of the task in which they have become interested. At sixth form level this is just as valid, and will increasingly be, in the form of independent research.

Modelling is not seen as merely the application process, through which students acquire higher level process skills by tackling real problems. It is also seen as the starting point for new learning. We are beginning to see the potential of combining modelling and problem solving with the structured acquisition of new concepts and skills. (This is the approach we are taking to the applied mathematics modules of the new SMP 16–19 course being developed by MAP and SMP). This is the challenging formula for our future curriculum development.

REFERENCES

Bell A, Brekke G, Swan M. (1987). Diagnostic Teaching 4 : Graphical Interpretation. *Mathematics in School*, Vol 11, **9**.

Bruner JS. (1966). *Towards a Theory of Instruction.* Harvard.

Cherouvim N, Kitchen A, Robbins P, Williams JS. *Practical Projects with Mathematics.* Cambridge University Press. (In press).

Cherouvim N, Kitchen A, Robbins P, Williams JS. (1989). Mechanics in Action: Modelling Bounciness. *Mathematics in School*, Vol 18, **2**.

Dienes ZP. (1960). *Building up Mathematics.* Hutchinson.

Mason J. (1988). Modelling: What do we really want pupils to learn?. In Pimm D (ed), *Mathematics, Teachers and Children*, Open University.

Savage MD, Williams JS. *Mechanics in Action.* Cambridge University Press. (In press).

Williams JS. (1989a). Real Problem Solving in Mechanics: The Role of Practical Work in Teaching Mathematical Modelling. In Blum W, Niss M and Huntley I (eds), *Modelling, Applications and Applied Problem Solving*, Ellis Horwood.

Williams JS. (1989b). Modelling Real Data with FGP and Makefil. *Micromaths* 2.3.

CHAPTER 29

Groundwater Modelling Courses at the Oberstufen-Kolleg

S Holz
University of Bielefeld, FR Germany

SUMMARY

The Oberstufen-Kolleg is an experimental college of the Land Nordrhein-Westfalen at the University of Bielefeld, combining the last three years of the Secondary School with the first two years of the university. One hundred and fifty to 200 students begin study each year. The volume of general education courses is about 50%. Every student chooses two of the 22 subjects which are offered as her or his future university discipline.

A sequence of courses is described in which mathematics work with students is based on general education courses on groundwater pollution problems. A report is given on three Facharbeiten (papers students are obliged to write during their last Oberstufen-Kolleg year) on groundwater questions, which are solved by partial differential equation techniques.

In the 1988 summer term I gave a course at the Oberstufen-Kolleg called 'groundwater pollution and its mathematical modelling' which was open to students from every year and with no particular mathematical background. Fortunately, all the students from the mathematics course I had been giving continuously since 1986, whom I had invited to join the groundwater course, took part. Thus there was an opportunity to combine a general education course on groundwater pollution with an analysis of quite advanced mathematical models describing groundwater seepage and pollution.

I want to emphasise this particular two-step combination of courses:

- a general education course on groundwater pollution problems containing maths *as a part of* general education,
- after that, more specialised work in a particular field of groundwater pollution, with a small group of maths students using differential equation techniques.

A short report on some of the differential equation techniques is given in section 2.

1. REPORT OF THE COURSES

In the summer of 1988 I gave a 72-hour course at the Oberstufen-Kolleg called 'groundwater pollution and its mathematical modelling'. The course was an option for students of all levels and from all disciplines and was intended as an introductory general education course containing the following subjects.

- The cycle of water and the arousal of groundwater
- Human impact on the quality of groundwater
  - leaking waste-disposal sites
  - agricultural impact: nitrogen and pesticides
  - acid rain depositing poisonous minerals in the soil
  - industrial impact: halogenized carbon hydrogenes seeping into the soil
  - sewage treatment by the communities.

The course contained several two to three hour visits to waste disposal sites, sewage treatment works and authorities controlling the quality of water.

The course was advertised as an introduction to secondary level maths techniques. Thus, I introduced the following concepts on the basis of ground and drinking water examples.

- Linear and quadratic functions
  (the amount of NO_3 washed out depends linearly (or quadratically) on the nitrogen fertilizer put on the fields)
- Exponential functions
  (between 1890 and 1980 the demand for drinking water in Bielefeld doubled every 20 years, and tends to be constant since 1980)
- First order differential equations
  Atrazin - a pesticide - is contained in the water of a drinking water storage lake near Dortmund, its content is much higher than it should be compared to the Common Market limit value which is the legal limit since October 1989. How long does it take to

obtain *clean* water in the lake if the agricultural industries respect the limit values? This is a first order inhomogeneous linear differential equation question – I did this separately with a small group of maths students taking part in this course.

A three–week project followed in which one group of students had closer contact with the local groundwater authorities. Another group of students studied seepage equations (this was a group of students studying physics), a third group built a $1.5m^2$ plant sewage cleaning model which was tested for one and a half weeks, in particular with respect to nitrogen (NO_3).

The 25 students who took part in the two general education courses just described came from nearly all subjects which are taught at the Obsestufen–Kolleg: history, linguistics, music, German, French and English, physics, mathematics and others – only seven of them had maths as their main subject, but even these didn't do specialised maths work except attending one additional ordinary differential equations lesson.

The kind of work *all* students in these courses did was as follows.

- Investigatory work (interviewing experts, visiting waste–disposal sites or sewage sites) and writing reports on that
- Experimental work (for example with the plant sewage cleaning model)
- Calculatory work (without higher maths) for example on pollution prognoses or cost of recycling.

After this general education and project period, I asked the seven maths students who had taken part in these general education courses if they wanted to improve their mathematics (in particular, analysis in several dimensions) by studying groundwater models. Three of them followed this offer and decided to do their obligatory Facharbeit (a paper of 30–40 pages students at the Oberstufen–Kolleg are obliged to write during their last year, within the branch of science which they want to study later) in this field. As usual, the students had three months time during their last winter term to do their Facharbeit. During this time each of them had a one and a half hour tutor session with me every fortnight. Between the sessions, they worked out their notes using the literature and met separately. Additionally, they met a local water engineer several times who gave them valuable hints. The degree of independence of their work was not very high, obviously, but high enough to increase their activity and strengthen their self–confidence.

Before I describe the Facharbeiten more precisely I will derive two of the seepage equations. This will be a summary of the first Facharbeit and shows from which point the other two started.

2. DIFFERENTIAL EQUATIONS

In 1856 H Darcy, a French engineer, planned the water supply for the city of Dijon. He carried out several experiments to determine a systematic way for placing wells around the city. His experiments helped him to formulate the following proportionality law called *Darcy's filter law* which says that the filter velocity v of seeping water is proportional to the piezometric head difference Δh stated along the horizontal length Δx:

$$v \sim \frac{\Delta h}{\Delta x} \quad \text{or, more precisely} \quad v = - k_f \frac{\delta h}{\delta x}$$

The constant k_f is called *permeability value*; it contains viscosity and density of the fluid, gravitation, and permeability of the soil (see Bear (1988) pages 119–125, where the range of validity of Darcy's law is discussed).

The *flow continuity equation* is derived by regarding a differential volume dV in the soil and taking into account that the amount of water entering this volume equals the amount of water leaving it. In precise terms – the flowrate Q is described as filter velocity times surface crossed by the velocity vector, that is dim Q = m^3/s.

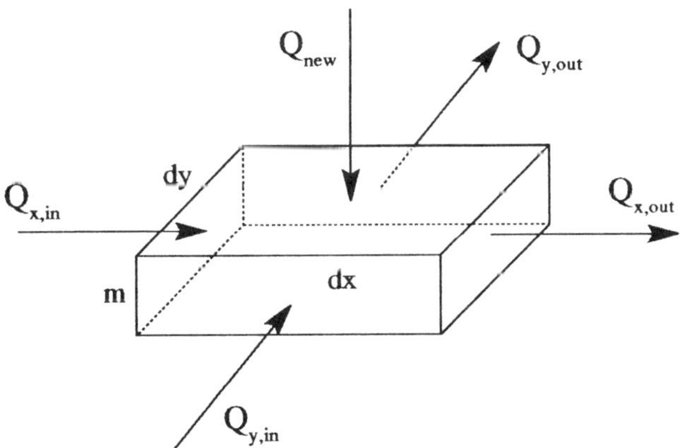

The figure shows that dV = dx . dy . m; m is the thickness of the groundwater–containing layer of soil. By the definition of the flow Q, we have the following three equations for the flow *entering* dV

$$Q_{x,in} = v_x \cdot m \cdot dy \quad \text{and} \quad Q_{y,in} = v_y \cdot m \cdot dx$$

Taking the groundwater renewal rate W into account (dim W = m/s) we get an additional term

$$Q_{new} = W \ . \ dx \ . \ dy$$

Rounding the Taylor approximation after the linear term we get for the flow *leaving* dV

$$Q_{x,\,out} = \left[v + \frac{\delta v_x}{\delta x}\, dx\right] \ . \ m \ . \ dy$$

and

$$Q_{y,\,out} = \left[v + \frac{\delta v_y}{\delta y}\, dy\right] \ . \ m \ . \ dx$$

The *flow continuity principle* says that the sum of the flows entering and leaving dV equals zero:

$$Q_{in} + Q_{new} - Q_{out} = 0$$

which implies that

$$-\left[\frac{\delta v_x}{\delta x} + \frac{\delta v_y}{\delta y}\right] + \frac{W}{m} = 0.$$

Inserting Darcy's law for v_x and v_y and supposing constant k_f we get

$$k_f\left[\frac{\delta}{\delta x}\left[\frac{\delta h}{\delta x}\right] + \frac{\delta}{\delta y}\left[\frac{\delta h}{\delta y}\right]\right] + \frac{W}{m} = 0$$

which is an elliptic second order partial differential equation (see Bear, 1988 page 196 equation 6.21).

In the case of a time-dependent situation, a similar argument yields

$$k_f\left[\frac{\delta^2}{\delta x^2}\, h + \frac{\delta^2}{\delta y^2}\, h\right] + \frac{W}{m} = S_s\,\frac{\delta h}{\delta t}$$

where S_s is the storage coefficient of the specific soil (see Bear, 1988 page 202 equation 6.2.37, page 207 equation 6.3.39 and page 214 equation 6.4.3).

Neglecting the rain term $^W/_m$ for a moment, we get

$$\frac{\delta^2}{\delta x^2}\, h + \frac{\delta^2}{\delta y^2}\, h = \frac{S_s}{k_f}\,\frac{\delta h}{\delta t} \qquad \text{(HDE)}$$

which is a differential equation equivalent to the heat diffusion equation

(HDE) being a linear (parabolic) second order partial differential equation (see Bear, 1988 page 204 equation 6.3.11).

Having prepared this background so far, I am now able to say more about the three Facharbeiten that the maths students Birgit Schrupp, Ingo Isensee and Thomas Kassens who had taken part in the general education groundwater course decided to write. Their areas are the following.

1. Physical, mathematical and engineering background for seepage equations (see Schrupp, 1989)
2. How to solve the heat diffusion equation analytically (given convenient initial and boundary conditions); the methods involved here are Fourier analysis, eigenvalues and eigenfunctions of a boundary value problem, and Bessel functions (see Isensee, 1989)
3. How to solve the equation (HDE) numerically (given less convenient initial and boundary conditions); the methods involved here are: numerical differentiation, finite elements, convergence of mesh refinement and iteration processes (see Kassens, 1989).

3. A REMARK ON THE PEDAGOGICAL PROCESS

The particular conditions for the small group of maths students who decided to specialise on groundwater modelling questions after having taken part in a general education course on groundwater pollution are the following.

- They had two preparatory courses on groundwater pollution in the early summer of 1988 which gave them, as *citizens*, the opportunity to learn several things about pollution in the region around Bielefeld. That these students, as *mathematicians*, worked in the same field a couple of months later was their own voluntary decision, which they made in October 1988. As a Facharbeit theme, they could have chosen any theme based on the maths courses they had had before. Probably those themes would have been easier.
- The *communication* between the general education groundwater course and the small group of maths students on the one hand, and local authorities controlling groundwater quality and water engineers working in our community on the other hand, was very good: we obtained data, literature or talking time from them whenever we needed it.
- The three students who have written their Facharbeit on groundwater models had done a lot of shared work during the past three years, so they knew one another quite well, and their *cooperation* was excellent.
- When the groundwater course started, I had *no* idea which of the modelling techniques would be the appropriate ones for groundwater

> flow and pollution. Later on, having discovered that partial differential equations might yield appropriate modelling techniques, I have to admit that I had never learned this material before, so, *I had to learn it together with the students.* Thus, they had the opportunity to watch a mathematician (being their teacher, at the same time) seeking his way in a field of mathematics which had been completely unknown to him.

Having finished their Facharbeiten the students were slightly disappointed that they had understood only the very beginning of the theory as, for example, a look into Bear (1988) shows. Additionally, they had not studied any pollution analysing or preventing models. Two of them will do that at the university.

These models ought to be studied with a next group of students at the Oberstufen-Kolleg.

REFERENCES

Bear J. (1988). *Dynamics of Fluids in Porous Media.* New York.

Blum W, Berry JS, Biehler P, Huntley ID, Kaiser-Messmer G, Profke L (eds). (1989). *Applications and Modelling in Learning and Teaching Mathematics.* Ellis Horwood, Chichester.

Darcy H. (1856). *Les fontaines publiques de la ville de Dijon.* Paris.

Effe-Stumpf G, Kemper A. The Role of Study Projects in Mathematical Education in Blum *et al,* 1989, p219-224.

Gerull K. Aspects of General Education in Mathematics in Blum *et al,* 1989, p225-229.

Holz S. (1989). Analytische Modelle, um Wasserverschmutzung zu beschreiben - Preprint Oberstufen-Kolleg, to appear in *Praxis der Mathematik* **6** (1990).

Isensee I. (1989). Analytische Lösung von mathematischen Modellen der Grundwasserströmung. Facharbeit in Mathematik, Oberstufen-Kolleg.

Kassens T. (1989). Numerische Lösungsverfahren zur Modellierung der Grundwasserströmung. Facharbeit in Mathematik, Oberstufen-Kolleg.

Kinzelbach W. (1986). *Groundwater Modelling - An introduction with sample programs in BASIC.* Elsevier Science Publishers BV, Amsterdam/Oxford.

Open University. (1972). *Initial Value Problems.* The Open University Press, London.

Smith G. (1969). *Numerical Solution of Partial Differential Equations.* London.

Schrupp B. (1989). Aufstellen von mathematischen Modellen der Grundwasserströmung. Facharbeit Oberstufen-Kolleg.

Schülert J, Vohmann D. A Sequence of Courses Dealing with Empirical Methods in the Social Sciences in Blum *et al,* 1989, p237–242.

Tolstow GP. (1976). *Fourier series.* London.

CHAPTER 30

Thinking Up Problems

R Eyre
College of St Paul and St Mary, Cheltenham, UK

SUMMARY

The one aspect of mathematical modelling that is often ignored is that of finding suitable problems for students to tackle.

Having justified why students should undertake the active aspects of mathematics, this paper illustrates, through some examples, how the author came to devise the problems quoted from recent assignments and examinations, and how students are encouraged to follow their own ideas for individual projects.

1. INTRODUCTION

The mathematics courses organised at the College of St Paul and St Mary for students intending to become teachers of mathematics are based on the premise that mathematics is a participatory activity. Our prime requirement of students is, therefore, that they must participate actively in the doing of mathematics.

In the content versus process debate within mathematics education in the UK, we have chosen to claim that a combination of both approaches must be close to an ideal situation, even if we stress the process or mathematical modelling approach on our courses.

Our perception is that students who have recently left school perceive mathematics as a fixed body of knowledge, which can be learned from books and easily conveyed to children in the same manner in which it was presented to them. It has been one of our aims to break this cycle by emphasising the active aspects of mathematics.

Making mathematics a participatory activity means using an area of mathematics labelled as problem solving. Since this label disguises a variety of activities, it is hoped that figure 1 shows how other aspects can be related to problem solving and yet exist under different names.

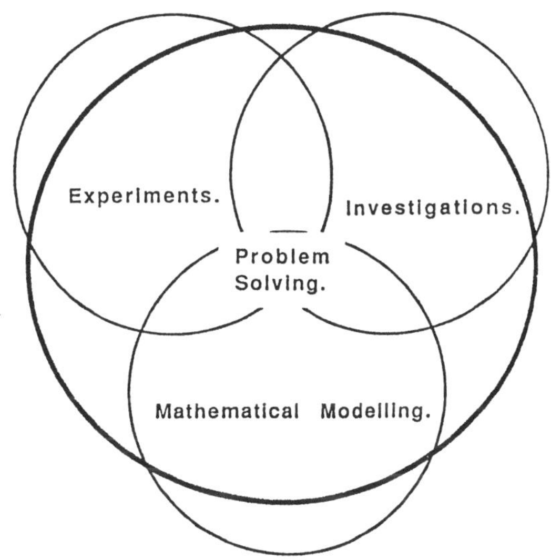

Figure 1 : The relationship between different types of active mathematics

2. MATHEMATICAL MODELLING

In justifying the use of mathematical modelling, we try to explain who might do such an activity and why. Figure 2 addresses this issue and aims to show how we use it and who might make the most use of it.

One of the reasons we, as teachers, use mathematical modelling is to introduce new areas of mathematics or new mathematical techniques via a situation. This is best typified by the Mathematics Applicable scheme, developed in conjunction with the Schools Council.

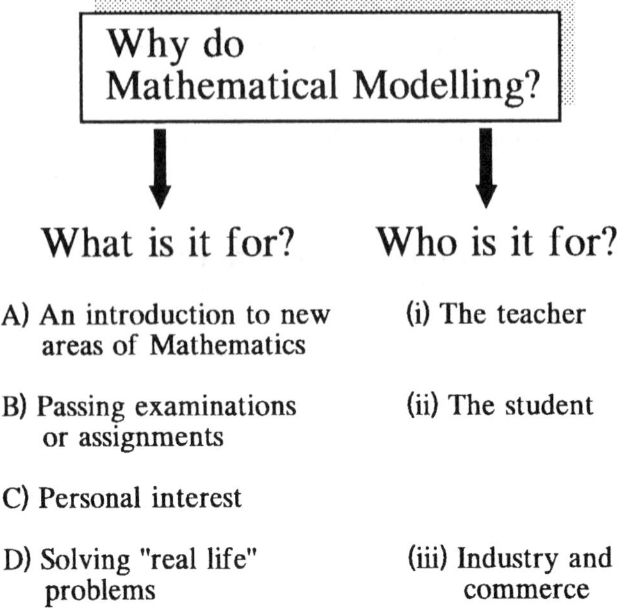

Figure 2 : Why do mathematical modelling?

From the students' point of view, participation is usually seen as a means of passing examinations or completing necessary assignments. We are hoping that students will want to use the approach to analyse problems that they find of personal interest.

The ultimate aim is that staff and students can attempt to solve real–life problems as they arise in the world of work outside education.

3. WHERE IDEAS FOR MODELS COME FROM

The next difficulty which arises is that of generating ideas for mathematical modelling. Figure 3 is an attempt at summarising how such ideas can be generated, but leads to the debate about whether it is modelling that is being taught or specific models.

If we are looking for source materials to set examination questions, then it seems reasonable to look at past examination papers and questions at the end of chapters in books. Journals and other publications can also be used to get ideas for assignments and investigations, especially when the authors have indicated the mathematics used in solving the problem they are presenting. Similarly, we can pick up ideas for problems, which may or may not have simple solutions, at conferences.

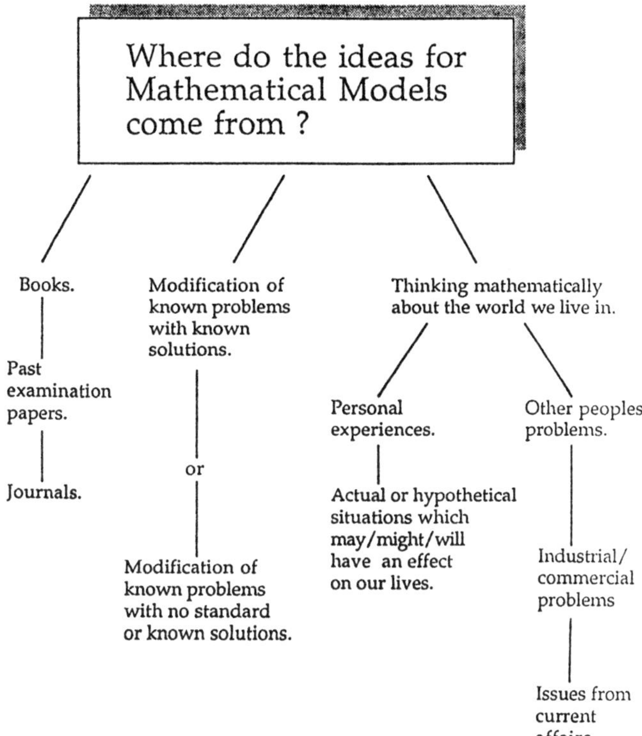

Figure 3 : Where ideas for models come from

The following problem was described by Mike Hamson of Thames Polytechnic at ICTMA-3 in Kassel. "A road is to be built joining points A and B in figure 4. The costs for building such a road are £10 000 per metre over ground type a, £20 000 over b and £15 000 over c. Plan a route that will minimise costs."

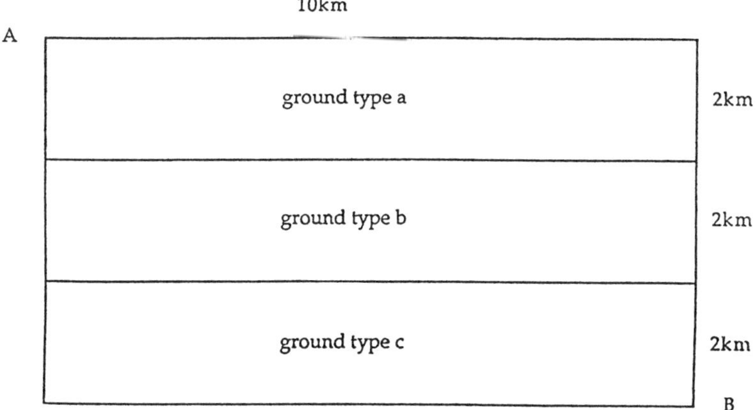

Figure 4 : Road building problem

I set this problem to our first year intake at the first lecture session, and asked if their solutions could be generalised.

Many students tackled the problem using a trial and error approach, which started with the section across ground type b as perpendicular to the 10km side. Some students gave this middle section a slight slope, and discovered cheaper routes. A few students tried a computer programming approach, one of which I have attached as Appendix 1. Another student drew a comparison with the refraction of light through various media, and formulated a solution based on this approach which also yielded the optimum solution. This is also attached, as Appendix 2.

This one example allowed me to outline the modelling process and highlight the essential nature of our course in terms of getting the students to discuss their work with others and building in a degree of evaluation and self–assessment of their work.

A similar problem, set to a previous year group, involved the design and manufacture of a wine box, to hold four or six bottles in their own compartments, from a single sheet of card. The extension of this was to consider optimum packaging, and was reported in my paper of 1988.

I try to use problems which are a result of thinking about the world mathematically, and yet are personal. For example, a van that I had bought suffered from a spattered rear window when driven in wet conditions. I therefore asked the students to consider the position and effectiveness of a window wiper. The initial work was all theoretical, and involved assumptions such as the windscreen being rectangular, the wiper blade being in a straight line with the wiper arm, and so on. The work then took the students out into the car park to measure car windscreens, and observe the various approaches manufacturers had adopted. They then worked individually to write up this work, and extended it by considering factors such as single arm versus multiple arm wipers, front versus rear wipers, linkage versus fixed attachment of the arm movement, and so on. One advantage of working in this way was that I could set a framework within which I wanted the students to work.

When students take on a project of their own, it allows us to see the effectiveness of our course. One student set herself the task of seeing if a piece of waste ground near her flat was worth developing into a car park. Her final report was in the form of a set of four recommendations to the local council which was summarised, within her report, on two sides of paper. Each recommendation was backed up in the report with the investigation of factors such as layouts, charges, residents' opinions, entrances and exits. We sent her report to the Clerk of the Council, and found that her work was very close to their own findings.

4. EXAMINATION PAPERS

I have tried to keep this approach even within our written examination conditions. I have attached an example paper from 1986 (Appendix 3) when there was a teachers' pay dispute involving performance related pay from which I was able to construct mathematical scenarios. The paper follows the sustained model theme, as practised by the Mathematics Applicable Group, and aims to cover the topics of polynomial, exponential and differential equation modelling for students whose second academic subject is mathematics. As can be seen from this example, there is a lot for the students to read, but the questions are all based on the same topic with the appropriate mathematics being used to explore different aspects of each situation.

Differential equation models are a feature of most of our courses and examples are frequently needed for examination questions. Examples have been based on the growth of rust on old cars, pain levels suffered as a result of a hockey injury, the cooling characteristics of one of my motor cycle engines, and the growth of a hole in a tooth requiring necessary but non urgent dental treatment.

Examples of ideas which can be modified for use as examination or assignment problems can be found in Berry (1989), Eyre (1983) and the forthcoming Mathematics Applicable booklet on Conjectural Modelling Scenarios.

5. CONCLUSION

Our course aims to encourage students to think mathematically about the world we live in. We hope that they can add a mathematical perspective to other courses, such as the aesthetic, moral and scientific perspectives that help them to become educated citizens. This means utilising their personal experiences, and seeing that mathematics may be applied to solving and/or setting problems. At a simple level, it might concern the arranging of funds for supplying personal transport, or the designing and ordering of materials for some home improvement. If teachers can convey enthusiasm for mathematics to their pupils, by introducing meaningful problems in this way, we can consider that our courses have been successful.

REFERENCES

Berry J, (1989). A Bank of Modelling Problems. Polytechnic South West, Plymouth.

Eyre R. (1983). Mathematical Modelling at the College of St Paul and St Mary. *Teaching Mathematics and its Applications*, Vol 2.

Eyre R. (1988). Keep on Wrapping. *Teaching Mathematics and its Applications*, Vol 7.

APPENDIX 1

To make it easier for the computer we reduced the number of variables to 2.

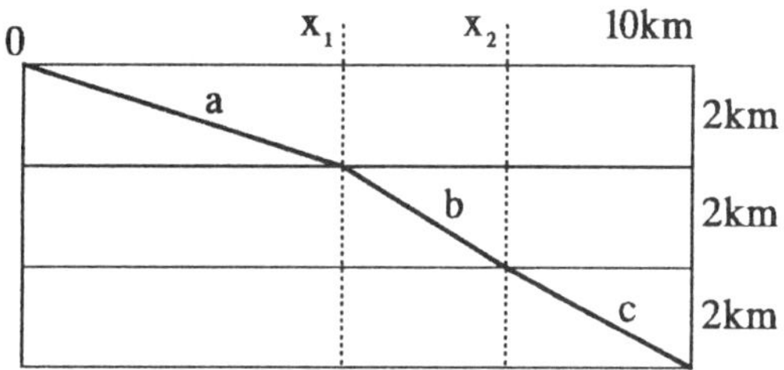

$$\text{Total cost} = 10\ 000\ 000a + 20\ 000\ 000b + 15\ 000\ 000c$$
$$= 5\times10^6\ [2(x_1{}^2+4)^{\frac{1}{2}} + 4((x_2-x_1)^2+4)^{\frac{1}{2}} + 3((10-x_2)^2+4))^{\frac{1}{2}}]$$

We then used a computer to find the minimum total cost.

APPENDIX 2

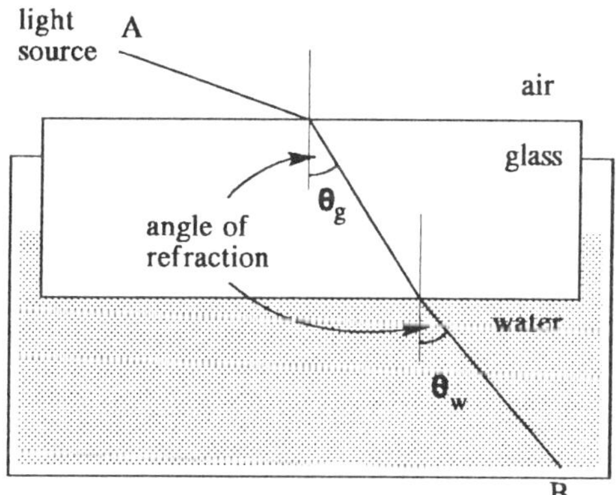

If we could reproduce $\mu_1 \equiv \mu_{air}\times2$ and $\mu_2 \equiv \mu_{air}\times1.5$ we could set up a model and direct light of one wavelength from point A towards point B, a scaled distance away in the same place. By tracing the path of the light, such as by flourescence, we could correlate the findings with the best route for our road.

However, if this is a reasonable facsimile, then there should also be a relationship between the 'angle of refraction' of our road and the cost m^{-1} of construction.

APPENDIX 3

Answer 3 of the 4 subsections in this Section

An education minister, Sephoj Thiek, has suggested that teachers' pay should be related to their performance as teachers. This means that assumptions are to be made relating teacher performance to the number of years spent teaching.

B1. One suggestion is that teachers' performance improves dramatically in the first few years but levels off after 12 years.

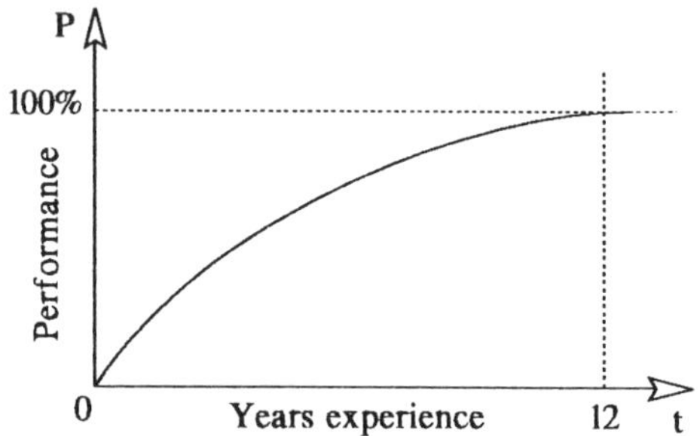

(a) Write down a mathematical sentence which relates performance, P, to the number of years experience, t, for $0 \leqslant t \leqslant 12$, assuming that it is a quadratic function.

The unions object to the notion that the rate of increase in performance is always decreasing.

(b) Explain *briefly* the unions' case.

The unions prefer to think that at the beginning of a teachers career, the rate of increase in performance level will be large, even though it might taper off later.

A new suggestion from the minister is as follows

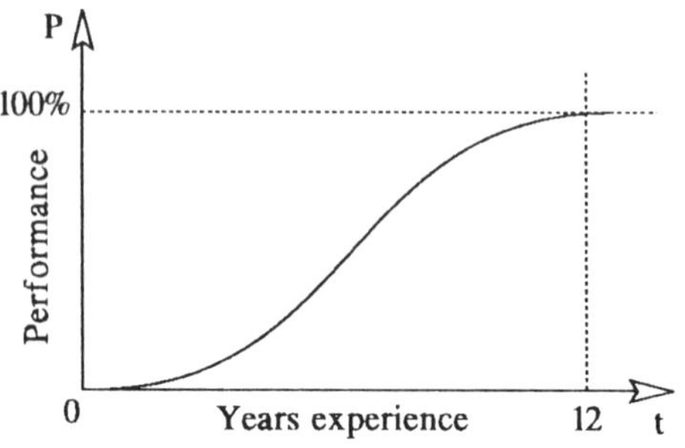

(c) Write down a mathematical sentence that describes this new model for $0 \leqslant t \leqslant 12$, assuming that it is a cubic function.

(d) After how many years teaching will the rate of increase in performance be at its maximum?

B2. The unions decide that they ought to base their case on a different basis, one which assumes that the longer a teacher teaches the greater is the performance level.

Sephoj recognises that the unions are opting for an exponential model, and says that this would mean that teachers performance would continue unbounded! The unions argue that their complicated model $P = 60(1-(0.8)^t) + 40$ means that teachers would never attain a perfect performance, but would still improve, hence the maximum pay should be awarded when a 95% performance figure is attained.

(a) Sketch a graph of the unions' mathematical sentence.
(b) Who is right, Sephoj or the unions?
(c) Is the union position reasonable?
(d) When would a performance level of 95% be attained?

B3. In an attempt to resolve the pay dispute, a new concept is introduced. It assumes that the teacher comes into teaching with a performance level P. At the end of the first year, this is increased by a half of the initial level of performance. At the end of the second year of teaching, the performance is increased by a third of the previous year's increase in performance. At the end of the third year by a quarter of the previous year's increase in performance, and so on.

(a) Write down a table showing the performance increase, and the total performance level, for each of the first five years of teaching.
(b) Write down an expression to describe the total amount of performance level, P, over an indefinite (ie infinite) amount of time.
(c) By factorising P out of this expression, you will be left with a series. Comparing this series with the expansion for exp(x), write down an expression for your series in terms of the number e.
(d) What is the limiting value of this expression (to 3 decimal places)?
(e) What is the expected percentage gain in performance level after a long teaching career, compared with the starting level of performance P?
(f) How does this compare with the total gain in performance level after five years?

(g) Does this seem reasonable?

B4. The only other alternative for the negotiators to consider is based on the knowledge, K, gained about what there is to know about teaching as time progresses.

The following statement is made. "The rate at which teaching knowledge, K, is gained over a period of time, T years, decreases with the amount of knowledge already known."

Let the rate of change in knowledge be $-cK$

(a) Write down a differential equation relating these facts.
(b) Derive an equation for K in terms of t.
(c) If teachers started with a 100 units of knowledge to be gained at the outset of their careers (ie when $t = 0$), find the value of the constant of integration.
(d) If the amount of knowledge to be gained after 10 years is 2 units, find the value for c.
(e) What would be the number of units of knowledge to be gained after 5 years?
(f) Does this seem a reasonable knowledge gain after 5 years teaching experience?

CHAPTER 31

Project Assessment for Mathematicians

CR Haines
City University, London, UK

SUMMARY

Assessment schemes in mathematics emphasise content and accuracy in seeking objectivity. The skills involved in mathematical modelling and project work include the ability to set the exercise in context, and to communicate the results to others in a written report. Models of realistic assessment schemes, the subjective judgements which are involved, and reference to two final year projects in applications and modelling, form the basis for discussion. The problem of making qualitative judgements on modelling projects in widely differing areas is discussed, in relation to their quantitative assessment. A criterion-based scheme is seen as a partial solution.

1. SEARCHING FOR COMMON GROUND

The inclusion of project work in undergraduate programmes, accompanied by an increasing interest in mathematical modelling, has highlighted the extreme difficulties which are to be faced in agreed assessment procedures. These procedures are drawn up in parallel with the aims and objectives of the project, and by comparing and contrasting the methods which are commonly available.

The nature of project work differs dramatically between institutions, across departments, within degree courses, and between supervisors. It is used to embrace interrelated ideas in undergraduate mathematics courses. Some projects might involve an extended essay or dissertation on a mathematical topic, which requires searching and expounding the relevant literature. It might be a piece of individual research or a group activity. Mathematical modelling courses have allowed some standardisation, to the extent that particular projects are given to classes of students to be attempted either on an individual basis or within a group. Data collection, applied statistical analysis and interpretation feature in a great many projects. These diverse practices are illustrated at the extremes by the second level course in mathematical methods and modelling presented by the Open University and the project component of the mathematical sciences honours degree programme of the City University. There are also examples of schemes which deal with both individual and group projects during the undergraduate course, as in Burley and Trowbridge (1983) and Gadian (1983).

The Open University course, Berry and Le Masurier (1983), contains a single project which is selected from a small pool of topics. The modelling is carried out on an individual basis by students, with advice and guidance available from a tutor who is usually remote.

The City University scheme deals with individual subjects for each final year student, according to staff interests and student choice. Amongst the projects chosen, a significant number are concerned with mathematical modelling; this paper gives a critical appraisal of two applications and modelling projects, as well as addressing the general question of project assessment.

2. A FRAMEWORK FOR ASSESSMENT

It is an ambitious task to lay down precise procedures which deal effectively with the assessment of every project scheme. However, detailed guidelines for such assessments should be considered and credible techniques must be developed.

Project work gives students the encouragement to think independently and the opportunity to show initiative, the experience of writing a report, and a training in the art of communication. In addition, applications and modelling projects develop those skills which enable the description and the solution of real problems.

Any assessment procedure must attempt to deal with the major ingredients, creativity, initiative and self-expression, as well as the more tangible area of content. In applications and modelling projects, considerable creativity is required in order to make realistic assumptions, to assign variables, and to take the first faltering steps towards a simple model. Each component is developed in a way which reflects the interaction between the student and the supervisor. Such support or advice as is given, and the level of mathematical expertise employed by the student, influence both the development of skills and the final mark given to the project.

The need to assess these qualities with integrity in the context of project work is vital, and during this process comparisons will be made with other constituent parts of the degree courses. Mathematicians are used to marking their examination questions using schemes in which each individual mark has a clearly defined purpose and is objectively assigned.

Hall (1983) has discussed the major issues of assessment and suggests clear directions to follow for the assessment of modelling projects. However, after making an effective partition of each project into content, presentation and drive, there is no specific indication of how the marks themselves should be assigned, although the question of how the marks for the three categories should be combined is addressed.

3. ASSESSING PROJECTS AT CITY UNIVERSITY

Projects, which are carried out in the first two terms of the final year of the mathematical science programme, are assessed under the following headings.

	to 1988	from 1989
Summary or abstract	10	10
Problem statement/introduction	20	16
Content	50	50
References/data sources	10	8
Presentation	10	16
	100	100

The role of the supervisor is stressed in the student guidelines which reinforce the importance of a meaningful dialogue and of working within agreed aims and objectives. The assessment, which does not include an oral presentation, is made by the supervisor, who prepares a short report. The report and the mark are subject to the scrutiny of a moderator, and a mark is agreed.

This scheme is similar to that used extensively elsewhere, UMTC (1988), and can take into account those qualities which are specific to applications and modelling projects. Clearly there are major deficiencies inherent in the method, which call into question the objectivity of the process. The wide band of marks available in each category leads to 'impression' marking and consequential subjective assessment. Whether this would be seen as a problem in other disciplines is by no means certain.

Some aspects of the process and the diverse nature of the modelling projects can be seen in two examples from 1989; the marks for each category appear at the end of section 4.

The probability modelling of a pinball machine

The bright lights of the arcade game pinball machines proved a strong attraction for one student. The features of the machines include impact with bumpers and tombstones, so that the scoring ball follows an unpredictable path which can be altered by the player operating flippers. The project was concerned with the modelling of tombstones, bumpers and flippers, so that an interesting and absorbing game could be played.

After an initial discussion period, weekly meetings with the supervisor laid the foundation for the investigation, which concentrated on a probability model. A passing acquaintance with simple Markov chains using 2×2 matrices led to some very good preliminary work, which soon dealt with simplifying a typical game. A model was developed which used an 8×8 matrix including absorption states and transient states, similar to that in a simple simulation package by Watson (1980). At this stage it became obvious that 'creativity' was very difficult to identify and to quantify in the assessment scheme, although the search leading to that package was keenly and intelligently carried out. The project then concentrated on the main constituents of an 'interesting game'.

The marks were apportioned with difficulty, even in the major area of content. Elementary errors were in the modelling assumptions for the probabilities of two particular states – there was no attempt to vary the main parameters in order to carry out a sensitivity analysis on the probabilities. Content achieved 60%, but of what? The goals of complete coverage were not defined in the scheme, but how could they be? Amongst the other defined areas, the abstract was well done, conveying to the reader exactly what had been attempted, and also gave the main results. This particular project grew with the student, and required clear criteria for the apportioning of marks, not given in the scheme. The modelling was assisted, as in many cases elsewhere, by a known and documented model.

Mathematics in music

In contrast, this project sought to order features of some compositions into well-known mathematical structures. Starting with the excellent stimulus of Hofstadter (1980) and a student pianist mathematician, early discussions with the supervisor established the extent of the literature, which resulted in a reasonable survey of the early attempts at a quantitative explanation of music. This formative stage showed some interesting ideas about representing scales by residue classes, and introduced an analysis of canons by means of groups; there were also ideas about fractal composition.

At this point the student found that she was not able to pursue the project with enough drive, and she submitted a disappointing report. She was able to achieve good marks on the introduction and the general presentation, but could not do justice to the subject. Some of this was due to motivation. Would another supervisor have been able to encourage her to be more positive? Was the algebra inherent in this project beyond the reach of the student? It should not have been. Where was the creativity and the drive? The former appeared briefly at the start, and the latter waned rapidly.

4. COMBINING THE MARKS

The aggregate mark is commonly determined by adding the constituent parts together, using a weighting factor as appropriate. Hall (1983) expresses the view that, especially for modelling projects, each constituent part is of crucial importance. Therefore, taking creativity, drive and content as the three ingredients, a zero mark in any one should lead to a zero mark overall. He proposed a product model, in which each constituent is raised to the power of the weight and the nth root taken, n being the sum of the weights.

A common property of both the sum and the product models is that a mark of, say, p% in each constituent part leads to an overall mark of p%. Each model is homogeneous in this respect. In general it is possible to devise homogeneous marking schemes in which the aggregate mark M is given by

$$M = S + P \qquad (*)$$

in which S is the mark obtained from a group of factors assessed using a sum model, and P is the mark obtained from those factors for which a product model is more appropriate. In practice, any scheme based on (*) is difficult to implement, especially if the number of factors is large. Hall (1983) established that the penalising effects in P are exaggerated, especially if the weight for a particular part is large or if the number of factors taken into consideration is increased.

Assessment at City University uses the sum rule, in which it is inappropriate to give a zero aggregate mark for the project if, say, zero was obtained for the abstract and non-zero scores were obtained for the other parts. The question arises as to whether the application of a product rule would indeed have given different results. Marks obtained by students over the past four years were analysed under this thesis, the five categories listed were accorded the appropriate weights, and the aggregate mark recalculated using the Hall algorithm.

The following table summarises the results and confirms, as expected, that the product rule leads to marks which are lower than those obtained by the sum rule. The differences are marginal, given that integer arithmetic is used for the final mark. The raw scores show, in each case, a depression of the original mark by 0, 1 or 2.

		Sum		Product	
	n	mean	st dev	mean	st dev
1986	11	71.45	13.13	70.55	13.28
1987	15	67.07	20.62	66.47	20.88
1988	27	74.52	18.20	74.00	18.01
1989	25	69.28	16.29	68.48	16.53

The specific changes in the marks for the projects discussed above are:

	Marks					Sum	Product
Pinball machine	8	12	30	4	10	64	63
Mathematics in music	4	10	24	2	10	50	49

5. CONCLUSIONS

This paper has reviewed procedures which are available for the assessment of projects. A major consideration is the allocation of marks to the constituent parts of the project, as has been demonstrated by reference to two modelling projects.

Clearly the assessment must be made in an objective manner, and it must be in line with precisely-stated aims and objectives. The assessment must therefore be criterion-based within well-defined categories. The scheme adopted could be set out in a similar manner to that described by Berry and Le Masurier (1983), but should not award marks in very wide bands. If, for example, ten marks are awarded for presentation, it is necessary to devise a ten-point scale of criteria which must be met in order to earn these marks. This task should not be under-estimated, but the payoff will be in the use of the complete scale for marks, rather than the narrow range which characterises impression marking.

The analysis of section four suggests that there is little difference between the sum and the product model. It suggests that the use of a product rule should not be instituted without very careful consideration, and even then with few categories. It is possible to devise a workable formula based on equation (*) which incorporates only two categories in P, perhaps creativity and drive, together with a range of options in S. Even so, this too will depend upon a satisfactory criterion-based assessment.

The experience gained at City University is such that project work has been beneficial to the student, without exception. Its worth has been accepted by even the most sceptical colleague, but the current discussion centres almost exclusively on assessment, which is moving towards a full criterion base. The City scheme involves projects in widely-differing areas, and the qualitative judgements so far made are not reflected in a satisfactorily quantitative manner. The need to assign a mark, for assimilation with the conventional parts of the degree courses, distorts the discussion, and a further realistic option involving a simple satisfactory or unsatisfactory grading has not been considered.

That a rigid criterion-based system should be used is not accepted universally. The view is often taken that a weak base should be taken, and that a strong system is only appropriate for gradings in which there is a clear division between categories (McLone, 1989). For applications and modelling projects in mathematics, and in other fields, this is not usually the case, and the techniques used by our colleagues in the humanities could be equally justified.

REFERENCES

Berry JS and Le Masurier D. (1983). OU Students do it by themselves. In JS Berry et al (eds). *Teaching and Applying Mathematical Modelling.* Chichester, Ellis Horwood, 48-85.

Burley DM and Trowbridge EA. (1983). Experiences of Mathematical Modelling at Sheffield University. In JS Berry et al (eds). *Teaching and Applying Mathematical Modelling.* Chichester, Ellis Horwood, 131-142.

Gadian AM. (1983) Experience with Team Projects in Mathematical Modelling. In JS Berry et al (eds). *Teaching and Applying Mathematical Modelling.* Chichester, Ellis Horwood, 143-148.

Hofstadter DR. (1980). *Gödel, Escher, Bach: An Eternal Golden Braid.* New York, Vintage.

McLone RR. (1989). Plenary lecture at Mathematics 16-20 Conference, Nottingham July 1989.

UMTC. (1988). The methodology of teaching and assessing project and investigative work in mathematics. University Mathematics Teaching Conference, Nottingham University September 1988.

Watson FR. (1980). A simple introduction to simulation. *Keele Mathematical Education publications,* Part 1.

CHAPTER 32

The Use of Mathematical Modelling in a Programme of Integrative Assignments

MJ Herring
Gloucestershire College of Arts and Technology, Cheltenham, UK

SUMMARY

The Business and Technician Education Council (BTEC) has instituted a Programme of Integrative Assignments (PIA) throughout their courses. A PIA is a component of a course consisting of assignments specifically designed so that students may develop what BTEC calls 'common skills'. These skills are commensurate with the skills which students should acquire when mathematical modelling. A description of how a Programme of Integrative Assignments has been implemented for the Higher National Diploma course in Mathematics, Statistics and Computering at Gloucestershire College of Arts and Technology is given, and a discussion of how the skills assessment may correspond with those skills displayed when modelling is included As the title Programme of Integrative Assignments suggests, subject areas in courses should be associated through the PIA, and assessments ought to be both realistic and relevant. These aims can be achieved through the vehicle of modelling. Examples of assignments are included in an Appendix.

1. INTRODUCTION

The Business and Technician Education Council (BTEC) was formed in 1983 from the merger of the Business Education Council and the Technician Education Council (BEC and TEC). The Council's fundamental aim is that students on BTEC courses should develop the necessary competence in their careers in their own and their employers' interest.

BTEC courses are made up of units, each defined in terms of learning support time and, therefore, each unit has a value depending on the course and the student. In these courses, core studies are emphasised, and the core typically includes the fundamental skills for the vocational area, and general skills needed in many areas – called *common skills*.

These skills are transferable skills, such as

self-development skills
communicating and working with others
problem-tackling, decision-making and investigating
information, quantitative and practical skills.

It is envisaged by BTEC that these common skills can be developed within each course unit and by specific assignments, which are grouped together to form a Programme of Integrative Assignments (PIA) which is operated in conjunction with the more vocationally-oriented course units. The PIA should not only develop a student's common skills, but also link the vocational areas of a course. The PIA should be developed by interdisciplinary teamwork. BTEC see a carefully planned PIA as an essential element in all their courses. The manner of implementation of a PIA will depend on the course structure, but should encompass all common skill areas in a vocational context and progress in difficulty and the independence required of the students.

It has generally been commented on that many students lack real problem-solving capability, show no experience of problem identification, construction or selection of models, identification of data requirements and methods of collection, model validation and review/criticisms of the results and the process used. From experience, it has been seen that students do not understand the underlying assumptions of a specific model, and rarely consider its appropriateness and the sensitivity of its assumptions.

Mathematical modelling courses have been designed to remedy these deficiencies. Thus, the development of such courses must encourage the development of the common skills as identified by BTEC. Conversely, a PIA should have a strong mathematical modelling element in it, which will allow realistic and relevant assignments to be set, and give excellent opportunity to develop the common skills.

2. PROGRAMMES OF INTEGRATIVE ASSIGNMENTS

A PIA can provide a vehicle through which course teams can develop assignments focussing on common skills. The keynotes are relevance – to vocational area, work tasks, the course as a whole, and the need to develop common skills

realism – in terms of situations, contexts and tasks involved in the assignments and contacts with interested parties.

The use of mathematical modelling assignments can develop the skills of collection, interpretation and presentation of information, problem identification and communication. It has the advantage that it can cover a wide range of topics, affect investigation of real–life problems, and introduce different sources and methods of obtaining information. The teamwork involved will help to strengthen group cohesion, identify issues, make decisions and evaluate solutions. A PIA can be implemented in two ways:

a unit allocated exclusively to the programme within each year of the course which has a distinct, timetabled activity additional to work on course units,

or

identified assignment work undertaken within the time allocated to a number of units or parts of the course.

The construction of the component assignments should cover the five main types: investigating and reporting, learning and evaluating, designing and making (although assignment may stop short at the design stage); work experience assignments; and industrial project. For those students who have work experience placements as a component of the course, the last two topics are suitable for assessment after their placement. In addition, interactive simulation and rôle play can form a greater part of assignments after students have completed their placements in industry. Examples of assignments which attempt to include the first three areas are given in Appendix 1.

For the development of common skills, all types of assignment should enable students to review their own strengths and weaknesses in major common skills, to plan and achieve objectives (in consultation with tutors) and to include team elements to develop interactive skills, group decision making and communication skills. Contact with interested parties outside the course should be made at information gathering, presentation and review stages.

3. OPERATION AND ASSESSMENT OF A PIA

The description below is that of a Programme which has been implemented for the HND in Mathematics, Statistics and Computing. The PIA is a distinct, timetabled activity since the course team saw this as giving greater flexibility in operation and assessment of the common skills. Three assignments of six to eight weeks duration are set per year, and students are expected to allocate three hours per week to an assignment. Ideally, teams of four are chosen, and a tutor allocated to each group. The overall management of an assignment is supervised by a moderator, who is a member of the course team. Students are given an Integrative Assignment Project Guide at the outset, which outlines the procedures to follow during an assignment and the common skills identification chart (Appendix 2) indicating the skills they need to demonstrate and develop. In devising the above programme, the course team has attempted to incorporate a wide range of interrelated assessment models such as the following.

Lecturer observation of group activity.
Student reports, which should include reflections on learning and on common skill strengths, weaknesses and development.
Peer assessment.
Group working papers, such as minutes and records of decisions, briefings to team members and checklists for interviews or investigations.
Assignment products such as reports and suchlike.
Process reviews, in which teams and tutors review the ways in which tasks were approached, team effectiveness and individual contributions.

Assessment methods will vary at different stages of the PIA, and they are used to build a picture of each student's development of common skills. Students' abilities and skills are expected to develop throughout the course. Thus the final grade for the PIA should not be based on an average grade, but should be weighted towards the level attained at the end of the unit.

4. CONCLUSIONS

Problems still arise in accepting a formal course in mathematical modelling into schemes. However, as the aims of mathematical modelling and the development of common skills are similar, then a PIA can be implemented with the underlying theme of modelling. From my experience, students have responded well to such a programme, and appreciate tackling what they regard as real-life problems. Also, their development of the common skills is enhanced by the monitoring process employed. They are pleased to have a contact, a tutor, to whom they can communicate their ideas as they progress through an assignment. The use of regular meetings, in the form of a committee meeting, allows the tutor to derive a better estimation of the development of a student's skills.

The development of modelling in schemes of PIAs for other disciplines, such as science and engineering, should be encouraged, as it offers a genuine opportunity for these students to apply the techniques of the core syllabus mathematics, which invariably form part of their course, to practical problems related to other core units and also to develop the common skills as discussed herein.

REFERENCES

Andrews J and McClone. (1976). *Mathematical Modelling*. Butterworth.

BTEC Common Skills and Core Themes – Illustrative Material and Advice. September 1986, 80–036–6.

BTEC Common Skills and Core Themes. General Guidelines, September 1986, 08–051–6.

Giordano F and Weir M. (1985). *A First Course in Mathematical Modelling*. Brooks/Cole.

APPENDIX 1

The following are general problem statements for assignments which may be used as components of a PIA. A task sheet may also be included with these statements in the early stages of a course.

TIMETABLING OF TRAINS

As a method of reducing costs, British Rail has introduced single track working on predominantly country routes as a way of reducing permanent way costs. One aspect of single track railways is the care that must be taken to avoid head-on collisions. A method employed by British Rail is the 'token' system, where, for each section of single track, the engine driver is provided with a large token, without which the driver is not permitted to proceed. The driver has to wait in a passing place, which is a short length of double track where one train can wait until the train travelling in the opposite direction has passed. The token is handed from one driver to the other, who will then travel in the opposite direction on that single track section.

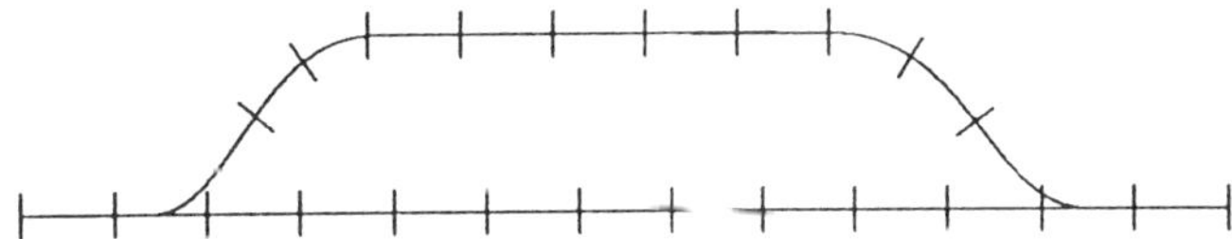

Although single track running saves on maintenance of the permanent way, it causes many other problems which arise from: introducing extra trains such as football specials, holiday extras, freight trains and so on; alterations to and retimetabling trains to take account of engineering work at a passing place; further economies reducing the number of trains used on a line, with the minimum disruption to existing or feeder services.

Your problem is to consider the single track rail system between Shrewsbury and Barmouth running through mid-Wales, with stops at Welshpool, Newtown, Caersws, Machynlleth and Dovey Junction. The whole journey takes about two and a half hours and, so unless some provision is made for trains travelling in opposite directions to pass each other, only one train could be used, which would result in a skeleton service with a gap of at least five hours. Stations are useful locations for passing places since the trains are scheduled to stop and thus no unnecessary delays are introduced by passing, but maintenance of extra buildings and platforms is required. With widely separated stations additional passing places are provided, to allow for greater flexibility in timetabling, but this increases the cost of the system since each passing place has to be controlled by a signal box.

You will devise and present a report to British Rail advising them of the operation of this line.

MINIMISATION OF SOUND DISTORTION PRODUCED BY TRACKING ERROR IN A RECORD PLAYER PICKUP ARM

The purpose of this assignment will be to investigate the way in which the stylus tracks the groove of a record, and make recommendations which can be implemented practically in a pickup-arm design. You may like to consider, but not investigate, what other features influence the quality of sound reproduction by a record player.

The stylus of a typical pickup-arm is attached to a cartridge by means of a thin rectangular strip of springy metal, and this cartridge is in turn attached to the pickup-arm. The two most common types of pickup-arm used are:

(a) *the straight-arm type*, in which the axis of the arm, cartridge and metal spring are all in line;
(b) *the offset-arm type*, in which the axis of the metal spring is aligned towards the centre of the turntable in relation to the pickup-arm axis, by offsetting the axis of the cartridge with respect to the cartridge.

You are required to investigate the straight-arm pickup with a view to minimising distortion produced by tracking and then, using these results, make recommendtions for the offset-arm type. The diagram illustrates a typical straight-arm pickup. The playing surface of the record has outer and inner radii of r_1 and r_2, the distance between the pivoted end of the arm and centre of the turntable is D, and the distance between the pivot and the stylus is L. If L is greater than D, the arm is said to be overhung.

One reason for an arm to be underhung or overhung is to avoid collision between the stylus and the centre spindle. Another reason is that analysis of the physical dynamics of the system shows that the distortion of the audio signal is directly proportional to α/r where α, the angle between the line of the pickup-arm and the tangent to the groove, is called the tracking error. Thus a good design will attempt to keep

α/r as small as possible for the whole of the playing surface.

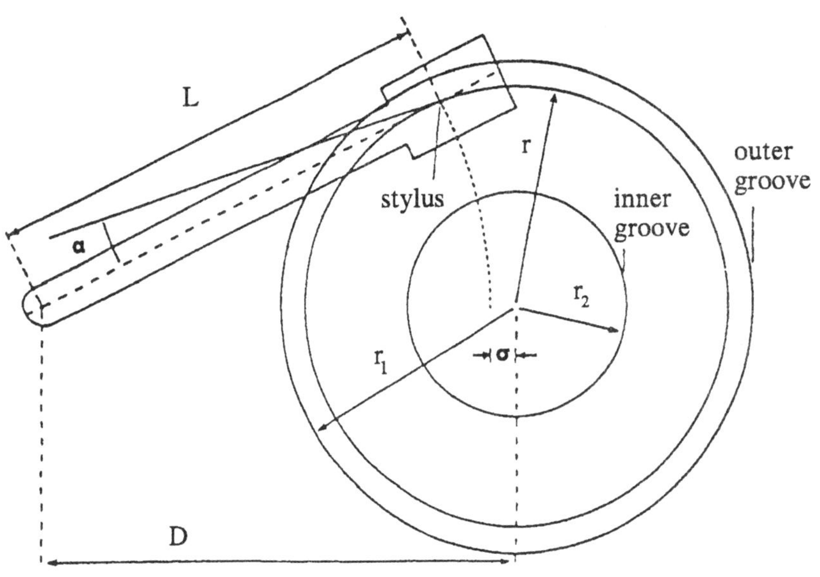

The geometry of a record player in the situation where the dimensions lead to underhang of the arm

Determine formulae which give α in terms of r, D and d (= D−L) for the cases of overhang and underhang. You should decide upon a set of typical data values for D, r_1, r_2 by actual measurements and, using these values, draw graphs of α/r against r for $r_2 \leqslant r \leqslant r_1$. Repeat this for several values of d. From your graphs and results, make recommendations on the design of straight pickup-arms. (It is suggested that a computer program is written to perform the computations of α/r).

You should now consider the offset pickup.

Discuss in general terms how the offset provides extra flexibility in controlling α/r and, based on the graphs drawn above, make suitable recommendations, if possible, regarding the design of offset pickup-arms.

Compare the results of your investigation and real-world record players, paying particular attention to the implementation of your model, and indicate ways in which your model may be developed further.

POSITIONING OF A REAR-SCREEN WIPER

The rear screens of hatchback cars often have a single windscreen wiper. To the casual observer, some cars seem to have so inadequate an arrangement that the wiper appears to have been included without any real regard to its effectiveness. Other cars have complex looking bent arms and off-centre pivot points, suggest that considerable thought has been given to the design.

Develop a model, which will help to determine the length of the wiper blade and its position, for a given size of rear screen. Collect relevant data for various makes of cars, which may be used to validate your model.

Produce a report for a car manufacturer advising the company on the design and position of the wiper blade. You must state carefully, with reasons, the criteria on which your calculations are based.

CONTAINERS FOR MILK

The design of milk bottles has changed over the past few years. Glass has traditionally been used, but more recently plastic and waterproofed cardboard have been used as alternatives.

A company is interested in minimising the cost of manufacture of one-pint glass milk bottles. Your team has to formulate and construct a mathematical model which will minimise the cost of manufacturing a milk bottle, whilst retaining the basic shape of the bottle (but allowing its dimensions to vary). Thus the problem will involve more than one variable. At present you may not possess the mathematical techniques to solve such problems analytically.

Therefore, in finding the solution to your problem, you should hold all variables except one fixed and solve your resulting equation(s). If necessary, you will have to resort to numerical techniques which will involve some computer programming. Any suitable programming language may be used. Having found the value of this variable, you should then hold this variable fixed at the calculated value and allow another of the variables to change until your criteria are satisfied. The above procedures should be repeated until you are satisfied that you have obtained the values which minimise the cost of production. Your design should be compared with those of existing milk bottles.

The implementation of a new design can only be made with the acceptance of the general public. You will need to gain information on the opinions of the public regarding changes in shape of bottles, and also the use of other materials for holding milk. You will also need to obtain information from those bodies who will have to use the new design, such as dairies and milkmen.

Some of the features of the milk container which should be considered are durability, ease of conveyance (both by hand and by vehicle), weight and shape of crates (where applicable), number of containers broken or not returned.

Your recommendations must be produced in a formal report, with an accompanying verbal presentation.

Structure diagrams, and hard copy listings, of any programs used should be given, together with results obtained from your programs and clear user instructions for running the programs.

APPENDIX 2

COMMON SKILLS IDENTIFICATION

1. Planning

a Recognises and meets deadlines
b Works to a plan
c Assesses what information is available and what is needed
d Identifies resources available

2. Communicating

a Listens, questions and responds in order to obtain information
b Uses technical terms appropriately
c Uses visual communications, graphs, charts, diagrams etc
d Writes effectively

3. Evaluating

a Assesses quality of information
b Distinguishes between fact and opinion
c Questions and justifies assumptions
d Analyses reasons for failure

4. Problem solving

a Recognises and defines problems
b Works out priorities
c Sets out aims and objectives
d Forms an understanding of different ways of tackling a problem

5. Implementing

a Manages time effectively
b Motivates self and others
c Adapts project schedule where necessary
d Reviews work as it is done and checks its accuracy

6. Working in a team

a Puts other team members at ease and helps them communicate
b Contributes constructively to the team
c Accepts cricitism within the team
d Shows sensitivity to the other members of the team and recognises their right to have different values and beliefs

7. Using Information Technology

a Sets up microcomputers with application packages to process data
b Uses computers and other devices as an aid to collecting, storing, analysing and disseminating technical and other information.
c Uses computer packages to help tackle technical and other problems

CHAPTER 33

Assessment of a Modelling and Applications Teaching Module

T Naylor
Edge Hill College of Higher Education, Liverpool, UK

SUMMARY

Discussion of the organisation and teaching of the module allows the place of mathematical modelling to be considered within the overall assessment pattern for mathematics at undergraduate level. A detailed analysis of coursework projects and examination questions is undertaken, through a review of the modelling process and the modelling sub-skills required to be developed or applied. Problems related to the assessment of mathematical modelling, including the difficulties of setting examination questions and coursework assignments, are identified, and consideration given to strategies for their minimisation.

1. MODELLING, APPLICATIONS AND ASSESSMENT AT EDGE HILL COLLEGE

Development

When planning courses and subsequent assessment of the content, it is paramount that the abilities students bring with them should receive major consideration. In this respect, each institution's situation is different. Similarly, the aims and objectives for each course and assessment programme will be unique. Provision by the subject area of mathematics is as a major subject within the BEd degree and a minor subject within the BA/BSc degree at Edge Hill College. In both cases, the minimum entry qualification is a good GCSE (General Certificate of Secondary Education) or equivalent.

Over the four years of the BEd and the three years of the BA/BSc courses, increasing emphasis is placed on modelling and applications, with all other areas facilitating this component. Course presentation takes the form of a one–hour lecture to a combined group of BEd/BA/BSc students, supported by a proportion of a two–hour workshop. Coursework consists solely of modelling activities – and the proportion this contributes to the yearly assessment grade has been the subject of much careful research and discussion. In 1973, within Part I and Part II of the course, the relative weighting assigned to examination and coursework was 60% and 40% respectively. However, it became clear that there was significant disparity between examination and coursework marks, the former being one grade (or sometimes two) lower than the latter. Consequently, while Part I assessment remained at 60:40, Part II moved to 80:20, resulting in a much closer correlation. As modelling is essentially a 'doing' activity this may well have some effect on the form of assessment. Coursework grades, therefore, may be seen as being essentially different from examination grades.

All assessed coursework is concerned with modelling, but the emphasis placed on the differing areas within such activity varies as the course develops. In Year 1, the six projects have two aims: firstly, development of the students' awareness of the modelling process and, secondly, their modelling sub-skills. The six–answer, three–hour examination paper has three of the ten questions presented devoted completely to modelling. Year 2 coursework consists of four 'situations approach' projects, and the examination paper includes four modelling questions. In Year 3, modelling accounts for half of the eight examination questions, and coursework constitutes one project.

2. ANALYSIS OF EXAMINATION QUESTIONS

Example 1

1. Let f(x) denote the population size of an urban place with x establishments of a particular service type. The following model has been empirically derived to show the relationship between population and number of establishments of a particular service type, using data for a collecton of urban places within a region.

$$f(x) = AB^x \qquad (*)$$

where A and B are constants, both greater than zero, and $x \geqslant 1$.

(a) Determine the domain and range of the function given by (*). (2 marks)

(b) Values for A and B are estimated for a particular service type, say jewellery stores, using the regional data. How would you use these estimates of A and B to determine the urban size that is necessary to sustain at least one jewellery store in a town? (2 marks)

(c) The following results have been obtained for two types of service function in the set of towns in the region.

Case 1	Petrol stations	: A = 145.19	B = 1.35
Case 2	Jewellery stores	: A = 253.7	B = 3.26

Sketch graphs of the function given by (*) for these two cases. (3 marks)

What do you conclude about the differences in the provision of these two services? (3 marks)

(d) A separate study is carried out in a different region, dealing only with the provision of petrol stations. The estimates of A and B in the two regions are as follows.

Region 1	:	A = 145.19	B = 1.35
Region 2	:	A = 190.18	B = 1.15

(3 marks)

What do you conclude about the differences in the provision of this service in the two regions? (4 marks)

2.

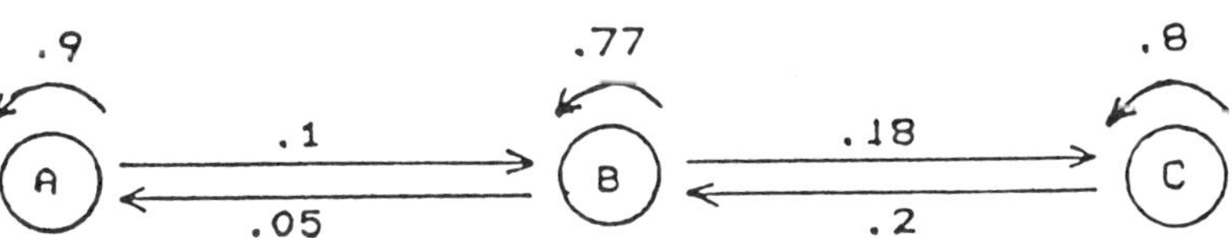

The transition diagram shows the movement of people between three sections of Blyth over a ten year period.

(a) State the transition matrix indicated by the diagram. (2 marks)

In 1951, 22% of the population lived in section A, 31% in section B, and 47% in section C. Over the period 1951–88, the population remained at about 35 000 and at the 1981 census was 34 553.

(b) Estimate the number of people in each section of Blyth in 1981. (4 marks)

(c) If the population continues to remain at about 35 000 in the future, what percentage of the population is likely to live in each section in the stable situation?
(7 marks)

(d) Discuss the relevance of this information.
(4 marks)

3. Discuss the cusp catastrophe and explain its use in social science.
(6,7 marks)
Illustrate your answer by considering the early stages of drug addiction.
(7 marks)

Example 2

1. Carefully remove the outer wrapper from a new toilet roll and file it away for later reference. Any relevant information on this wrapper may *only* be used for validating a model and *not* for building a model. A single sheet from a similar roll may be used for obtaining any measurements required.

   (i) Without unrolling the toilet roll develop models to find the length of paper on the roll.

   (ii) Within the specifications defined above, construct a model to find the total area of paper contained in the roll.

2. Analyse the conduct of a historical battle and comment on your observations.

Year 1

Part (a) of question 1 (example 1) checks the students' understanding of mathematical terms, while simple use of the given model is required in part (b). The skills required for drawing graphs of $y = AB^x$ are assessed in part (c), along with the ability to compare two graphs. Here, and in part (d), the question attempts to deal with real-world situations, and requires presentation of a written argument. Part (d) follows from (c), but requires a deeper level of understanding in order to advance the argument, hence the four marks allocated to its interpretation.

Analysis of the mark scheme suggests thresholds at 4, 10 and 17 corresponding to question elements which may be categorised as introduction, seen and unseen. The raw marks may be too high but, on the other hand, the demands of the grades may be different. The overriding consideration here is whether or not the threshold is at the correct standard for students progressing into Year 2. Part I assessment is, in effect, a 'gate-keeping' exercise.

The situations presented in this question are very real to our students, for most of whom geography or social science is their major subject. Is then this type of question a natural extension of mathematical modelling, or should some other kind of question be presented in Year 1? Are different skills to those assessed during coursework being tested, or should testing supplement coursework assessment?

Question 1 assumes a student's understanding of certain mathematical modelling sub–skills. The generation of variables, generation of relationships and modelling of situations are implicit in the function $f(x) = AB^x$ and the situation presented. Specific questions are identified, for example what increase of population is required to sustain a further petrol station, while estimating and validating are essential in constructing arguments in parts (c) and (d). The question thus supplements and complements the skills required for project 1 (see example 2). In essence it involves solution and interpretation. Is this what most examination questions require? Is this a good idea?

Year 2

Question 2 (example 1) has a simple introduction, with the situation explained in pictorial terms, which the students use in part (a) to formulate a model. Solution of the model is undertaken in part (b), and the situation predicted by using the model could be checked against data recorded by the local authority, as an application of the model. The stable situation, obtained in part (c), forms the basis of the report required in part (d), which necessitates presentation and interpretation of the solution.

This final stage corresponds to the unseen or 'acid test' elements of examination questions. It is difficult for students working under examination conditions to present a written report, based on their geography and social science backgrounds, incorporating interpretation of the stable situation that the population is moving towards in this local authority. Nevertheless, the usefulness criterion receives a positive rating from students, since an ability to interpret trends is vital to social scientists (service provision), historians (the implications of similar previous trends) and geographers (industrial and agricultural location).

Year 3

After the initial discussion, which might be considered similar to the bookwork/proof of a theorem, question 3 (example 1) is of a rather open nature. The students must define precisely any terms stated (for example Intention to Buy, Social Normative Factors, Attitude Factors) and then proceed to define their own problems. Following this, they are concerned with the presentation of a solution. Thus, the whole modelling process is embodied in the question.

However, following the discussion, what is the 'acid test' if such a question is used? Is there one, in fact, or do the component parts add up to something different from the acid test idea? Are we here more interested in the usefulness of the question, which for social scientists is of very high level? The increase in sophistication required between Year 1 and Year 3 is emphasised by the degree of understanding necessary to come to terms with the nature of the problem, and the nature of the arguments, involved. Such understanding is vital if students are to be aware of how to look at similar situations.

3. ANALYSIS OF COURSEWORK PROJECTS

Year 1

The value and relevance of an activity must be considered in terms of the aims set out for coursework. Project 1 (example 2) is the students' first piece of assessed coursework. The situation is presented in the form of written information and, by implication, is one where a consumer will ultimately be looking for the best buy. The situation is, therefore, real and relevant to the student. In parts (i) and (ii), the tutor has defined the problem to be solved under the conditions previously laid down. Within these constraints, the student must build mathematical models (for example equations). After making appropriate measurements, these values are fed into the model in order to give a solution for the specific case under consideration. Use of the wrapper to validate the model allows interpretation of the results produced in terms of any assumptions made during the building of the model, for example that the paper is wrapped in concentric circles. The aim of making students aware of the modelling process has thus been achieved, and consideration must now be given to the modelling sub-skills employed.

When representing parameters of the toilet roll by letters, either by definition or on a diagram, students are generating variables which they can then investigate for relationships. Initially, however, general relationships might be considered, for example

weight of toilet roll = weight of paper + weight of cardboard tube.

Estimating the weight of one sheet involves the strategy of averaging, as does finding the radius (diameter) and height of the roll. The validating skill is facilitated by using the information on the wrapper to build an accurate model. Not only are students able to consider any weaknesses in their assumptions and measurement techniques, but also to discuss the meaning of the phrase 'average number of sheets' displayed on the wrapper.

In addition to allowing the student to experience all stages of the modelling process, this project provides the opportunity to develop modelling sub-skills, a most vital requirement during initial modelling experiences. However, as the student progresses, emphasis on sub-skills decreases as these become fully integrated into the modelling process, which itself becomes of paramount importance.

Year 2

As in Year 1, a problem is set and defined by the question, but in a much wider context, with projects posing an individual problem for each student. The situation presented in project 2 (example 2) is far removed from the direct problem of project 1. Students must seek out unique information on which to base their models. This introduces the students to a kind of research, the information from which must be analysed in terms of available techniques such as Lanchester's square law. Alternative strategies must then be considered in view of such questions as "What would happen if part of the force were held in reserve?". This and other questions need to be posed by the student and the consequences investigated. Analysis and evaluation of such investigations will form the basis of a serious piece of writing.

In Year 2 the emphasis is firmly on the modelling process, particularly the ability to comment critically on models constructed. This requires an analysis of the available techniques on the basis of known information, which must also receive detailed discussion. By defining their own problems and by considering various strategies, students are moving nearer the open situation of true modelling.

Year 3

For their third-year coursework assessment, students are placed in a totally open-ended situation. Considering a situation which they find of particular interest, and posing questions to themselves relative to that situation, leads to the students formulating their own problems, for which they build appropriate models. Following validation, the results are interpreted and, along with all the other stages, discussed in a written report.

The quality and quantity of work produced by a student who considered social policy in Salford went well beyond the demands of this course. Changes in the Government's social policy were considered in terms of catastrophe theory, with the move from the Social Security Benefit system to the Income Support Grant forming the focus of the investigation. Different groups and sub-groups were identified and defined with extreme care, for example 'scrounger'. Before making statements in relation to these groups, the student was careful to provide evidence for her claims, and this required a considerable amount of genuine research. In the second section, the student considered carefully the amalgamation of two schools. Then, in an examination of extremely high quality, she investigated the profitability of different shopping areas of the city. Five zones were identified, and a production–constrained model developed to establish the movement of shopping expenditure between each zone. The student based her thesis on each zone's income, distances between zones, and the floor space of shopping centres (source – official statistics for 1984 and 1986). With this information she modelled the flow of shopping expenditure between zones. As a final focus for social policy, the student looked at health care, where she felt a cut–back in the provision of hospital beds would be most damaging. Death rates were used to measure the level of availability required, and the degree of correlation between death rate and available beds investigated.

4. CONCLUSIONS

At the end of their degree course it is necessary for students to be classified as I, II(i) and so on. We must, therefore, ask ourselves what we expect from a student in order to gain a particular classification, and to what extent our assessment procedure allows an appropriate grade to be achieved.

In the light of this analysis, what criteria determine a I classification? For most mathematical examinations, such quality is demonstrated by complete answers rather than a collection of fragmentary parts, and we at Edge Hill would subscribe to this view. As a key essential, therefore, we expect candidates to answer the unseen part of the question regularly. However, in both Year 2 and Year 3, the requirement that students should be able to integrate knowledge from their major subject into the interpretation of the model was found to be a barrier, and their inability to do this with any consistency led us to believe that we might have identified the correct criteria.

The attributes for an excellent piece of coursework are more difficult to identify, and usually show in the student's originality. Research of relevant information is meticulous, and forms the basis of a model which attempts to incorporate all the information. Finally, written analysis gives due weight to successful aspects of the model and, at the same time, indicates the need for caution where evidence is less strong.

A II(i) classification will, in the examination situation, result from evidence of some complete solutions. Coursework, whilst sound in its implementation of the modelling cycle, will lack originality. The II(ii) student will display competence in examination questions, up to the stage of our interpretation of the unseen part, and coursework will be open to criticism of some aspects of the modelling cycle.

It is obvious from tutorials with students that they derive great enjoyment from the coursework projects, which is also reflected in written submissions. This is a positive and satisfying outcome but, unfortunately, generates the problem of students devoting an inordinate amount of time to the investigation. This, they say, detracts from their efforts in other areas of the degree course. Consequently, and not surprisingly, they feel quite strongly that coursework assessment should form a much higher proportion of the overall grade for the unit.

CHAPTER 34

Mathematical Modelling for Engineering Undergraduates

LR Mustoe
Loughborough University of Technology, UK

SUMMARY

This paper describes the place of case studies and modelling in an undergraduate engineering course. Experience shows that students respond positively to actual case studies which illustrate the usefulness of modelling with mathematics. The first mathematics lecture is devoted to examples of modelling problems relevant to the particular discipline, which will be discussed later in the course. Shared lectures are used to present case studies; an example is given as an illustration. A typical tutorial exercise in modelling is presented and a modelling problem demanding a greater timescale is described. Student reaction to the modelling component of their course is discussed, and conclusions are drawn.

1. BACKGROUND

The lecturer of mathematics to an undergraduate engineering course in the United Kingdom faces an increasingly difficult task. There is pressure from the serviced department to reduce contact hours; there are requests to include new material in the syllabus, although every topic currently present is considered either essential by the mathematics lecturer or, at the least, very desirable by one or more of the engineering staff (they helped to draft the original syllabus, of course). The student intake is varied both in quality, and in range and depth, of mathematical experience.

It is against this background that the mathematics lecturer, if s/he is an advocate of modelling in the syllabus, must seek to find some room for this activity. S/he cannot afford to neglect the basic mathematical techniques which the students require in order to cope with their other engineering subjects, and by which s/he will be judged by engineering colleagues. The author has a total contact allowance of 165 hours for lectures and problem classes, supplemented by a weekly 'surgery' hour in which to impart the mathematics course to his civil engineering undergraduates. These hours are split into two roughly equal parts in the first and second years of the course. If the activity described in this paper seems modest, it must be judged in this context.

Among many articles dealing with modelling for engineers those by Murthy and Page (1981) and James and Wilson (1985) are especially worthy of note.

2. FIRST STEPS

On their induction course, the freshmen civil engineers received a series of ten-minutes talks from each of their subject lecturers. This allows the author an opportunity to introduce the idea that modelling will be a key feature of the mathematics component.

In the first lecture proper, the concept of mathematical modelling is explained using the bending of a simply-supported beam as an example. The assumptions on which the formula for the deflection of the mid-point, $5wL^4/(384\ EI)$, is produced are discussed. How thin is a 'thin' beam, what would happen if the beam were not homogeneous or uniform, and how small is 'small' deflection, are questions which are posed. It is emphasised that, before the formula was produced, certain simplifying assumptions had to be made, and therefore any predictions which follow from the application of the formula are only as reliable as these assumptions.

To conclude the lecture the students are given a hand-out entitled Models in Engineering, which comprises several examples of engineering problems requiring the use of mathematics for their solution. Some are relatively easy to tackle, some have very little information and require considerable thought before the application of any mathematical technique, some are quite vague and open-ended. In all cases, it is stressed that a modelling approach has to be undertaken. These examples will be discussed at suitable points later in the course. The modelling process is now discussed via a flow diagram.

3. RAINFALL AND RUNOFF: A SHARED LECTURE

In their second term the civil engineering freshmen are given the methods of separation of variables and integrating factor in a package on ordinary differential equations. Part of this package is a shared lecture in the form of a dialogue between the author (M) and one of his civil engineering colleagues (C).

C explains the need to relate the rainfall on a catchment area to the resulting surface run-off, in order to predict the likely outcome of a storm and take any necessary preventative action. M tries to model the catchment via a single leaking tank. C is not impressed when he compares the predicted outflow with the outflow from an actual case study, see figure 1(a). Having used the method of variable separation in the simple model, M tries a more sophisticated model with one tank sending its outflow into a second identical tank which is initially empty. This time, the integrating factor method is required. The graph of outflow from the second tank resembles the actual outflow of figure 1(b), but is not quite accurate. C asks what would happen if the two tanks were augmented by a third, a fourth and so on. M shows that it is possible, with the help of mathematical induction, to obtain a formula for the outflow from the last tank in a cascade of n tanks

$$Q_n = h_0 A k^n \frac{t^{n-1}}{(n-1)!} \qquad (1)$$

where h_0 was the initial depth of water in the first tank, A the cross-section area of each tank and k is a constant relating to discharge characteristics. The question of how to determine the 'best' values for n and k is discussed; remember that h_0A is the initial amount of water entering the system. It is agreed to take first and second moments of area about the t = 0 axis of both the graph of actual outflow and a 'theoretical' graph.

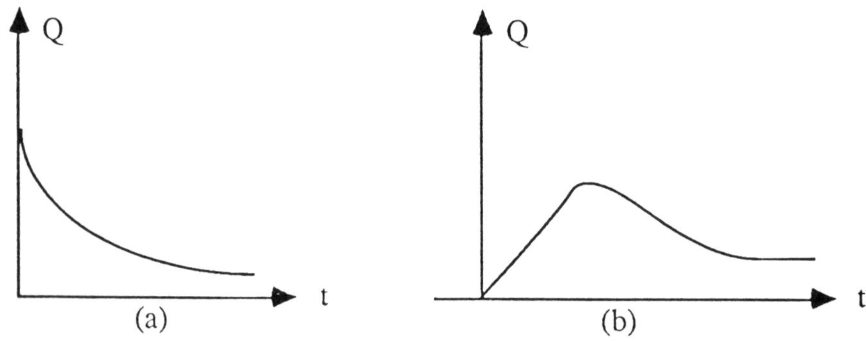

Figure 1

The nature of the dialogue so far has been mostly even-tempered, but it now takes a stormier path. C is highly critical of the model, which has assumed an initial amount of water dumped instantaneously into the system; this is not how a real rainstorm behaves. In any event there is no such thing as a standard rainstorm, and seasonal climatic factors may have an effect. M explains that no model is perfect, and it is better to build up gradually to more sophisticated models; after all, if a simple model turns out to be adequate for the purposes for which it was designed, why use a complicated one?

The serious problem to address is the modelling of the nature of the storm, its finite duration and variable intensity. To act as a baseline, Ah_0 is taken as 1 and the resulting outflow graph is named the Instantaneous Unit Hydrograph. The idea of convolution is developed by M; the intensity-time graph is broken up into vertical strips and each strip is combined with a suitably delayed IUH to obtain the outflow formula

$$Q(t) = \int_0^{t_f} I(\tau)u(t-\tau)d\tau \qquad (2)$$

where t_f is the duration of the storm, I(t) is the intensity of the storm and u(t) is the IUH starting at t = 0. C mentions that in industry he did a similar exercise using a tabular method; M states that provided the accuracy obtained was acceptable then a tabular method was satisfactory, but did C understand what he was doing?

C shows a diagram of the computer-based model used by the Welsh Water Authority which incorporated many interlinked systems of leaking tanks feeding into each other. He explains how the model is used in practice, and M is able to point out the merits of employing mathematical models to solve real-life problems.

4. RESERVOIR CONTAMINATION: A TUTORIAL SESSION

Immediately following the package of differential equations, the first year civil engineers are given the following problem in a tutorial session.

> A reservoir of 10^6 litres has been contaminated by effluent. The degree of contamination is 0.02% by weight. The average daily rate of consumption of water is 2×10^4 litres and this is continuously replaced by pure water. How long will it be before the concentration of contaminant drops to the safe level of $10^{-5}\%$?

The students are asked first to comment on the information provided. Is it sufficient? Is it in a suitable form? Is any of it redundant? What does degree of contamination mean: average, or maximum or what?

The class is encouraged to suggest ways of solving the problem by setting up a model. The expected response is a differential equation, although there is some disquiet about applying a continuous-time model to the information given. Generally, the students have to be restrained from diving into the production of a model equation before they have considered fully what they hope to achieve and what variables they will use in their model. It is recognised that mass of contaminant is an easier variable to use than is concentration. Taking δt as one day, the model obtained is simplified to

$$\frac{dc}{dt} = -\frac{c}{50}$$

where c is the contaminant concentration at time t. If c_0 is its initial value and c_f the desired safe level, it is straightforward to obtain the solution that the required time to achieve c_f is

$$t = 50\ \ell n(c_0/c_f) \cong 380 \text{ days.}$$

It is asked whether anything can be done to improve the accuracy of the solution. When it is suggested that δt be reduced, say to half a day, it is a disappointment, and a surprise, to the students to find that t = 760 (half-days). No improvement can be made by reducing δt. Why not?

Whilst this question is being considered, the author suggests that a different approach can be taken. Since the information is given on a day-by-day basis, why not let m_n be the mass of contaminant in the reservoir at the end of day n and try to relate m_n to m_{n-1} and hence to m_0? It is readily seen that

$$m_n = (49/50)m_{n-1} = (49/50)^n\ m_0$$

and that

$$n = \frac{\ell n(m_0/m_f)}{\ell n(50/49)}$$

where m_0 and m_f are the initial and final mass of contaminant. Noting that $m_0/m_f = c_0/c_f = 2000$ then n = 376 (days). This value compares favourably with 380 from the previous model. Which is correct? Which is the more accurate? Is the model optimistic or pessimistic? What advice would you give to senior engineers who want to know what action they should take? What other factors might decide their course of action?

To conclude the session the author comments on the need to formulate models carefully, in the light of information provided. In this instance, a simple discrete approach gave an almost identical result to a continuous-based model, yet neither could be considered really satisfactory.

5. THE FURNACE BUMPER PROBLEM: A LONGER TIME SCALE EXERCISE

Part of the process of producing steel plate consists of reheating steel slabs in a furnace, and discharging them onto a roller table which transports them to a plate mill. It is necessary to slow down the slab after it leaves the furnace. A method used in practice was to allow the slab, of typical mass 18 tonnes, to hit a bumper, of typical mass 2 tonnes. The bumper was attached to six steel cables, which passed through slits on the roller table to a lower concrete slab of mass 40 tonnes, and to six more cables attached to a higher identical slab; each slab rested on a rough slope inclined at 45° to the horizontal. The difficulty was that the cables broke much earlier in their working lives than had been expected, and it was required to model the system and suggest suitable modifications.

This exercise is given to mathematical engineering freshmen in term 3. The students are divided into groups of approximate size 6, each with a spread of abilities, and are taken through the exercise over a period of three weeks, spending six class contact hours in addition to their own time. The problem is described, and the students come to a tutorial one week later to hold initial discussions on the problem. The author acts as a fairly passive chairman Their is a need to ensure that progress is reasonably on schedule, and this has to be balanced against the desirability of as little intervention as possible. At the end of each session progress is reviewed, and the exercise is concluded by a full plenary session in which the overall progress of the groups is discussed and general points highlighted. A computer program has been written to solve the model equations; this is given to the students to obtain some results about the output in the form of graphs, which are presented and discussed at a later stage in the term. The conclusions reached when the modelling was originally carried out in industry are described.

6. STUDENT REACTION

Feedback has been obtained both in the form of verbal comments and responses to short questionnaires. The shared lecture was well received, and the dialogue approach retained the student's interest: the fact that the models discussed formed the basis of a real-life application was appreciated.

The reservoir problem proved less popular. It was the novelty of having to work things out for themselves that was resented at the time by many, although a subsequent questionnaire indicated that, by the end of their first year, the students recognised the value of the exercise.

The furnace bumper exercise proved surprisingly satisfactory. The majority of students accepted that learning to model was a difficult task, but a crucial one in their mathematical development. They accepted the need at this stage of their careers to have occasional help and prodding. Surprise was expressed at the simple level of mathematics that had been used in constructing the model. There was acceptance of the need to learn a few extra techniques to effect the solution.

7. CONCLUSION

The progression from case study to active modelling is crucial, and must be handled by introducing the latter by models of increasing complexity, the guidance provided being gradually reduced.

One problem confronting the author is the size of group being taught. The furnace bumper problem is ideal for 18 mathematical engineering freshmen; it is impracticable for 80 civil engineers. It would be a bonus if engineering staff were able to lead some of the subgroups, but pressures on UK academics discourage them from devoting more time to such teaching activities.

A further difficulty is the restriction of mathematics to the first two years of the engineering courses at Loughborough. Moves are in progress to introduce a modelling course as one of the final year options, with lecturers from both engineering and mathematics staff. In the meantime, efforts will be made to find whatever time can be spared, without sacrificing too much of the teaching of concepts and techniques, which are also a vital part of the undergraduate engineers' training and which are the yardstick by which most mathematics courses are judged.

REFERENCES

Murthy DNP and Page NW. (1981). Teaching Mathematical Modelling to Undergraduate Engineering Students. Int J Math Educ Sci Technol, 12 (2), 234-243.

James DJG and Wilson MA. (1985). How Should the Mathematical Teaching of an Engineering Undergraduate be Conducted? Int J Math Educ Sci Technol, 6 (12), 163-171.

CHAPTER 35

Use of a Consultancy Project to Teach Mathematical Modelling to Students of Engineering

CE Wilsdon
Brighton Polytechnic, UK

SUMMARY

A method of teaching mathematical modelling techniques to engineers is described which used a consultancy project as a 'really real' introduction to problems of modelling, but followed this up with a simplified version which was used to introduce solution methods. There was then the opportunity to move back towards the original problem, with more sophisticated computer techniques. The driving force behind this approach was to sell the idea of mathematical modelling in a situation where the students have varied backgrounds, have limited time available, and need convincing of its relevance. In this respect it was found to be very successful.

1. INTRODUCTION

In September 1988 I was asked to teach the initial part of Mathematics and Modelling to students studying for an MEng degree at Brighton Polytechnic. The MEng course is an interdisciplinary extension to the three year BEng courses in Civil, Electrical and Mechanical Engineering, and is a relatively new development in the UK. The emphasis of the MEng is on developing the managerial skills of the professional engineer.

There were a number of factors to consider when deciding on the best approach to teaching the mathematics and modelling component of the degree.

(1) The mathematical background of the students was not uniform, due to differences in the specialist BEng courses.

(2) There was a need to demonstrate that the course was relevant to engineers.

(3) The course was allocated only one hour of teaching per week, and there were heavy competing demands on the students' time from project work and case studies.

A project-based approach (see Cornwall and Schmithals, 1976) was rejected, in spite of its obvious attractions in terms of relevance and giving the students some measure of autonomy. Project work is very time-consuming for students, particularly when there are several, unrelated activities of this sort going on simultaneously.

The traditional taught-course method was also not felt to be suitable. Real applications are often not included and the process can have the appearance of 'solutions in search of a problem'.

2. AN ALTERNATIVE APPROACH

An alternative approach was to accept that there would have to be a large element of prescription, due to the time constraints of the course and its relationship to the other courses, but to achieve relevance by basing the course on a real consultancy project of which the institution had experience. The original problem would be used to spark off thinking about modelling, and ideas would then be channelled in a predetermined direction. This removes student autonomy, but is not totally artificial because the direction taken would follow one that had already been developed by the consultants on the job. Having explored the problem and developed ideas about how to formulate it mathematically, it could be replaced with a smaller, simplified version which could be used to teach particular techniques. It would then be possible to show the limitations of the approach, and to add extra features which demand more sophisticated models.

3. INITIAL STEP : GENERAL DISCUSSION OF MODELLING

It could be expected that engineering students would have sufficient background to be able to discuss general ideas about modelling, and transfer these ideas to mathematical modelling in particular. A fairly brief introduction to general ideas was given, which emphasised that modelling was not an abstract activity but had to be related to a well-defined purpose. In terms of business operations, the purpose suggested was to aid decision-making towards a corporate goal. Further discussion took place about the structure of a mathematical model of a business activity. The emphasis was on an optimising model, and the concepts of objectives and constraints.

4. THE ORIGINATING PROJECT

The second stage of the course was to present a real problem for consideration. The project used was one that had been carried out the previous year, for the Sussex Division of the Southern Water Authority, by two members of the Mathematical Sciences Department and an industrial training student (Gostick et al, 1989). Briefly, the project had been to produce a computerised tool to assist in the planning of facilities to dispose of residues from the treatment of sewage waste.

The Sussex Division of the Southern Water Authority is responsible for sewage disposal throughout a large area of southern England (figure 1). Sewage is disposed of in two ways: in some coastal towns it is discharged through outfalls to sea after preliminary treatment; in all other areas it is collected in sewage treatment works of which there are 184 in the Sussex Division. These treat the sewage in a variety of ways to produce clean water and a residue of sludge (a suspension of up to 7% solid material) which requires further processing. However, processing equipment in the Sussex Division is located at only 23 of the 184 treatment works, and sludge from the other 161 satellite works has to be pumped on to road tankers and transported to these treatment centres.

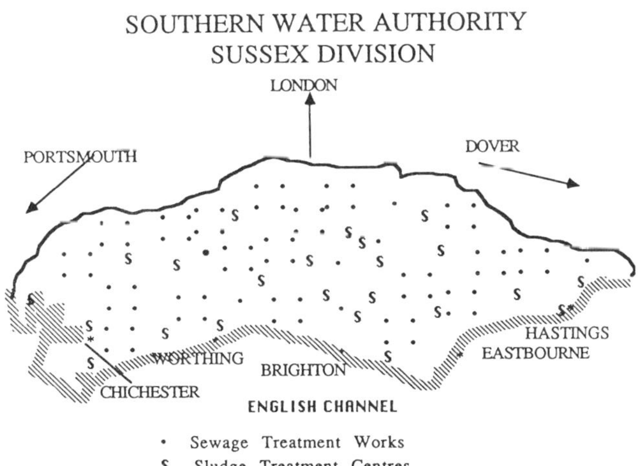

Figure 1

At the treatment centres there are a number of different ways that the sludge is processed, some of which may impose constraints on the quality of the sludge which they treat. In addition, there are limits on the proportions of industrial wastes that can be treated, and overall capacity limitations on the plant. Tankering costs depend on the mileages involved, the labour agreements, the size of tanker; in addition, there are the operating costs of the treatment centres, which vary considerably from centre to centre. A final stage is the disposal of the treated, dried sludge. Some is offered to farmers for land fertilisation, and some is stored on the Authority's own land close to the sludge treatment works. The availability of disposal sites, and their cost, is a significant factor in the operation.

The aim of the Division was to minimise the costs of sludge treatment and disposal, within the constraints that existed. It already had a plan which identified where the sludge from each satellite works should be taken, but this took several months to draw up by hand. What the Division wanted was a tool to optimise the present arrangements, and provide a means of planning for the future.

5. INITIAL STUDENT ACTIVITIES

An outline of the problem on the lines of the previous section was presented to the students, and they were given a week to think about it. At the next lecture, a brain-storming discussion took place about how such a problem could be tackled. There was considerable insecurity on the part of the students, who had never been faced with anything like this within the area of mathematics. However, in the course of the discussion it was possible to identify the objective, the key parameters, and the constraints, and to write down equations expressing these algebraically. The difficult part was not the algebra, but the dissection of the original problem.

The starting point was the need for a schedule that coped with tankers of a given size, based at a limited number of depots, and driven by drivers who had working agreements that governed mileage, hours of work and bonus rates.

The first simplification was to treat the transfer of sludge as if it were a continuous process through pipelines. Having done that, it was hoped to return to consider the problem of scheduling the individual tankers as a perturbation of the original solution.

The second simplification was to include only the constraints that were to do with the capacities of the treatment works and the production levels from the different satellites.

A third assumption, agreed with the students, was to treat the costs as linear with respect to the quantities transferred. The set of equations that were agreed were the following.

Minimise

$$\text{Cost} = \sum_{i=1}^{23} \sum_{j=1}^{184} c_{ij} x_{ij}$$

where c_{ij} is the unit cost of transporting sludge from satellite works j to sludge treatment centre i, subject to the constraints

$$\sum_{j=1}^{184} x_{ij} \leq k_i \qquad 1 \leq i \leq 23$$

$$\sum_{i=1}^{23} x_{ij} = p_j \qquad 1 \leq j \leq 184$$

where k_i is the capacity of sludge treatment works i and p_j is the production level of satellite j.

No further steps were taken at this stage to solve the originating problem.

6. A TEACHING VERSION OF THE PROBLEM

We then turned to a much smaller and simplified model problem, involving sludge from seven districts being tankered to three treatment works, and a dump which could be used to take any excess. Capacities of the various plants were known, and it was possible to calculate the production of sludge from the population figures for each district.

These are shown in the table below, where 50g of sludge were assumed to be produced per person per day, (in fact 75g would have been a more accurate figure).

Brighton Sewage

District	Population	Sludge Treatment Centre	Capacity (kg/day)	Treatment Cost (£/kg)
A	30000			
B	20000			
C	40000	1	4000	1
D	60000	2	2000	2
E	34000	3	4000	1
F	10000	dump	as required	0
G	16000			

The transport costs (£/kg) for transporting sludge between the satellite works and the treatment centres are shown below.

	A	B	C	D	E	F	G
Plant 1	5	3	4	6	7	3	2
Plant 2	3	1	4	6	9	3	6
Plant 3	6	5	6	3	4	2	8
Dump	6	6	6	6	6	6	6

Students were provided with copies of the tableau (figure 2) and this was used to demonstrate how a minimum-cost solution could be obtained using the well-known transportation algorithm (for example Anderson et al, 1988). The model was then used to investigate the effect of changes in the problem, just as might happen in practice. For example, if a detour became necessary on a particular route, and this put up the tankering costs, what effect would this have on the optimum solution? How much was it worth spending on enlarging a particular treatment works in terms of the effect on the costs of tankering, treatment and disposal?

	A	B	C	D	E	F	G	sludge capacity(kg)
1	6	4	5	7	8	4	3	4000
2	5	3	6	8	11	5	8	2000
3	7	6	7	4	5	3	9	4000
DUMP	6	6	6	6	6	6	6	500
sludge production (kg)								

Figure 2

6.1 Student assessment

Assessment of this part of the course took the form of a management report, in which students were asked to examine a range of planning options and use the model to investigate them. This form of assessment set the problem in context, and balanced the modelling aspects with the need to communicate with users.

6.2 Extension of the model

Other changes can now be introduced into the simplified model. For example, it may be necessary to use small expensive tankers to service one satellite, where local residents have objected to large vehicles. This could lead to trans–shipment from small to large tankers at another satellite works. Constraints on the quality of the incoming sludge may now be brought back into the problem. These constraints are a convenient way to introduce linear programming as a more general technique. How this is dealt with depends on whether it is intended to pursue hand–calculated solutions, for example using the simplex method, or to rely on a computer package. In practice, it has been found better to go straight to the computer at this stage, since the output of a linear–programming package provides plenty of scope for interpretation of the model, whereas the mechanics of solving the linear programme will tend to distract from the original problem.

6.3 Assessment of the success of the approach

Student response to this way of introducing modelling was very favourable. It was seen as obviously relevant to engineers and the combination of real problems with smaller ones that students can handle was seen as helpful. The students did flounder at the initial exposure to the problem, but this is probably no bad thing provided they are not left in that state for too long! There can be problems in maintaining the consistency of the approach, but no fundamental difficulties were encountered.

7. CONCLUSION

Whilst the approach described above derives from a particular set of teaching constraints and the availability of a suitable case study, the principles are more widely applicable. Clearly the number of different techniques that can be dealt with by this approach, in a single course, is limited, but this is more than made up for by the integration that is achieved, and by the increased motivation of the students.

REFERENCES

Cornwall M and Schmithals F. (Eds). (1976). What is project–orientation? – an overview. *Proceedings of the International Seminar on Project–orientation in Higher Education in Science.* University of Bremen.

Gostick C, Parramore K, and Wilsdon CE. (1989). A problem with sludge – A transportaton method becomes an effective planning tool for Southern Water. *OR Insight*, Vol 2 Number 3, July–September, 3–5.

Anderson DR, Sweeney DJ, and Williams TA. (1988). *An Introduction to Management Science.* West Publishing Company, 5th Ed.

CHAPTER 36

Industrial Enhancement of Group Activity

JR Usher, DG Simmonds and SE Earl
Robert Gordon's Institute of Technology, Aberdeen, UK

SUMMARY

RGIT's successful tender for Enterprise Initiative Funding (EIF) from the Government's Training Agency, in August 1988, is intended to facilitate the development of student enterprise skills on all courses in the Institute over a five year programme. In the first year of the programme five courses were targetted; one of these courses was the BSc in Mathematical Sciences with Computing Degree (MSC). In particular, the specific targetted areas on the degree were the Mathematical Models and Methods (MMM) course and a new innovatory Honours Industrial Group Project (HIGP).

It was fortuitous that, in the academic session before the EIF took place (1987/88), the course team had been developing an Honours extension to the degree with an Industrial Group Project as a central part of the course, being a natural development from the MMM component of the degree course. Thus, it was quite natural to embed this development in RGIT's tender to the Training Agency. Also, many of the group modelling exercises (GMEs) given to students on the MMM course had arisen from consultancy projects undertaken by members of the course team, and the EIF provided an opportunity for direct employer involvement in the design, delivery and assessment of the GMEs as well as the new HIGP.

1. HISTORICAL PERSPECTIVE

The BSc Degree in Mathematical Sciences at RGIT was initially approved in April 1974, with the first student intake to the course in October of that year.

In 1984, student group activity was introduced into the degree via a new Mathematical Models and Methods (MMM) course, which contained both a lecturing component and a Group Modelling Exercise (GME) component. The initial development of this course was described in a paper presented at the Second International Conference on the Teaching of Mathematical Modelling (Usher and Earl, 1987). Subsequently, an Honours extension was approved in 1988 together with a change in title of the degree to Mathematical Sciences with Computing (MSC). The focal point in the Honours extension is the Honours Industrial Group Project (HIGP) which is seen as a natural development of the MMM course. The major innovation of the HIGP is the full involvement by industry in *all* aspects of the operation of this component of the degree – from the joint preparation of material, through briefing of students on the unresolved problems, to monitoring student progress and joint final assessment.

2. DEVELOPMENT OF SKILLS FOR AN EFFECTIVE CONSULTANT MATHEMATICAL SCIENTIST

An effective consultant mathematical scientist should possess some technical knowledge of all areas in the mathematical sciences – mathematics, computing, computational methods, numerical analysis, statistics and operational research – together with a deeper knowledge in selected areas. A mathematical scientist must be able to apply this knowledge to solving problems in order to obtain practical solutions, that is possess modelling skills. In particular, nowadays, a consultant mathematical scientist must be able to operate in a group situation, where group dynamics and communication skills are of paramount importance.

RIGT staff have developed the MSC Degree with the primary intention of producing an effective consultant mathematical scientist.

We now focus our attention on the MMM course and HIGP components (see figure 1).

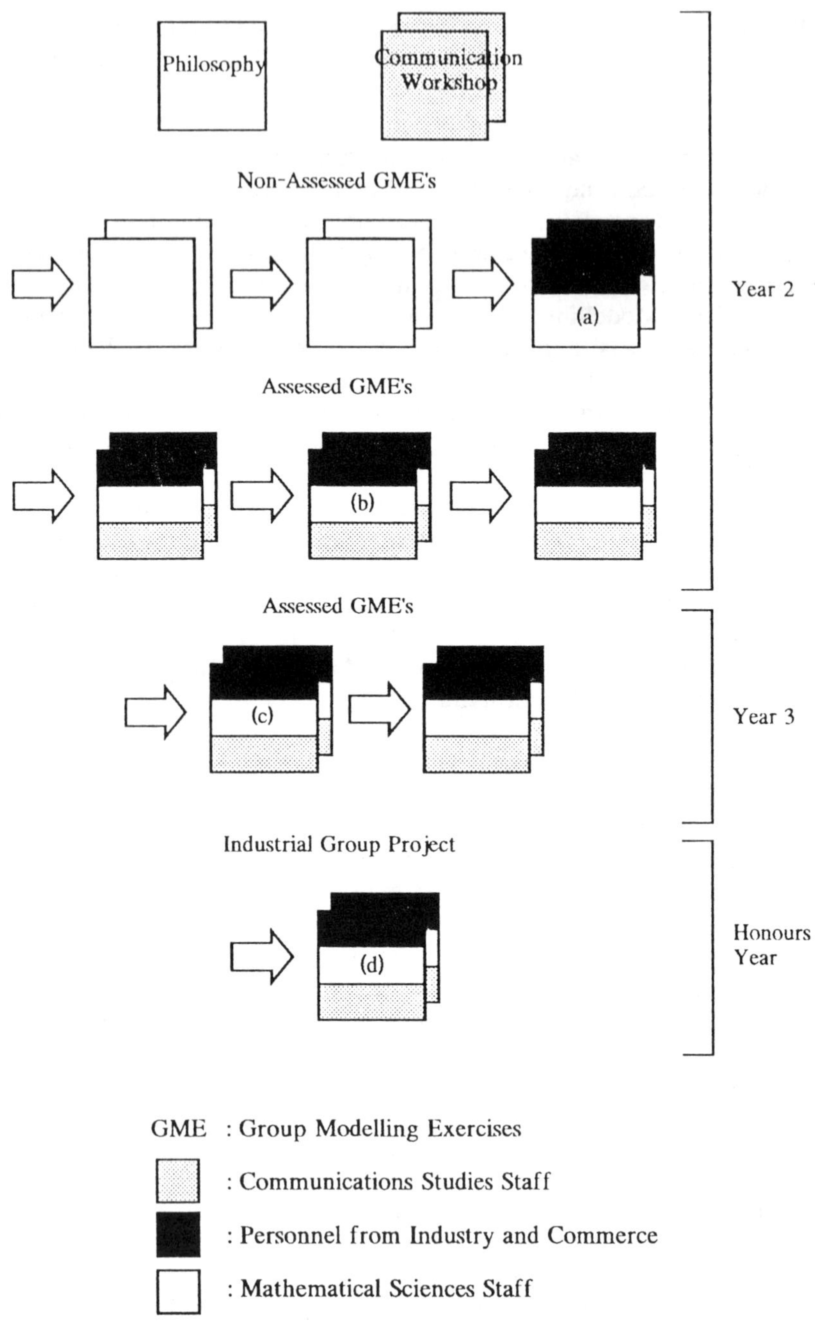

Figure 1
Team-teaching of MMM and Industrial Group Project

(a) MMM course

The MMM course focusses on the development of interpersonal, modelling, oral and written presentation skills. The operation of the course involves team teaching. Until 1988 the team consisted of staff from the School of Computer and Mathematical Sciences, Communications staff and RGIT library staff. For details of the course operation see Usher and Earl, 1987. Since October 1988, with the aid of EIF, industrialists are also members of this team. Fuller details of the industrial involvement will be presented in section 4.

(b) Honours Industrial Group Project

- In years two and three of the degree there is an emphasis on group problem solving, together with the fostering of communication skills, within the MMM course.
- The main aim of the Industrial Group Project in the Honours year is to provide the student with an opportunity to apply the skills of modelling and group problem solving to *real* problems arising in industry and commerce in order to achieve practical solutions.
- Suitable projects are generated by personnel from commercial and industrial organisations.

The HIGP is spread over 16 weeks.

- Each project comes from a customer in industry who is closely involved in the setting-up, operation and assessment of the project.
- Each group has a staff supervisor who monitors and may direct the work of the group.
- Each group produces three written reports during the 16 weeks.
- Each individual submits a written report at the end of the project (as well as the final report of the whole group).

3. ASSESSMENT OF GROUP WORK

There are four assessment categories

(i) Group activity (GA)
(ii) Oral presentation(s) of the group (OP)
(iii) Written report(s) of the group (WR)
(iv) Written report of the individual, Hon year only (IR).

Assessment is by consensus of the supervisory staff, including consultants from industry, who assign base marks for the group in categories (i), (ii) and (iii) and individual marks in category (iv). Individuals in the group are given marks by their peers in categories (i) and (iii). By combining peer marks with base marks, individual marks for categories (i) and (iii) are obtained. Thus, in all but (ii), an individual mark is obtained. For fuller details of the assessment scheme see Usher (in press).

4. INDUSTRIAL INVOLVEMENT IN GROUP ACTIVITY

RGIT's five year programme under EIF is made up of the following three-element model.

(i) Development of transferable skills

- interpersonal skills, including working in teams
- communication skills
- task management skills, including planning, organising, decision-making, monitoring and evaluation
- problem-solving skills and creative thinking

(ii) Work applications (ie application of transferable skills)

Here transferable skills are applied within the framework of problem-based learning projects - for example casestudies and projects developed in consultation with industry with *employer-related data* being adapted to achieve particular learning outcomes.

(iii) Work experience

Here enterprise skills are developed and applied to live industrial projects.

The targetted activities on the MSC degree focus on the second and third elements described above. In years two and three the GMEs focus on element (ii) while in the Honours year the HIGP focuses on (iii).

A description of industrial involvement in years two, three and four of the degree for the academic session (1988/89) is now given (see figure 1). In all cases, except where stated, the industrial sponsors of the problems acted as clients, providing appropriate data and background information for the students, and then subsequently attended the students' oral presentations and were involved in assessment.

(a) Year 2 GMEs

Introductory GMEs : Consultant from an international software company presented training videos on Report Writing and Oral Presentations before students were given exercises developed by academic staff.

GME 1 : Engineer from Roads Department, Grampian Regional Council introduced students to Preparation of a Proposal Document, before presenting them with various problems.

GME 2 : Technical Manager from an international papermaking company took students on a conducted tour of company premises before presenting them with various problems.

GME 3 : Engineer from a company involved in the use of sonar and radar equipment for subsea survey work provided problems and data which academic staff, involved in consultancy work with this company, adapted for student use; the academic staff then took on the role of clients.

(b) Year 3 GMEs

GME 1 : A Human Relations Manager and the Cost and Accounts Manager from an international oil company were fully involved with the students in the simulation of the operation of an oil company using a software package (the Bruce Oil Management Simulation) developed in-house. (Details of this package are available form the authors).

GME 2 : Students *collectively* selected one problem from a portfolio of problems provided by various companies; the originator of the problem selected by the students then acted as the client.

(c) Honours Industrial Group Project

Portfolio of problems provided by various companies; those from BP and DAFS - Marine Laboratory were selected by teaching staff.

The Head of Management Informations Systems, BP, and the Head of Sonar Section, DAFS, attended the sessions and were fully involved in

- initial negotiation with course team
- preparation of material
- briefing meetings with course team
- briefing of students

- supplying data
- accessible as client throughout
- monitoring student progress
- final assessment.

Both companies expected to implement solutions or suggestions proffered by the student groups.

5. EXAMPLES OF GMEs AND HIGP

The problems that follow illustrate how industrial involvement develops through the course. They are taken from four consecutive stages (see figure 1).

(a) Road Survey problems : Year 2

The non-assessed GMEs are intended as practice problems and are of a relatively short duration, some of them being designed to last an hour or two only. The last of these non-assessed GMEs is a two-week exercise carried out in collaboration with the Roads Department of Grampian Regional Council. In this exercise groups are provided with problems associated with the collection of road-survey data. These problems have included the need to assess the levels of traffic flow in a city centre or through a rural town requiring a by-pass, with the minimum of cost and disruption to the public. The students have to recommend methods for obtaining the data. Each problem concerns a survey which either has been done or is about to be done by the Council. An introductory talk by an engineer from the Roads Department precedes the exercises and the same engineer listens to the group's presentations and helps with the subsequent debriefing.

Each group was given a map of the area to be surveyed for traffic usage and a budget to cover survey costs. There are a number of ways of doing a survey.

- Hiring a policeman and a car for a day and using him to stop some of the traffic.
- Using recorders (human) to note number plate codes.
- Handing out a questionnaire.

The modelling process involved developing a survey proposal which would, in the group's view, provide the information needed, involve minimum disruption to the public, and fall within budget.
The involvement of an engineer from outside RGIT providing easily accessible problems has resulted in much higher levels of motivation and presentation in the groups. In future years it is intended that,

at a later stage in the second year, groups will be involved in the actual collection and analysis of road–survey data, again in collaboration with the Regional Council. The engineer also presented sample tenders, so students realised the standard expected.

(b) Quality Control Problems : Year 2

The second set of assessed group exercises in year two of the course is devoted to statistical problems. When first instigated these exercises usually arose out of staff experience in consultancy work for industry, but in recent years more direct industrial involvement in the actual presentation and assessment of the problems has been sought. Wiggins–Teape, a local papermaking company, has provided problems associated with the quality control of paper manufacture, problems of maintaining acceptable consistency in paper weight, thickness, and colour. Managers and engineers from the company act as clients, supplying production data for analysis, and assessing student presentations and reports. Before the start of the exercise the whole class is taken on a visit to the company to observe the manufacturing process and absorb technical background to the problems.

Four specific problems were involved as follows.

- Investigation of the quality of paper based on 180 samples, size 5 of grammage (g/m^2). The actual industrial specification limits were supplied. Control charts of means and ranges were used in the solution.
- Comparison of strength of paper before and after summer shutdown, data provided for burst strength of 12 different paper types (pre– and post summer), hypotheses testing (parametric and non–parametric) used.
- Investigation of relationship between downtime on paper machines and effluent flow–rate, on–line process control information being supplied for five machines over 26 weeks. Time series and regression analysis used.
- Test the effectiveness of a formation calibration meter, samples of paper with corresponding meter readings are supplied. Samples assessed by ranking with respect to formation (blotchiness) using number of 'judges' and results compared to meter readings using rank correlation, coefficient of concordance.

Each group produces a written report on its findings and also makes an oral presentation. When assessing the presentations, representatives of the company and academic staff subject students to rigorous questioning. At this stage it does not matter if the answers are incomplete. The aim is to familiarise students with systems of appraisal used by commercial companies.

(c) Bruce Oil Management Simulation : Year 3

In the first of the GMEs in year three all groups are involved in a computer-based simulation, developed at RGIT, of the management of an offshore oil field. All groups start with the same amount of capital and the same number of developed wells. The aim of the simulation is to develop the best strategies for the bringing-on of new wells, the refining and delivery of existing oil production, borrowing, investment, and so on. Groups effectively compete with each other to finish with the highest level of reserves.

The students are provided with a detailed scenario of production, transportation and processing of oil and gas from the North Sea. They are encouraged to use spreadsheets to perform the economic calculations related to costs and transportation logistics before participating in the simulation; they are thus well armed with data necessary for decision taking when simulating running their own oil company.

For the last two years a Budgets and Costs Supervisor from Marathon Oil has been involved with this exercise, giving students some background to the oil industry, helping the groups get started on the simulation, and assessing presentations. Marathon personnel say they enjoy the overview of the industry, and academic staff and industrialists have found that their levels of expectation are similar. They home in on the same weaknesses in students' arguments, and they rarely demur over the acceptability of particular points of presentation style.

(d) Database Problems : Honours Year 4

Students in the Honours year of the MSC devote one afternoon per week for sixteen weeks to their HIGP. In small groups they tackle a problem from a portfolio sponsored by local industries.

In the HIGP the students must organise weekly meetings, properly chaired and minuted, and make interim and final reports and oral presentations. A member of the academic staff acts as a supervisor to each group, but the clients are from industry and as such they play a major part in the exercise. They have attended some of the students' meetings, they obviously take part in the assessment of written and oral reports, they are available to receive visits from the groups on their own premises, they provide resource data and they answer telephone queries. Students learn to respect their clients' workload and to value time.

One of the problems presented usually involves the design of a database. For example, BP, an oil major employing a large number of contract workers needed a database to hold information on these contractees.

A data model was set up showing the four tables containing
- agency details
- contractee CVs
- follow-up and interview details
- work or contract details

and how they were related.

This model was then transformed into a relational database model and complemented using INGRES.

The groups involved met regularly with representatives from the company. After the initial meeting the group produced an initial user specification which was modified at later meetings. Their initial functional design was presented to the BP representatives in the form of an interim report and presentation five weeks after the initial meeting. The system design was then finalised and implementation started. The group did not complete all the implementation in the time available, but were judged to have done a good piece of work which the company could refine.

It is intended to continue with industry-sponsored database problems. At the time of writing, another oil major is sponsoring a database design problem.

6. CONCLUSIONS

In general terms, the development of student group activity and the industrial enhancement of this activity on the MMM course and the HIGP has been successful in stimulating both students and staff. A major problem encountered in the first year of operation of the HIGP was that the students spent so much time in trying to achieve a highly polished final product for their industrial clients that their other studies were affected in the half of the session whilst the HIGP was in operation. The course team have been investigating ways of ameliorating this situation for the future.

The success of the group activity on the MSC Degree has convinced RIGT that academics and industrialists can work together on courses and has led to the inclusion of a similar component in the Computer Science Degree that the School operates and also in a new Advanced Diploma in

Computing which started in October 1989.

7. ACKNOWLEDGEMENTS

The authors wish to thank all members of the teaching team who have helped to develop and to operate successfully the MMM course and the HIGP at RGIT, together with the industrialists who have willingly provided their time and effort this year.

REFERENCES

Usher JR and Earl SE. (1987). In Berry JS *et al* (eds), *Mathematical Modelling Courses*. Ellis Horwood Ltd.

Usher JR. (in press). Development of a Staff and Peer Assessment Scheme for Group Work in Mathematical Modelling. *IMA Journal of Teaching Mathematics and its Applications*.

CHAPTER 37

The Integrated Microcomputer Software Environment for the Support of Mathematical-modelling Teaching

RR Clements
University of Bristol, UK

SUMMARY

In the teaching of mathematical modelling, we often incidentally place students in the position of needing to undertake extensive algebraic manipulations and numerical computations. Modern microcomputers offer a powerful resource to carry out these functions in support of the teaching of mathematical modelling (and indeed of other areas of mathematics). This paper attempts to establish the basic requirements of a software environment for the support of modelling and other mathematics teaching, comments on the extent to which currently available software meets these requirements, and suggests directions for future developments.

1. COMPUTATIONAL SUPPORT FOR THE TEACHING OF MATHEMATICAL MODELLING

The author's experience of teaching mathematical modelling includes many year's use of the case study exercises which are described in, for instance, Clements (1978a, 1978b, 1982, 1984a). In these exercises students must extract, from a description of a real-world problem, a mathematical formulation of that problem, solve the mathematical problem, re-interpret the mathematical solution into real-world terms and then evaluate the solution in those terms. The essential modelling activities are those contained in the first, third and fourth stages of this process - the solution of the mathematical problem is not really an exercise in modelling. Nonetheless, if students fail, totally or partially, at this stage they will usually obtain less experience, and certainly less satisfactory experience, in the third and fourth modelling stages. The author's observation is that, for a majority of students, their lack of facility in algebraic manipulation, numerical analysis and computer programming often presents a significant obstacle to satisfactory completion of the second stage of the modelling process, and therefore to their learning from the casestudy exercises.

One possible solution to this problem is to base mathematical-modelling teaching around exercises which are sufficiently restricted in scope that the models created are simpler to solve. Such an approach is less than entirely satisfactory, because the problems are often thereby rendered unrealistic or unconvincing. If sufficient assistance with the mathematical skills is available, it is not necessary to restrict the scope of the problems to be modelled; the modern microcomputer can offer this assistance. When the casestudy course was first developed and used, in 1978, microcomputers had barely been conceived. Such numerical work as was necessary had to be done by programming in BASIC (and latterly, in Pascal) on a mainframe computer. The introduction, over the years, of a variety of readily-available, microcomputer-based, computational aids has greatly enhanced the students' capabilities and possibilities. Microcomputers can now offer support facilities which enable students to undertake more complicated numerical and algebraic manipulations quickly and accurately and, through these facilities, microcomputers can significantly enhance the teaching of mathematical modelling. Of course, the advantages which microcomputers offer in assisting with quick and accurate numerical and algebraic manipulation can also be applied to the teaching of mathematics in other areas.

This paper is largely devoted to a discussion of the support facilities which are desirable for both mathematical modelling courses and other courses in applicable mathematics, and to a review of the microcomputer systems and software which are available, and how they meet the requirements.

2. HOW SHOULD COMPUTERS SUPPORT THE TEACHING OF MATHEMATICS AND MODELLING?

The traditional role of computing in mathematics courses is as an adjunct to the teaching of numerical analysis. In order to achieve a proper understanding of the theoretical concepts, students need experience of practical numerical problem solving. Such appropriate, practical experience has, in the past, needed the power of at least minicomputers, and the use of conventional high-level programming languages such as Pascal, FORTRAN or BASIC. Mastering such languages requires significant time and effort from the students, and their lack of skill and maturity in programming usually becomes a significant obstacle to the full realisation of the learning potential of computational approaches to numerical analysis. An easily-learnt programming system operating at a higher level than conventional high-level languages would, therefore, be a considerable asset to mathematics lecturers.

In many areas of mathematics, understanding of the theoretical ideas can be considerably enhanced by the use of pictorial or diagrammatic approaches. Teaching the techniques of curve sketching is partly motivated by the need for mathematics students to be able, quickly, easily and reliably, to construct sketches of functions and other mathematical entities. The development, as computer terminal devices, of visual display units with graphical capabilities, and the parallel development of graphical programming aids such as the Gino library and similar systems, has enabled students to use computer assistance in the construction of graphical representations of mathematical ideas.

The construction of mathematical models of dynamical systems often gives rise to differential equations. Considerable enhancements of the understanding of such models can be achieved if students have a qualitative grasp of the behaviour of the solutions of differential equations of various types. Again, such understanding can be enhanced by the use of computational assistance with the solution of differential equations. If this makes the simple and rapid solution of a wide variety of differential equations possible, students can gain broad experience and an intuitive feel for the properties of various classes of differential equations. Such assistance can be provided by use of computer-programming skills, allied with the use of graphical means of displaying the solutions.

Much of mathematics teaching and learning relies on a certain basic manipulative skill in the learner. It is a common experience of mathematics teachers that the students' lack of manipulative skill and experience is an obstacle to their understanding of basic mathematical ideas at a deep level. The unskilled learner is handicapped by the need to concentrate on the mechanics of the manipulation, this need distracting attention from the understanding of the underlying mathematical ideas. The development of the hand-held calculator has helped to alleviate a similar problem in respect of basic numerical computations. If routine algebraic manipulation could be similarly delegated to a simple and accurate mechanical aid, attention could be directed more fully to understanding the underlying mathematical ideas.

In summary, the teaching and learning of mathematics, and particularly of the ideas of mathematical modelling, can be significantly enhanced if students have available to them four basic computing aids.

(a) A simple and high-level system for carrying out numerical computation
(b) A system for presenting graphical representations of functions and other mathematical entities
(c) A system for solving differential equations with minimum computational effort
(d) The algebraic equivalent of the pocket calculator, a simple-to-use, accurate system for carrying out routine algebraic manipulation

3. WHAT SORT OF SOFTWARE CAN FULFIL THESE NEEDS?

Traditionally the facilities identified above have been provided by teaching a high-level computing language on time-shared mainframe or minicomputers, together with appropriate use of software backup such as the Gino library for graphical operations and the NAG library for some numerical procedures. The cost, however, in terms of effort on the part of the students, is often considerable, and difficulties with the computer language (and also, often, the computer's operating system) tend to obscure the actual mathematical objective of the programming activities. Are there better ways of providing the facilities to back up mathematics teaching than through a conventional high level language? The author's experience suggests that indeed there are.

The advent of desktop microcomputers and the rapid growth in their processor speeds, memory sizes and other capabilities, has meant that many of the facilities previously provided on a time-shared basis can now be provided by microcomputers on a single-user basis. At their best, single-user microcomputers are much easier to access and more user-friendly than time-shared systems. The programmer-support environment provided by microcomputer implementations of common high-level computer language compilers and interpreters is often enormously superior to that provided on mainframe computers. The Turbo Pascal compiler (which, in its latest versions, is more like a simple IPSE than a mere compiler) is a good example of the sort of comprehensive programming-support environment which modern microcomputer systems can provide.

The provision of more advanced facilities for traditional high-level language programming is not the only advance which has come in the wake of the desktop microcomputer. Mainframes and minicomputers have provided some packages partially meeting requirements (c) and (d) above. Simulation languages, such as CSMP, CSSL IV, ACSL, and Dynamo, have provided systems for solving differential equations more readily than by programming in high-level languages. Mainframe systems, however, lacked the flexibility and interactive capabilities which have arrived with the development of equivalent languages for microcomputers. The advent of the TUTSIM system, on the Apple microcomputer and later on MS-DOS computers, and the BCSSP system, for the BBC microcomputer and for MS-DOS computers, has brought cheap, interactive simulation with graphical capability within the reach of microcomputer users. Moscardini, Cross and Prior (1984) and Clements (1986a, 1986b) have described how simulation systems can be integrated into the support of the teaching of mathematical modelling.

The requirements of (d) above are met, at least partially, by symbolic manipulation systems such as Macsyma, Reduce, SMP and Maple, available on mainframe computers. These systems, with the exception of the Maple program, were not designed primarily for student use, and leave something to be desired in the user interface. However, with the growth in power of microcomputers, symbolic manipulation systems designed specifically for these machines have arrived, and these show much more promise for smaller scale educational use. Barrozzi and Clements (1987) have commented on the desirability of introducing computer algebra systems into the education of engineering students.

Whereas systems meeting requirement (c) and (d) were available on mainframes, and the advent of microcomputing has merely made these sytems more user friendly and more accessible, it is in area (a) that microcomputers have brought about a real revolution. The microcomputer stimulated the development of the spreadsheet - originally targeted at accountants and business analysts, but now recognised by many other types of user as offering considerable advantages (for instance Orvis (1987) pointed out their potential for scientists and engineers). In mathematics they have great potential as a form of high-level programming system for numerical computations; they can readily be harnessed to provide the quick and easy numerical computing environment which is needed for the teaching of numerical analysis and the exploration of many mathematical models.

Only in the case of requirement (b) have systems dedicated to meeting the perceived need not arisen. This is not entirely surprising, when it is recognised that the graphical facilities provided by most simulation languages and spreadsheets can be used to meet the requirement, and that most microcomputer symbolic manipulation systems also provide graph plotting as a subsidiary facility.

Thus, we see that the requirements identified in section two may be met as follows.

(a) spreadsheets
(b) spreadsheets, simulation systems and, usually, computer algebra systems
(c) continuous system simulation languages
(d) symbolic manipulation (or computer-algebra) systems

4. WHAT SORT OF MICROCOMPUTERS SHOULD BE USED?

The author's first attempt to design and teach an undergraduate course in which use of a computer package was an integral part, was a course in systems modelling described in Clements (1986a). The course was originally intended to be based around the CSSL IV mainframe simulation language, but it was quickly apparent from pilot usage that the language, designed with industrial applications in mind, was too complicated and the whole system too unreliable to be practical for undergraduate use. The problem alluded to above, that students' unfamiliarity with high-level languages obscures the main learning objective, was equally apparent with the rather complex CSSL IV language and run-time system.

When the TUTSIM system appeared, completely interactive, offering a straightforward user interface, good graphical output capabilities and based on the Apple microcomputer, it was immediately apparent that it was better oriented to the needs of undergraduate students. Unfortunately the author's university had just standardised on the BBC microcomputer for its first generation of teaching microcomputers, so he designed and wrote a similar simulation system, BCSSP (Clements 1984b, 1985), to run on that machine. The combination of interactive operation, carefully-designed user interface, graphical output and microcomputer basis has made the package highly acceptable to students, and very successful in practice.

The systems modelling course has made use of the BCSSP package and the BBC microcomputer for a number of years. The experience of the author, and the other academic staff involved, has been that students are freed, by using a package which is quick to learn and easy to use, from undue concern with the means and mechanisms of simulation, and are therefore able to concentrate more readily on the mathematical issues which the simulations are designed to illustrate.

The introduction of the BBC micro as a low-cost graphics-terminal substitute meant that the author's Department was able to commission a small microcomputer laboratory in 1983 and gradually to enhance it subsequently. The experience, and the success of the systems modelling course, encouraged the Department to consider other uses of microcomputers. The ideas for the use of spreadsheets in support of numerical analysis and other courses were developed, but it was soon apparent that the BBC micro's spreadsheet program was not really powerful enough for such use.

Experience has emphasised how important it is that, if computational support for mathematics teaching of all kinds is to be a reality, the facilities provided should be adequate for the number of students involved. If facilities are too scarce students rapidly become demotivated, standards of work on courses requiring computational backup fall, students avoid taking option courses which involve using the computing facilities, and the general level of student dissatisfaction with the course as a whole rises. The author's Department has a ratio of one microcomputer to every six students on its degree course. As the ambitions of the Department in the field of microcomputing grew, and the limitations of the BBC microcomputers became more restrictive, thought was given to their replacement. The possibilities, in early 1988, included the following.

Atari ST
Commodore Amiga
MS-DOS machines of PC/XT type
MS-DOS machines of PC/AT type
Apple MacIntosh
Acorn Archimedes
Sun workstations

The requirement, in the light of the perceptions outlined in sections two and three of this paper, was to obtain machines which would

act as a terminal emulator for access to mainframes,
host a Pascal compiler/interpreter,
host a spreadsheet,
host a simulation package,
host a computer-algebra package,

whilst, at the same time, providing as many student places as possible at as small a cost as possible.

The price range of the hardware mentioned covers from around £600 to more than £4000 and the software covers a similarly-wide price range. MS-DOS machines have the considerable advantage that a very wide range of software is available, much of which is simultaneously of good quality and reasonable cost. The Amiga and Atari options were discounted, partly because of the restricted range of options for software and partly because of the lack of local expertise for their support and maintenance. The MacIntosh is a very attractive machine with considerable power, the MacIntosh II being in the workstation class. It would, however, have been roughly double the cost of the low-end MS-DOS machines, and low-cost software is less readily available. The Acorn Archimedes is a powerful machine, again in the workstation class, but software is less readily available. A laboratory of Sun microcomputers, networked together and running the Unix operating system, would have been an ideal option, but would have meant settling for perhaps a quarter of the number of machines which could be obtained by adopting a cheaper solution.

In the event, the advantages of the *de facto* MS-DOS standard determined the choice. This was an 8MHz 8086-based MS-DOS microcomputer with 1Mb memory, Hercules monochrome graphics board, 8087 floating point coprocessor, 30Mb hard disc and single 5.25" floppy drive, at around £800. The possibility of an 80286-based machine with a similar specification, except for EGA colour graphics, at a cost of around £1200 was the most favoured alternative considered. The main argument in favour of this option was 'futureproofness', particularly the ability to run OS/2 (should that operating system ever become widely accepted!). However, in order to run OS/2, VGA graphics, considerably more memory (informed opinion suggests at least 2Mb and preferably 4Mb) and a larger hard disc would be needed. Futureproofness would therefore have cost considerably more, and restricted the number of machines obtained to an unacceptable degree. It was decided that a larger number of more basic machines would best fulfil the immediate needs. By the time VGA graphics, 80386 processors and OS/2 become necessary the cost of machines offering these will have fallen considerably, and the current machines will be ready for replacement.

In summary then, the choice of the hardware was governed to a considerable extent by availability of software. Formerly, choosing microcomputers for the support of mathematics teaching might well have involved ascertaining that a good FORTRAN, Pascal or BASIC compiler was available, and then selecting the most powerful microcomputer available. Having determined, however, that several categories of software are required complicates the decision considerably. The selection of MS-DOS machines was conditioned primarily by the perception of the wide range of software available in each of the categories. Choosing MS-DOS machines inevitably meant that raw computing power was probably going to be sacrificed, but this was acceptable in order readily to obtain the software. Having picked MS-DOS, there was still a wide range of machines, both in terms of manufacturer and level of machine, within any manufacturer's range. Our choice was to maximise the number of machines provided within a fixed budget, this being achieved by picking machines with just sufficient capability to run the software we had selected.

5. WHAT SOFTWARE IS AVAILABLE?

The best known spreadsheet is probably Lotus 1-2-3. This was not the first on the market, nor is it necessarily the best of the spreadsheet packages. It has, however, established a *de facto* standard which other spreadsheets have subsequently built upon. The major packages, like Lotus 1-2-3, Microsoft Excel and Borland Quattro, offer a vast range of arcane features, and support for a wide variety of displays and printers, at prices from £150 to more than £400 per copy. More recently, some companies have offered special deals for educational users - presumably on the theory, first developed by drug dealers, that if they can get students hooked on their products during their formative years, they will remain addicts for ever after! Whilst such deals are, for hard-pressed and cash-starved higher education institutions, a move in the right direction, they only alleviate the problem partially. For instance, Lotus 1-2-3, £395 to a commercial user, may be obtained for only £180 per copy by a UK university. Other companies, seeking a market niche, have stepped in and produced look-alike products, albeit with reduced capabilities and support for a more restricted range of graphics systems and printers, but still conforming to the *de facto* standard, at a much lower price. In the UK ExpressCalc and AsEasyAs are two such examples. AsEasyAs offers most of the functionality of Lotus 1-2-3 at 20% of the cost. For most educational use, such packages are perfectly adequate.

The range of possibilities in the market is much smaller when it comes to continuous system simulation languages. All the best-known packages in this field are mainframe packages, though some are beginning to appear in PC versions. Conversions, however, never seem to be quite as satisfactory as systems written *ab initio* for PCs. The basic, mainframe-oriented philosophy of these packages is often manifest in the PC equivalent as an awkward or unfriendly user interface.

The classic computer-algebra packages were also written originally for large mainframe computers. The pattern by which simulation packages, and other mainframe software, have been ported to microcomputers is typically linked to the continually growing processor speed and memory capacity of microcomputers. The Sinclair ZX81, the first microcomputer to gain any widespread currency in the UK, was equipped with 1Kb of RAM. The BBC microcomputer had 32Kb. The first IBM PCs had 128Kb of RAM and a processor speed of 4.66MHz. The current top end of the MS-DOS microcomputer market offers machines with 8Mb or more of RAM and processor speeds of up to 30MHz. Microcomputer versions of each type of mainframe package have tended to appear at such time as the memory sizes and processor speeds of microcomputers have reached realistic levels for the operation of the package concerned. Symbolic manipulation packages are traditionally recognised as memory and power hungry entities. It is for this reason that, whilst packages in other areas have been available in microcomputer versions for some time, the major computer algebra packages, Macsyma, Maple and Reduce, have not yet become so available. Reduce, it is true, has been ported down to Sun and Archimedes workstations but not to less capable machines. The newer Mathematica package has been developed, from its inception, for a wide range of computers, from near workstations like the Apple MacIntosh, through true workstations and right up to supercomputers. There is an MS-DOS version of Mathematica, but it has a much reduced graphical capability when compared with the implementations for the MacIntosh and the Sun machines and, even so, requires an 80386-based machine with at least 3Mb of RAM running, realistically speaking, at 20MHz or more. Such machines are more expensive than the basic £800 MS-DOS machines, mentioned in the last section, by a factor of two or more. The symbolic manipulation packages which have, so far, become available on lower end MS-DOS machines are the MuMath/MuSimp system and its successor, Derive. The capabilities of Derive are considerably less, both in terms of functionality and size of problem which can be handled, than those of Mathematica. On the other hand Mathematica for MS-DOS machines costs $700 or more, whereas Derive costs £105. For introducing students to the basic ideas of symbolic manipulation, and familiarising them with what can and cannot be achieved with these systems, packages at the level of Derive are, for the most part, perfectly satisfactory.

Summarising then, in many areas and particularly in the areas of interest to us here, there is a wide range of software offering equally-wide ranges of capabilities and facilities at considerably differing prices. It has been argued that the requirements of educational usage do not necessarily require the most elaborate features available, and wise choice of lower-priced software with adequate and appropriate facilities will often provide a cost-effective solution allowing larger numbers of microcomputers to be equipped within restricted budgets. For the author's microcomputer laboratory, the AsEasyAs spreadsheet, the in-house BCSSP simulation package and the Derive computer algebra package have been chosen. Additionally the computers have been equipped with Borland's Turbo Pascal and the Kermit terminal emulator. The total software cost is around £200 per machine, 25% of the hardware cost.

6. WHAT WILL THE FUTURE BRING?

It has been suggested that one possible solution to the problem of provision of computer resources on a large scale is that, on certain courses at least, students might be required to purchase their own microcomputers. Following discussions of this issue with colleagues from Eindhoven Technical University (Netherlands), the author undertook a survey of all second-year engineering students at Bristol University, reported in Clements (1989). This survey provides strong evidence that, were a single university to introduce such a requirement, the vast majority of students would chose to study at other universities, or to study other subjects, in order to avoid the expense of such a requirement. This possible solution seems, therefore, not to be practical in current circumstances, at least in the UK.

One of the drawbacks of the approach to the mathematical support environment which has been outlined in this paper is that it rests on a number of different packages. This presents a considerable information load for students as they learn to use the various packages. There are, of course, some elements of commonality; for instance, the panel-style menus of Lotus 1-2-3 are a fairly-widespread motif in MS-DOS software. In our case, the AsEasyAs spreadsheet and the Derive package share this style of menu presentation, though they are not entirely consistent in the finer details of their operation. This information load will limit the rate at which the whole environment can be introduced to the students. It would obviously be desirable to integrate the four elements of the support environment into a single package, with a completely consistent interface. Inevitably such a monolithic package would be quite complex, and take considerable time and effort to become completely familiar and competent with, but it should be an improvement over three separate packages. The Mathematica package does, perhaps, go some way towards providing this integrated environment.

What can be predicted, with some confidence, is that the power and memory size of microcomputers will continue, for some time to come, to increase rapidly and the price/performance index will continue to fall. Whilst the Mathematica package, and the machinery needed to run it, is currently beyond the scope of most UK universities to provide in quantity, it might be anticipated that Mathematica, or more likely one or more Mathematica clones will, within the next five years, fall in price to a point where they will become affordable. It is unlikely that the price of basic-level MS-DOS microcomputers will fall very far, but far more likely that the performance and facilities provided by machines at around the £1000 level will increase considerably. As memory sizes and processor speeds increase, it seems likely to the author that Unix rather than OS/2 will become the *de facto* operating system standard. This will bring about convergence, at least from the point of view of the user, between the microcomputer, the workstation and the multi-user minicomputer. For our students this must be a good thing.

The other factor which can be predicted, with a depressing degree of confidence, is that higher education institutions in the UK will continue to be starved of the resources which they need in order to operate to best advantage. The emphasis which, in this paper, has been placed on issues related to obtaining the largest number of student microcomputing places within a restricted budget will continue to be necessary and appropriate.

REFERENCES

Barrozzi G and Clements RR. (1987). The potential users of computer algebra systems in the mathematical education of engineers. *Int J Math Educ Sci Technol*, Vol 18, 681-683.

Clements LS and Clements RR. (1978). The objectives and creation of a course of simulations/case studies for the teaching of Engineering Mathematics. *Int J Math Educ Sci Technol*, Vol 9, 97.

Clements RR. (1978). The role of simulations/case studies in teaching the practical application of mathematics. *Bull IMA*, Vol 14, 295.

Clements RR. (1982). Initial experience in the use of simulations/case studies in the teaching of Engineering Mathematics. *Int J Math Educ Sci Technol*, Vol 13, 111–116.

Clements RR. (1984a). The use of the simulation and case study technique in the teaching of mathematical modelling. In Berry JS *et al* (Eeds), *Teaching and Applying Mathematical Modelling*, Ellis Horwood.

Clements RR. (1984b). BCSSP – User Manual. Internal Report, Department of Engineering Mathematics, University of Bristol.

Clements RR. (1985). BSETR – User Manual. Internal Report, Department of Engineering Mathematics, University of Bristol.

Clements RR. (1986a). The role of system simulation programs in teaching applicable mathematics. *Int J Math Educ Sci Technol*, Vol 17, 553–560.

Clements RR. (1986b). Mathematical modelling using dynamic simulation. In Berry JS *et al* (eds), *Mathematical Modelling Methodology, Models and Micros*, Ellis Horwood.

Clements RR. (1989). A survey of microcomputer ownership and attitudes amongst second year engineering students. University of Bristol, Department of Engineering Mathematics, Report EM/GR/5/89.

Moscardini AO, Cross M and Prior DE. (1984). On the use of simulation software in higher educational courses. In Berry JS *et al* (eds), *Teaching and Applying Mathematical Modelling*, Ellis Horwood.

Orvis WJ. (1987). *1–2–3 for scientists and engineers.* Sybex.

CHAPTER 38

Mathematical Models in Water Quality Control and Educational Aspects

B Barabas
Technical University, Budapest, Hungary

SUMMARY

The aim of this paper is to outline a few important stochastic models which play a basic role in water quality control. The first part will concern random variable models, and different methods to observe the changing water quality will be introduced. The second part will outline briefly our experience with stochastic process models, and educational aspects will be dealt with.

1. INTRODUCTION

A few years ago I was approached by certain engineers, whose special field is water management, and asked whether the water quality of the Danube river is better or worse than it was.

The engineers produced a great deal of data including water samples taken every day or every second day, from various points along the river. The concentrations of 15–20 ingredients were laboratory analysed, and included ammonium and nitrate. The chemical oxygen demand (COD) and biological oxygen demand (BOD) were also obtained. This amount of data enabled the question 'How can we decide from this data whether the water quality has changed or not?' to be asked.

The question is difficult and complex, with many problems such as how to draw the sample, and when, where and how many times a day or week.

We need to know the ideal state of the water – if the concentration of a certain chemical is too high it turns the water poisonous, but if it is chemical free (distilled water) then fishes die. A further problem is that we don't know the cumulative effect of different chemicals. If, for example, the concentration of one chemical increases and another decreases, is the water quality unchanged? It is also important to know about the changing water quality along the river, for we know the river is able to clean itself but we don't know all the conditions of this process.

2. MODELLING

These are examples of the many problems, but I would like to tackle one question to illustrate our methods. The question concerns how to observe changes in the concentration of characteristic components – a question which plays a basic role in water quality control. Different mathematical models will be presented and summarised via their educational aspects.

We have, for example, data from observations at the same place and at different times – weekly means of ammonium concentration for a given year – and we would like to know if the average concentration has increased, decreased or remained unchanged over the year. We must decide whether we want to use a deterministic or stochastic model; in this case stochastic models seem to be appropriate.

3. RANDOM VARIABLE MODELS I

It is convenient to consider the chemical concentration as a random variable, and we can therefore utilise statistical methods to detect the change of water quality. The basic condition of every statistical method is the data from a sample, in the sense that they are independent, identically-distributed random variables (IDRV) but, unfortunately, in most cases this condition does not hold. If the time span between two observations is short, we can make a strong connection between the data; increasing the interval means the dependence will decrease. We could justify constructing an empirical distribution function if the data were independent. In spite of this fact the examination of water quality is frequently based on empirical distribution functions. However, in many cases this procedure is acceptable, because we can draw correct conclusions provided the dependence is weak.

As an example, let us consider the monthly means of chemical oxygen demand (COD). It is easy to verify that the dependence between the monthly means is weak, and thus if we want to know whether this parameter remained unchanged or not in the last two years, we compare the data with those from the previous three years. Figure 1 shows the empirical distribution functions using real data of monthly means of COD for the years 1976, 1977, 1978 (function F) relative to the years 1979, 1980 (function G) for the Danube river at Baja city at the southern border of Hungary.

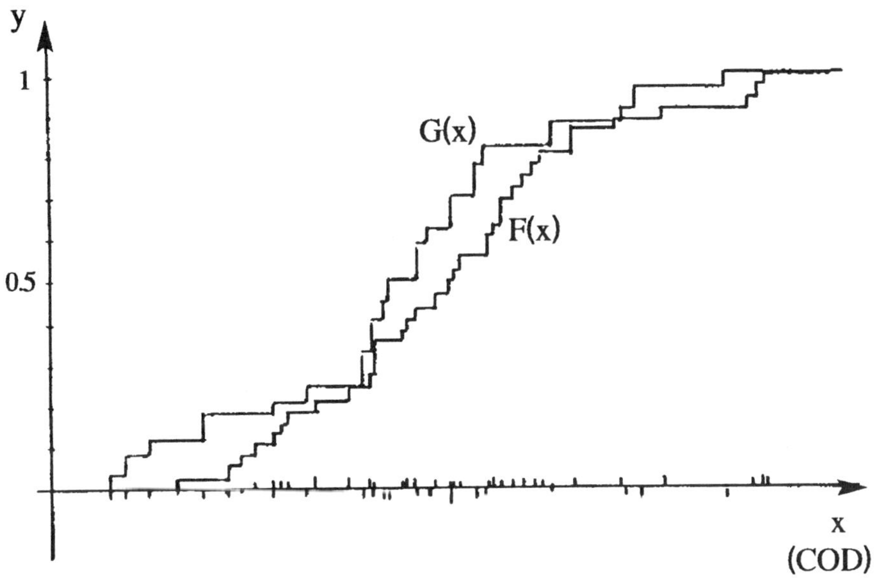

Figure 1

The x axis stands for the data. The graph of the function F(x) has an increment $^1/_{36}$ at every x_i (i=1,2,3,...,36) (indicated with marks 'above' the x-axis). Similarly the graph of the function G(x) has an increment $^1/_{24}$ at every y_j (j=1,2,...,24) (marks 'below' the x-axis). The data for the first three years are

x_1 = 19.87	x_2 = 19.87	x_3 = 24.77	x_4 = 23.55
x_5 = 22.88	x_6 = 15.00	x_7 = 18.33	x_8 = 19.66
x_9 = 16.88	x_{10} = 16.25	x_{11} = 20.55	x_{12} = 23.00
x_{13} = 29.22	x_{14} = 24.00	x_{15} = 17.55	x_{16} = 23.87

$x_{17} = 20.66$ $x_{18} = 21.66$ $x_{19} = 22.00$ $x_{20} = 21.33$

$x_{21} = 22.66$ $x_{22} = 29.50$ $x_{23} = 25.66$ $x_{24} = 22.66$

$x_{25} = 24.77$ $x_{26} = 29.25$ $x_{27} = 23.00$ $x_{28} = 20.87$

$x_{29} = 19.22$ $x_{30} = 16.55$ $x_{31} = 17.33$ $x_{32} = 21.88$

$x_{33} = 23.25$ $x_{34} = 17.66$ $x_{35} = 19.88$ $x_{36} = 27.00$

$\mu = 21.72$
$\sigma = 3.61$

The data for the second two years are

$y_1 = 26.33$ $y_2 = 28.50$ $y_3 = 24.33$ $y_4 = 22.62$

$y_5 = 20.20$ $y_6 = 19.55$ $y_7 = 13.66$ $y_8 = 22.44$

$y_9 = 26.00$ $y_{10} = 19.88$ $y_{11} = 21.66$ $y_{12} = 18.14$

$y_{13} = 20.11$ $y_{14} = 19.66$ $y_{15} = 22.44$ $7_{16} = 20.87$

$y_{17} = 19.62$ $y_{18} = 15.55$ $y_{19} = 14.33$ $y_{20} = 13.37$

$y_{21} = 20.88$ $y_{22} = 17.33$ $y_{23} = 21.12$ $y_{24} = 21.66$

$\mu = 20.42$
$\sigma = 3.74$

If the component did not change during this time the two empirical distribution functions $F_n(x)$ and $G_m(x)$ must be close to the same theoretical distribution function. Thus they must be close to each other as well. To measure whether they are close or not we use the Smirnov theorem – or, more exactly, a generalised version of the Smirnov theorem, because $m \neq n$. We found there was no reason to reject the hypothesis $F(x) = G(x)$ at a 90% level. However, one can see from figure 1 that the graph of the function $G(x)$ is to the left of $F(x)$. It is difficult to believe that this happened by chance so, in addition, we used a homogeniety test. This was a combinatorial method as follows.

Denote by

I. $x_1, x_2, \ldots, x_n$
the ordered sample which arises from the observations for the first period and

II. $y_1, y_2, \ldots y_m$
the ordered sample according to the observations for the second period.

Let the ordered sample of the union of I and II be denoted by

III. z_1, z_2, z_{n+m}

We also introduce the notation

$$V_i = \begin{cases} +1 & \text{if } z_i \in \text{I} \\ -1 & \text{if } z_i \in \text{II} \end{cases}$$

and

$$S_i = V_1 + V_2 + \ldots + V_i \qquad (i=1,2,\ldots,m+n)$$

The partial sum of V_i is illustrated in Figure 2. It is calculated in the following way. Starting from the origin, take one step in the direction of (1,1) if $z_i \in$ I, or take one step in the direction of (1,−1) if $z_i \in$ II ($i=1,2,\ldots,m+n$). For example, using the given data we get a zig-zag graph as shown in the figure. Thus, the x-axis represents the variable i, and the y-axis represents the partial sum S_i. Every graph has its starting point at the origin, and its terminal point at $(m+n, S_{m+n})$.

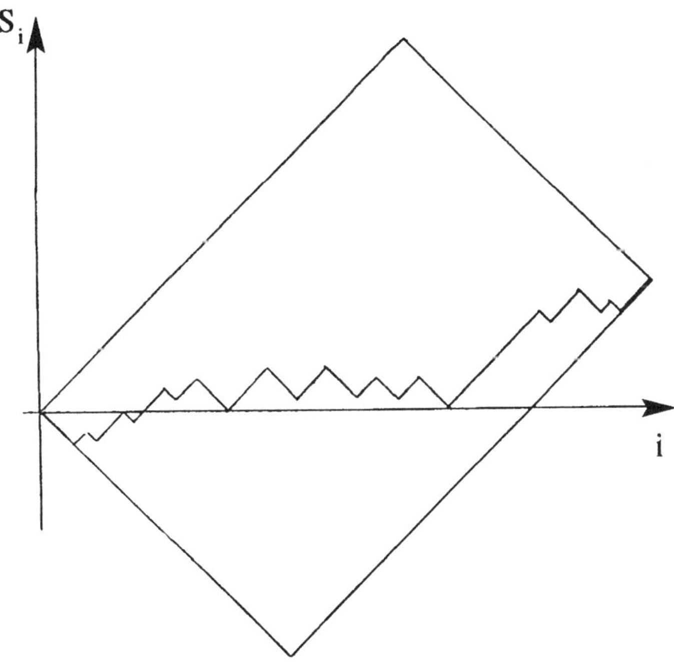

Figure 2

Every possible graph of S_i takes its course in the rectangle. Roughly speaking the graph remains in the vicinity of the diagonal if the hypothesis $F(x) = G(x)$ is true and the probability is small concerning the event that the graph exceeds a certain high level k. Testing the hypothesis $F(x) = G(x)$ at a level $1-\varepsilon$ we determine the proper value of k by a combinatorial method.

The method appeared to work well but actually did not. Although the survey of the graph shows a change in COD concentration – it runs on one side of the diagonal – the test fails to demonstrate the nonhomogeneity. Therefore, after many trials we took the Wilcoxon test. It is a rank statistic to test the hypothesis $F(x) = G(x)$. It is very efficient if the counter hypothesis is $G(x) = F(x+c)$, and we found the Wilcoxon test to be sufficiently sensitive to show the change in this case.

4. RANDOM VARIABLE MODELS II

I would like to mention one more aspect of this random variable model. Suppose an industrial or agricultural firm pollute the river uniformly over a given period. The engineers are interested to know how large a quantity of pollutant implies a significant increase in the average of the concentrations. They know, of course, how many litres of water flow in the river during a particular period.

To find the answer we could use either the Student 2–sample t–test or the Welch test (if the variances were different); in this case we must check that the parameters have a Gaussian distribution. According to our experience the t–test to detect the changes is not as sensitive as the Wilcoxon test.

Until this point we planned to model the concentration as a random variable, despite the fact that in most cases the data were not independent. We thought that a better model could be constructed if the concentration were considered as a stochastic process – even though a stochastic process is more difficult to handle than a random variable – to observe the change of water quality enough to detect the change the distribution of these random variables.

It is of particular interest to examine the excess of a high level of a certain pollution, and the duration of this excess. Let us consider the excess as a random variable denoted by Z in respect to level C, and let Y be the duration above level C.

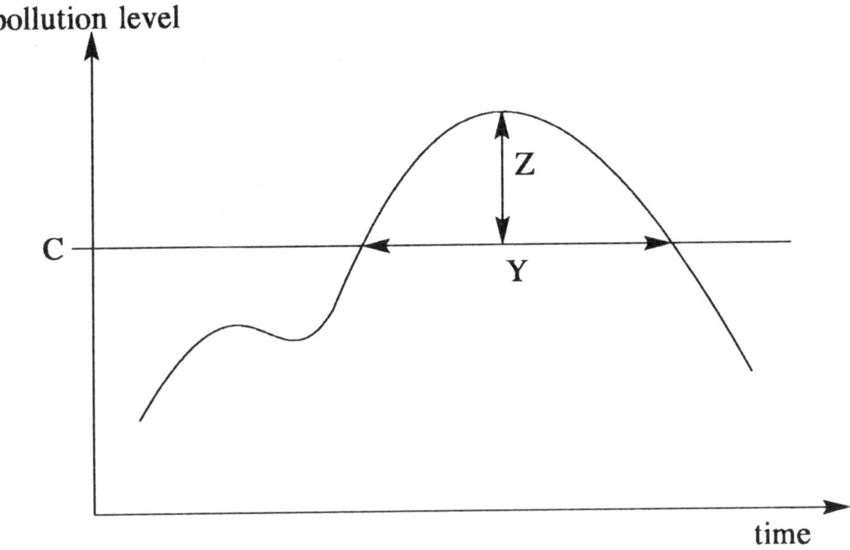

Figure 3

The distribution of Y and Z and their joint distribution H(y,z) can be determined by the sample (or estimated).

We found the joint distribution function H(y,z) fits the following type of bivariate distribution function

$$H(y,z) = \alpha \min [F(y),G(z)] + (1-\alpha) F(y) G(z)$$

where $0 \leqslant \alpha \ll 1$ and α can be estimated by the correlation coefficient.

The last model I should like to mention is a time-series model, the ARMA process:

$$x_t = r_1 x_{t-1} + r_2 x_{t-2} + \dots + r_p x_{t-p} + a_t - s_1 a_{t-1} - \dots - s_q a_{t-q}$$

Since the first and second order process already yield a high variety we restricted our attention to them ($0 \leqslant p \leqslant 2$, $0 \leqslant q \leqslant 2$). This is an easy model to use. We have the mathematical model and can therefore produce forecasts, for example by computer simulation. We can determine a band where the realisations of the process must lie, with a high probability. If we find that the real process transgresses this band, we can infer a sudden change in water quality. According to our experience this method is effective, especially in short term forecasting.

5. EDUCATIONAL REMARKS

In summarising the educational aspects of this paper, we understand how important it is to teach modelling in addition to basic mathematics. Our department began organising optional subjects to teach modelling in both the graduate and postgraduate courses, and we expected the students to be familiar with the use of computers.

The theoretical parts of the methods were discussed first; some of these were known and others were new to the students. We found the students could analyse the data without any help, but we needed to get together again to discuss the results and make plans for the next step.

At an early stage it became clear to all participants that there is no such thing as a single true model of the reality. Different mathematical models can be made of the same object; one can be better than another, but there is no best model.

An important point which is always stressed to students is to try to produce mathematical models of reality. The model exists; we don't need a model which exists but does not operate. We must find a model which exists and which also operates. Never the opposite - fixing a model and finding an example for it, or changing the data to fit the model - the result may be a model which operates but does not exist.

REFERENCES

Barabas B, Császár J, Reimann J. (1988). Mathematical methods in water quality control. *Periodica Polytechnica*, **32**.

Box GEP, Jenkins GM. (1976). *Time series analysis, forecasting and control*. Holden-Day Inc.

Blomqvist N. (1950). On a measure of dependence between two random variables. *Am Math Statist*, **21**.

Cramer H, Leadbetter MR. (1976). *Stationary and related stochastic processes*. John Wiley and Sons.

Feller W. (1957). *An introduction to probability theory and its application*. John Wiley and Sons.

Todorov I, Zelenhasic E. (1970). A stochastic model for flood analysis. *Water Research Pub*.

Yevjevich V. (1972). Stochastic processes in hydrology. *Water Research Pub*, Colorado, USA.

CHAPTER 39

Computer Modelling of Epidemic Spread

J Cunningham, LRT Gardner and GA Gardner
University College of North Wales, Bangor, UK

SUMMARY

This paper describes four case histories of short projects used as part of a postgraduate MSc course in Applied Mathematics. All four projects are linked to a specific research activity of the UCNW Biological Simulations Group, namely finite difference and finite element deterministic and stochastic modelling of the spatial spread of rabies by foxes and its control. We discuss the philosophy of using a major research programme to provide short projects. Each of the four case histories is outlined and placed in context, but detailed discussion is given for one case only, namely a simulation of rabies spread in Anglesey, an island region in the British Isles.

1. BACKGROUND

The one-year postgraduate MSc/Diploma course provides *either* a preparation for research degrees or, in the case of the majority of the students who come from overseas, a rounded preparation for teaching (applied) mathematics in their home universities in developing countries. The major aims of the course are to teach techniques of mathematics and of numerical analysis, to provide *modelling experience*, and to emphasise both written and oral communication skills. Traditional lecture courses, reading courses, tutorials and continuously-assessed assignments are used for the first six months. The selection of courses is tailored to student's individual needs, but invariably includes mathematical methods, FORTRAN and Pascal programming, and numerical analysis. In the cases to be described students also followed a reading course in mathematical biology based on Bailey (1975) and Frauenthal (1980). After completion of the continuously-assessed part, all MSc students proceed to gain modelling experience by undertaking a short (three month) project which is then written up as a dissertation. Students are also encouraged to prepare a conference paper for presentation at a suitable meeting; the Colleges of the University of Wales have an annual colloquium at Gregynog, which sometimes provides a suitable forum for a talk previously rehearsed within the School. Topics for the dissertation, which are linked with ongoing research projects in the School, are particularly suitable because the student feels that s/he is making a real contribution rather than carrying out an academic exercise. Furthermore, while students may develop their own computer codes, large-scale programs cannot be developed in the time available. We have, therefore, encouraged, not simply permitted, the use of packages already developed within the research group – this places an emphasis on actual investigation, since the techniques to be used should be well digested, and suitable software is available for use or modification. Since students have varying backgrounds of computer provision, all computation is carried out using codes which will run on IBM-compatible PCs, even though some actual large-scale problems may require transfer to a mainframe due to memory restrictions.

2. CASE 1: RABIES – 1-D FINITE DIFFERENCE MODEL

In 1985 an MSc student set out to model rabies spread in foxes, in one dimension; reference material included the work of Arcuri (1984, 1985), who visited our research group and spoke with the student, and the first task was to learn something about fox ecology and rabies spread. The student's literature search included Anderson et al (1981), Kallen (1984), Kallen et al (1985), Murray et al (1988), Toma and Andral (1977) and a visit to the South Kensington Science Museum (1985) Louis Pasteur Centenary Exhibition.

The project catalogued biological facts about rabies, and from them made assumptions to justify the terms in model equations, by and large, following previous authors. It was decided to set up a reaction-diffusion model of rabies spread among foxes (normally territorial animals) the main vector in Europe (though bats, thought to have been introduced to Denmark in fruit ships, are now also becoming significant). The main departure from earlier work was to add logistic growth terms because, although existing models could predict epidemic waves with speeds consistent with European data, without them the observed periodic recurrences of rabies were not predicted. Our simulations produced good agreement with the observed European data.

Fascinated by the story of Pasteur and vaccines against rabies, this student concentrated his study on disease control. The idea pursued was very simple: either reduce the effectiveness of transmission by vaccinating foxes, or reduce the effective carrying capacity of a habitat by means of a slaughter policy to keep fox populations low. Some detailed predictions are discussed in Case 3.

3. CASE 2: RABIES - 1-D FINITE ELEMENT MODEL

The simulations group decided to develop a model with realistic geography and ecology, and a hybrid finite-element/finite-difference code was written for spread in one-dimension. At this stage a second MSc rabies modelling project was commenced, with full access to the previous work, and the newly-developed finite element code. Since considerable groundwork had been done, a critical appraisal of the existing UCNW model was undertaken. No account had been taken of the fact that rabies has a very long and very variable incubation period, although cognisance of this had been taken in arguments assigning a value to a contact parameter α. Moreover, a leading authority on foxes and rabies, author of a work on fox ecology (Lloyd, 1980), suggested strongly that α was impossible to estimate accurately. He believed that seasonal migrations of young dog foxes in a latent state (infected but not yet infectious) were very important in rabies spread and that rabid meanderings were very local. Experiments with rabid foxes, released carrying radio transmitters, produced dead foxes very close to the point of release, suggesting that many rabid animals die within their own territory (death within a week after symptoms appear is the invariable outcome of rabies). Also, 50% of all foxes fail to survive their first year of life, producing large seasonal variations in fox populations. Perhaps this seasonal variation could drive the observed four-yearly recurrence phenomenon, explained in the UCNW model by the natural logistic (essentially seasonally-averaged) growth law.

Modification of existing software enabled the student to investigate the effect of including a latent class, and the superimposition of sinusoidal population fluctuations (see Cunningham et al, 1987). It was found that *average* behaviour was no better predicted by the more complex models, and it was concluded that no modifications of the basic model were desirable. Even if this were true, and recent private discussions with the urban fox experiment group in Bristol suggest otherwise (see Harris, 1986), it was of little real importance to the group project, because in a computational approach there is little difficulty in incorporating such modifications. However, from the theoretical viewpoint, the simpler the model the better the intuitive understanding of the gross behaviour. Of research significance, and valuable to the student's experience of modelling, was the clear appraisal of the underlying biological facts, and the gaining of an understanding of how one should assign parameter values in the UCNW model. Mathematically the dissertation contains an analysis of numerical stability (see Gardner et al, 1987) and interesting simulations, which show that the progress of the epidemic depends crucially on the initial healthy population density levels. In Case 1 disease had always been introduced into a healthy population at the carrying capacity of the habitat and, in practice, it is rare to find every ecological niche initially full.

4. CASE 3: RABIES - 2-D FINITE ELEMENT MODEL

Britain is rabies free but, in continental Europe, the disease has spread during the past 50 years from the East at a rate of 30-60 kms per year, with lesser recurrent outbreaks every three to four years. The main vector is known to be the red fox, (*vulpes-vulpes*) and the main danger to man is when infection in wildlife is passed to domestic animals such as dogs. In Europe, typical densities of foxes are one fox per square kilometre but, in the UK, both rural and urban densities are much higher. There is considerable interest in the possible invasion of Britain by rabies with the opening of the Channel tunnel, although we believe that the real threat is already present, posed by the smuggling of pets through any port. We assigned the modelling of rabies spread, in a two-dimensional region with irregular boundaries and realistic internal geography, as a research topic to a PhD student. In his preliminary year (1987-88) this student pursued the MSc course, and the group provided him with a two-dimensional package modelling rabies spread in a rectangular uniform region using rectangular finite elements (see Gardner et al, 1988 and also Cunningham et al, 1988) with the brief to see if this package could reproduce the agreement with observation already achieved in the one-dimensional pilot scheme (Cases 1 and 2 above) and to model the spread of rabies introduced into a realistic island region.

The diffusion mechanism assumes that rabies is spread by contact (a bite passing infected saliva into susceptible's blood stream) when infective animals lose their senses of territoriality due to rabid madness or seasonal dispersal. It was decided also to incorporate population growth because, although existing models could predict wave speeds consistent with European data, without it they did not account for the observed periodic recurrences of rabies.

In suitably-scaled space and time units, the basic model equations are

$$\frac{dS}{dt} = - p_1 IS + bS(p_2 - S - I) \qquad (1)$$

$$\frac{dI}{dt} = p_1 IS - rI - bI(S+I) + \nabla^2 I \qquad (2)$$

where S and I are respectively the number densities of susceptible and infective individuals; b, r are non negative parameters and, for the moment, $p_1 = p_2 = 1$. The biology is condensed into essentially two parameters and reflects the following facts.

(a) Typical fox densities are one fox/km^2 with a territory size of five square kilometres
(b) Rabies is invariably fatal; on the average, symptoms appear about 30 days after infection, and last for about five days, before death
(c) Healthy fox families meet every four or five days, more often when rabid, but not all contacts transmit disease (making the contact parameter α referred to in Case 2 hard to estimate reliably)
(d) Rabid migrations occur at the outset of symptoms, and migration of latents about six months after birth

With these assumptions we set $r = 0.456$ and $b = 0.022$.

In the long run only two outcomes are possible, corresponding to stable equilibrium states

$$S = 1, \qquad I = 0, \qquad \text{if } 1 \leqslant b + r \qquad (3)$$

$$S = br + b^2 + r, \qquad I = b - br - b^2, \qquad \text{if } 1 \geqslant b + r \qquad (4)$$

When condition (3) holds, $I \to 0$ and $S \to 1$ as $t \to \infty$.

When (4) holds, an epidemic wave propagates, followed by lesser trailing waves of decreasing amplitude, and an endemic equilibrium is attained via damped oscillations of S and I.

The theorctical propagation speed is estimated at $2(1-r-b)^{\frac{1}{2}}$, which is converted to physical units by a conversion factor of 37.3. The passage of the epidemic wave causes a crash of the healthy fox population to about 20% of the natural carrying capacity and, in our simulations, a level around 18% is predicted. When condition (4) holds, simulations produce good agreement with both the mathematical theory and the reported European rabies data in respect of speed of propagation of the epidemic wave, periodicity of recurrent epidemics, and proportion of infective foxes in the endemic state. Some results are given in Table 1.

5. VALIDATION AND SIMULATION OF THE 2-D MODEL

Apart from the numerical analysis of the method and reviewing the earlier work, the project fell into two parts, validation of the model, and application to a semi-realistic island region. To validate the model, disease spread was considered in a long thin region two elements wide and, apart from some end-effect differences, gave excellent agreement with the one-dimensional model as shown by Table 1. By introducing plane polar coordinates it was proved that, in a uniform region remote from the introduction point and from any boundaries, the theory predicted a radial disease spread with precisely the one-dimensional theoretical velocity.

	Theory	Field Data	Simulations			
			Finite Diff 1D	Finite Element 1D	Hybrid Code 2D	Space Time 2D
Pulse Speed Km/yr	54	30-60	46	51	51	51
Post Epidemic S minimum %	≥ 15	20	19	18	18	20
$\frac{100*I}{(I + S)}$ %	2.4	3-7	2.4	2.3	2.3	2.4
Period yrs	3.8	3-4	3.8	4	3.8	3.9

Table 1 : Simulations compared with theoretical and field data

The island region chosen for the simulation was the Isle of Anglesey, an approximately rectangular island measuring about 30×32.5 km^2 lying off the North Wales coast. Likely introduction sites for rabies into Anglesey were identified as the port of Holyhead, at its North Western extremity, and Menai Bridge on the South coast, where the island is connected to the mainland.

Simulations for Anglesey showed the region to be too small for the theoretical travelling wave to develop properly, and almost immediately the damped oscillatory behaviour described above was established. In well under a year, even with modest fox population densities of one fox/km^2, disease introduced at Holyhead had swept all the way across the island, and recurred with the predicted periodicity of about four years. The assumed density of one fox/km^2 is almost certainly an under estimate, giving about 750 foxes in Anglesey, and the actual population could be four times this number, with even more rapid disease spread!

Graphical output, produced during the computer runs, dynamically displays rabies spread by means of continuously-updated contour maps of the region, showing the populations levels of both infective and susceptible foxes. Data stored selectively during runs is used to produce contour map snapshots of the epidemic region at various times (see figure 1). This information can also be used to produce plots showing infective and susceptible levels (see figure 2), for all times, at predetermined sampling stations.

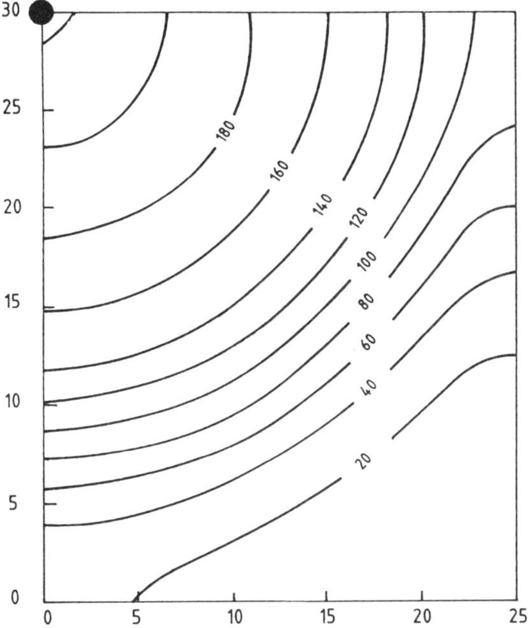

Figure 1: Infective contours eight months after rabies introduced at Holyhead

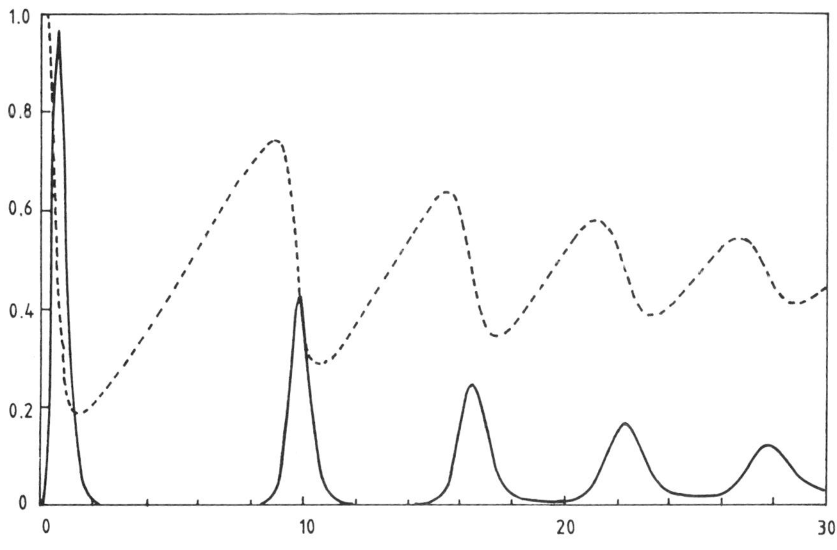

Figure 2: Evolution of susceptible (dotted) and infective (×5) densities

When condition (3) holds, the system cannot support a travelling epidemic wave, and the disease tends to fade out. This is the basis of the control studies initiated in Case 1. If, in some regions, foxes were vaccinated, then the effective contact parameter p_1 would be reduced in equations (1) and (2) from the value $p_1 = 1$ to a fraction, $0 < p_1 < 1$; the condition (4), necessary for the system to support a travelling epidemic wave, becomes $p_1 > b + r$, so that p_1 must simply be chosen small enough; the rabies propagation as a travelling wave ceases, and the disease tends to extinction. Control is therefore achieved, provided extinction of disease occurs more rapidly than the time taken for the wave to cross the vaccination barrier region. The snag was that the extinction occurs as an exponential decay, and so takes infinitely long! The problem was resolved by choosing an arbitrary cut-off level, below which disease was deemed to be extinct. With 65% vaccination protection and a cut-off at around 10^{-5} infectives/km^2 (giving a massive safety margin!) simulations predicted barrier widths of order 60 kms, which was consistent with North American experimental vaccination schemes. Modifying the effective carrying capacity of a barrier region, by means of a slaughter policy of continuous culling, can be modelled by choosing $0 < p_2 < 1$ in equation (1), with similar results (see Cunnigham et al, 1986).

It was noted that the characteristic dimension of Anglesey was of the same order as the control barrier widths found above, and it was concluded that halting epidemic spread in Anglesey was not practicable. For rabies introduced at Holyhead, the best one could hope for would be to prevent spread to the mainland, and attempt to reduce fox populations to a very low level by an intensive extermination programme to inhibit the recurrent epidemics.

Speeds of spread in different parts of the island were much influenced by the local boundary shape. Figure 3 shows the actual shape of the Isle of Anglesey, with sampling stations (mainly population centres) indicated: M = Menai Bridge (where the island is linked to the mainland), L = Llangefni, H = Holyhead, A = Amlwch and C = Carmel Head. Figure 4 shows contour lines of susceptibles and infectives about eight months after introduction of rabies from the mainland at Menai Bridge, from a more recent study by the authors (Gardner et al, not yet published) which improves on the rectangular boundary approximation of the MSc project.

Consider the market town of Llangefni (L in figure 3). Initially the susceptible density is 1.0. When the disease reaches the sampling point, the susceptible density falls as the infected density increases. When the infective density attains its maximum of about 12%, the susceptible population has been reduced to about 60%. The susceptible population is then too small to sustain the epidemic at this site and, as the susceptible density falls further to about 25%, the infected density also falls away, apparently to zero in the manner of figure 2. The expected recurrent phase of disease then occurs. Since the island is so small, the standard-sized infective pulse does not develop, but most of the features of the progress of the disease described above are observed. The fox population increases, by natural growth, until it reaches such a size as to be able to maintain an epidemic once more, at which time an infective pulse, smaller in magnitude than the original one, arises at the initial site (the Menai Bridge) and travels once more northwards across the Island. This process is repeated again and again in following years, and the endemic ratio of 1.9% is approached with a period of 3.9 years.

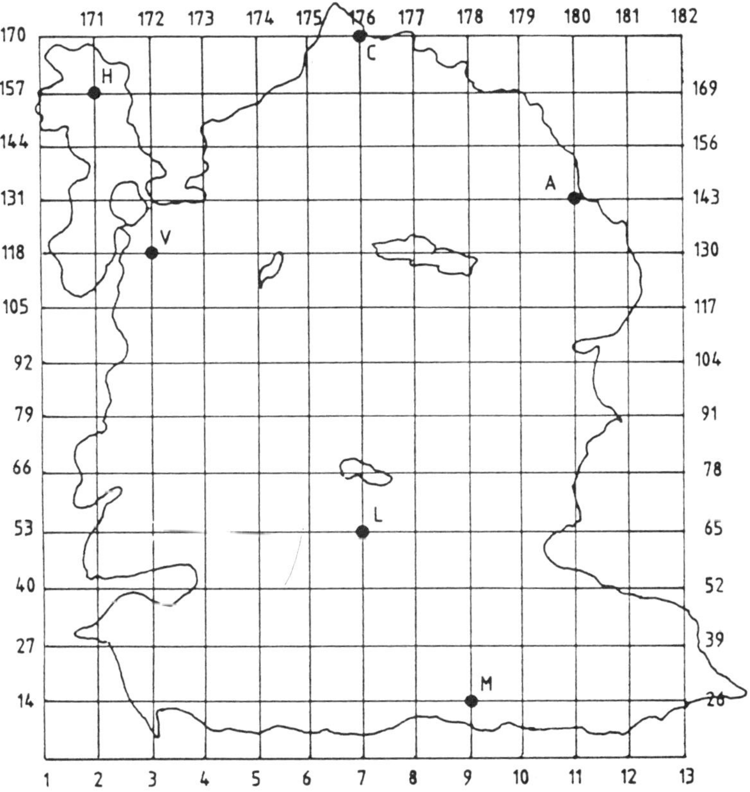

Figure 3: A map of Anglesey showing its division into 12×13 finite elements each 2.5 km square. Node numbers and sampling points are shown

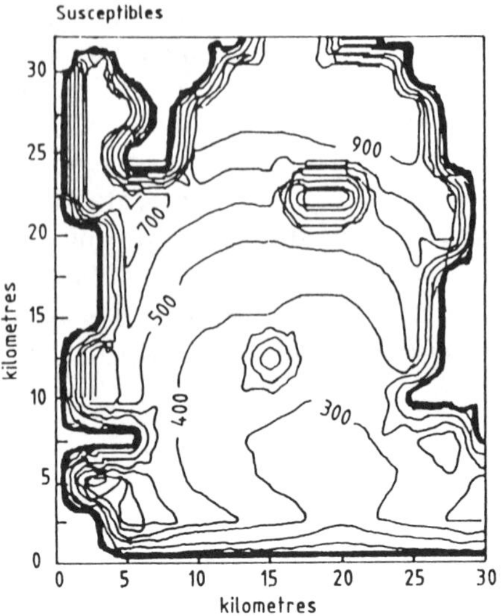

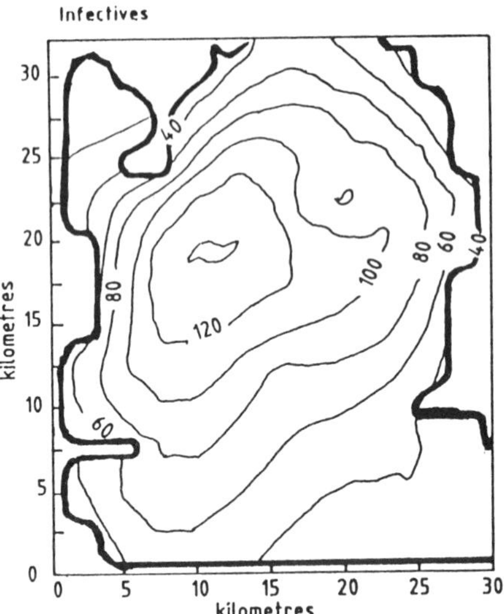

Figure 4: Contour lines showing the density distribution (×1000) of susceptible and infective fox population over Anglesey, about eight months after the infection was introduced via the Menai Bridge

At the other sampling stations (see figure 3) similar behaviour is observed, with identical endemic periods but with final constant endemic levels of susceptibles S_∞ and infectives I_∞ of differing magnitudes. The results are shown in Table 2. The theoretical result shown is that reproduced in one-dimensional simulations, and validation runs with two-dimensional codes for essentially one-dimensional problems.

Station	S_∞	I_∞	$I_\infty/(I_\infty+S_\infty)$
M	60%	0.8%	1.3%
L	52%	1%	1.9%
V	69%	0.6%	0.9%
H	87%	0.3%	0.3%
A	66%	0.7%	1%
C	59%	0.9%	1.5%
Theory	48%	1.1%	2.4%

Table 2: Simulated endemic levels at various sampling stations

The anomalous values at H, Holyhead, arise because of the difficult access for diseased animals. Variations between the sampling stations may be caused by edge effects, due to the small size of the island and its irregular boundary. Such non-uniformity was not observed with the previous model, which made no attempt to approximate the boundaries.

6. CASE 4: RABIES - STOCHASTIC MODEL

The control of disease needs the study of very small populations - extinction of infectives is of crucial interest. Also, in our rabies studies, the level of infective density is exceedingly small in the immediate aftermath of the leading epidemic pulse (see figure 2), and the sort of cut-off levels necessary in the vaccination and slaughter control studies might, without intervention, operate to wipe out the disease, thereby preventing the simulation producing recurrent outbreaks. This has motivated a current MSc project, not specifically linked into rabies, of building a stochastic model of disease spread in a logistically growing population in a single territory. Details of this study cannot be given here, but the evolution of absolute numbers of infectives and susceptibles given deterministically by a model similar to that described by equations (1) and (2) is modelled stochastically by a set of partial differential equations for

$$\frac{dP_{i,s}}{dt}$$

where $P_{i,s}(t)$ is the probability that at time t there are i infectives and s susceptibles; whereas previously there were two equations, now there are a double infinity. To make comparisons, stochastic averages are taken. Suitable parameter choices make the average equations identical with their deterministic analogues. This does not mean that the stochastic averages evolve deterministically, because the equations are nonlinear and the variances and covariance are not, in general, zero or small. Thus, the stochastic predictions differ even on the average from the deterministic ones, with a marked tendency for small populations to become extinct. Stochastic modelling is very instructive in emphasising that many outcomes are possible in real situations.

The essential difference between deterministic and stochastic theories of rabies in a *single* region is that, in the former, when the susceptible population has recovered sufficiently, there *can* be a resurgence of disease from a very low level, while in the latter, the probability of disease extinction *never* decreases. In the deterministic theory, when the disease level first falls, it does so to a level far below that of any reasonable cut-off, and the stochastic investigation fails to provide any justification for the deterministic explanation of recurrent epidemics. In dealing with disease spread in a single, closed region, where typically a territory of 5 km^2 contains five foxes, a cut-off *should* be used.

However, in spatial spread *many* regions are chained together, and population densities are used - very low densities should be interpreted as disease present in a few territories which, when susceptible levels recover sufficiently, allow disease to erupt from these isolated sources. In dealing with disease spread through n regions, densities *should not* be cut off at any level above 1/n, representing a single infected fox in one territory only. This aspect has not been exhaustively investigated.

7. CONCLUSION

Research group studies can provide short modelling projects for taught courses, to the psychological and educational advantage of students and with real benefit to the group. Projects may exploit the existence of relevant sophisticated packages and expert knowledge of the subject but, as we discovered in Case 3, care must be taken not to pre-empt the student's necessary initial struggle with model formulation through the supervisor's prejudicial familiarity with the problem.

REFERENCES

Anderson RM, Jackson HC, May RM and Smith AM. (1981). Population dynamics of fox rabies in Europe. *Nature*, 289, 765-771.

Arcuri P. (1984). Model Mechanisms for Biological Pattern Formation. D Phil Thesis, Oxford.

Arcuri P. (1985). A Simple Model for the Spatial Spread of Rabies. University of Wales Applied Mathematics Colloquium, Gregynog.

Bailey NTJ. (1975). *The Mathematical Theory of Infectious Diseases and Applications*. Griffin, London.

Cunningham J, Gardner LRT, Gardner GA, and Shaoul HA. (1986). Epidemiology of rabies in a logistic population. *Proc Int AMSE Conf 'Modelling and Simulation'*, Sorrento (Italy), Vol 4.2, 3-12.

Cunningham J, Gardner LRT, and Gardner GA. (1988). The Spatial Containment of Epidemics. *Proc Int AMSE Conf, 'Modelling and Simulation'*, Istanbul (Turkey), Vol 4A, 25-34.

Frauenthal JC. (1980). *Mathematical Modelling in Epidemiology*. Springer Verlag, Berlin.

Gardner LRT, Gardner GA, Cunningham J and Kanwati A. (1987). A finite element model for the spatial spread of rabies. *Proc Int AMSE Conf, 'Modelling and Simulation'*, Karlsruhe (West Germany), Vol 4, 13-24.

Gardner GA, Gardner LRT, and Cunningham J. (1988). A finite element model for the two-dimensional spread of rabies. *Proc Int AMSE Conf, 'Modelling and Simulation'*, Istanbul (Turkey), Vol 4A, 11-24.

Harris S. (1986). *Urban Foxes*. Whittet, London.

Kallen A. (1984). Thresholds and travelling waves in an epidemic model for rabies. *Non-linear analysis, Theory, Methods and Applications*, 8.8, 851-856.

Lloyd HG. (1980). *The Red Fox*. Batsford, London.

Murray JD, Stanley EA and Brown DL. (1988). On the spatial spread of rabies among foxes. *Proc R Soc Lond, B*, 299, 111-150.

Toma B and Andral L. (1977). Epidemiology of fox rabies. *Ad Virus Res*, 21, 1-36.

South Kensington Science Museum Exhibition. (1985). Louis Pasteur and Rabies, London.

CHAPTER 40

Good and Bad Mathematical Modelling in Mechanics

W Kliem
The Technical University of Denmark, Lyngby, Denmark

SUMMARY

Although mechanics has been a popular field of mathematical modelling for a long time, there are still pitfalls hidden in this subject.

Should we avoid fictitious forces or not? We shall look in particular at the difficulty of using those forces in the dynamics of rigid bodies and in rotor dynamics.

Can we drop damping in simple modelling? We will explain, by a few examples, how crucial damping can be in the modelling of many mechanical systems.

Finally, mathematicians will be encouraged to model in mechanics under strict consideration of physical facts; this is essential for good mathematical modelling.

1. INTRODUCTION

Beginning with Galileo and Newton, mechanics has been an outstanding field of mathematical modelling for over 300 years. Names like Bernoulli, Euler, d'Alembert and Lagrange are connected with a development, which led to the concept of a mechanics-inspired determinism, most clearly formulated by Laplace.

So, since classical mechanics is by no means a new science, one would imagine that there are no hidden pitfalls remaining in the subject. I will demonstrate that this is not the case by three themes.

Modelling with or without fictitious forces
Modelling with or without damping
Modelling with or without consideration of physical facts

All three topics are of great interest when teaching mechanics at university level.

2. MODELLING WITH OR WITHOUT FICTITIOUS FORCES

To understand the problem, the reader should know a little about the concepts of absolute and relative motion.

By definition, the *absolute* motion of a particle is measured with respect to a fixed inertial system. However, there are problems for which the analysis of motion is simplified by using measurements with respect to a moving reference system; this is known as a *relative* motion analysis. If $\mathbf{F}$ is the external force on a particle with mass m, these two approaches lead to different expressions for Newton's 2nd law:

$$m\,\mathbf{a}_{abs} = \mathbf{F} \qquad (1)$$

and

$$m\,\mathbf{a}_{rel} = \mathbf{F} - m\,\mathbf{a}_{rf} - m\,\mathbf{a}_{Cor} \qquad (2)$$

Here $\mathbf{a}_{rf}$ means – roughly speaking – the acceleration of the moving reference system with respect to the inertial system. The so-called Coriolis acceleration $\mathbf{a}_{Cor}$ is due to the rotation of the moving system. The connection between (1) and (2) is simply the analysis

$$\mathbf{a}_{abs} = \mathbf{a}_{rel} + \mathbf{a}_{rf} + \mathbf{a}_{Cor} \qquad (3)$$

Equation (2) is often connected with the name d'Alembert's principle and the terms $- m\,\mathbf{a}_{rf}$ and $- m\,\mathbf{a}_{Cor}$ are called *fictitious forces* or inertia forces. The first term includes a centrifugal force, and the second is known as the Coriolis force. Whereas the analysis of (3) is of vital importance in kinematics, it is doubtful whether the use of (2) will give any advantage. Some textbooks, mainly American, refute the use of fictitious forces. I quote from Meriam and Kraige (1987): Equation (3) "introduces no simplification and adds non-existent forces to the diagram".

Nevertheless, it is my opinion that there are problems where fictitious forces are convenient and, also, these forces are used frequently in the European engineering tradition. So I think we have to discuss the problems arising from this approach.

Besides common pedagogical difficulties in teaching students to use fictitious forces in a correct way, I can see two main problems in the modelling procedure.

(a) Extending the dynamics from one particle to a rigid body, fictitious forces act on all elements Δm_i of the body. Then the question arises, whether all these forces are equivalent to the fictitious forces acting on the centre of mass, where the whole mass is assumed to be concentrated. Equivalence means the same torque about an arbitrary point. This equivalence would simplify modelling, but is only valid in certain cases (Jorgensen and Kliem, 1975). Normally it is necessary to integrate all torques over the whole body.

I shall give two simple examples:

A ball is rolling on a rotating disk (figure 1). In an analysis relative to the disk, the centrifugal forces can be collected in the centre of mass C. However, the Coriolis forces may not be collected in C, and their torque about C demands quite a cumbersome computation.

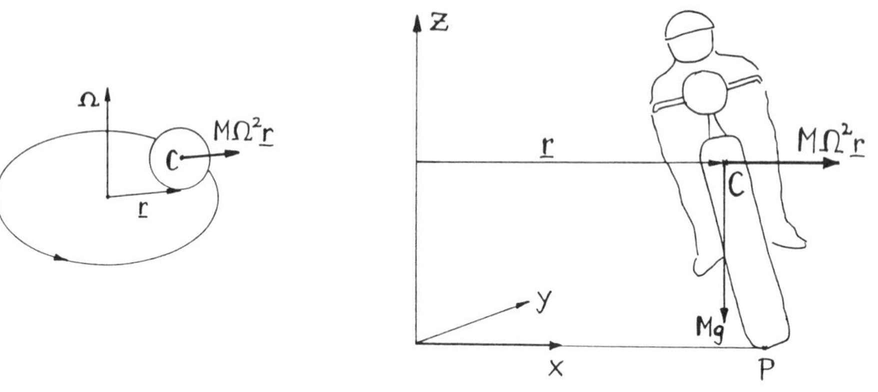

Figure 1 **Figure 2**

Another example is a motor cyclist on a curve (figure 2). In respect to a rotating reference system the rigid body is in equilibrium, and one might be tempted to collect the centrifugal forces in the centre of mass C, and then base the moment equation on P. This is actually wrong.

The answer to this kind of problem was given 200 years ago, is widely used in gyrodynamics, but is often overlooked. For rigid bodies an absolute motion analysis is preferable, where the physical quantities are measured from an inertial system, but are *expressed* in coordinates of an appropriate moving system. This modelling procedure is used, for example, to establish the famous Euler equations, and should be central in teaching space dynamics.

(b) My second point concerns rotor dynamics in particular. We want to model the oscillations of an elastic shaft with a disk (figure 3).

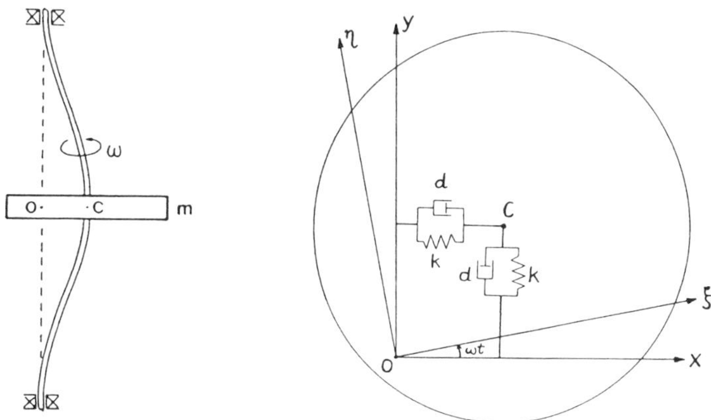

Figure 3

Using a rotating reference system $\xi\eta$, the matrix differential equation for the centre of mass C becomes

$$\begin{bmatrix} m & 0 \\ 0 & m \end{bmatrix} \begin{bmatrix} \ddot{\xi} \\ \ddot{\eta} \end{bmatrix} + \begin{bmatrix} d & -2m\omega \\ 2m\omega & d \end{bmatrix} \begin{bmatrix} \dot{\xi} \\ \dot{\eta} \end{bmatrix} + \begin{bmatrix} k-m\omega^2 & -d\omega \\ d\omega & k-m\omega^2 \end{bmatrix} \begin{bmatrix} \xi \\ \eta \end{bmatrix} = \begin{bmatrix} 0 \\ 0 \end{bmatrix} \tag{4}$$

where k and d are coefficients of elasticity and of external damping. (4) is a so-called non-conservative system, with asymmetric system matrices.

We can, however, model a fixed referencc system xy much more easily.

$$\begin{bmatrix} m & 0 \\ 0 & m \end{bmatrix} \begin{bmatrix} \ddot{x} \\ \ddot{y} \end{bmatrix} + \begin{bmatrix} d & 0 \\ 0 & d \end{bmatrix} \begin{bmatrix} \dot{x} \\ \dot{y} \end{bmatrix} + \begin{bmatrix} k & 0 \\ 0 & k \end{bmatrix} \begin{bmatrix} x \\ y \end{bmatrix} = \begin{bmatrix} 0 \\ 0 \end{bmatrix} \tag{5}$$

Needless to say, one should prefer the uncoupled system (5), but, even in papers, (4) appears with all its disadvantages.

Conclusiion: Fictitious forces are not easy to handle. In particular, we have seen that they do not simplify the analysis of the dynamics of rigid bodies. In fact they can even complicate modelling and lead to bad models. If possible, avoid fictitious forces.

3. MODELLING WITH OR WITHOUT DAMPING

Mathematicians often think that one can neglect damping in the modelling process, because it only slightly reduces movements and oscillations and does not change the qualitative behaviour of the solutions. Problems exist, however, where modelling without damping leads to paradoxical results.

Take a rope or chain with mass μ per unit length, lying rolled up on the floor (figure 4). What is the force F required to give the end of the rope a constant velocity v? Neglect gravity.

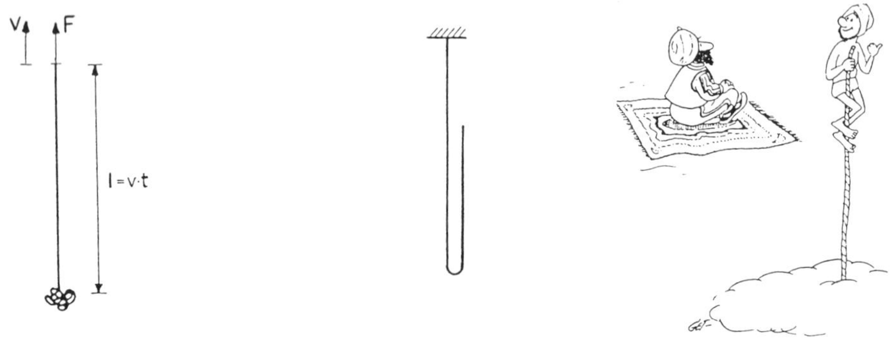

Figure 4 Figure 5 Figure 6

Assuming no energy loss, F's work equals the change in kinetic energy.

$$dA = F d\ell = \frac{1}{2} (\mu d\ell) v^2 \qquad \text{or} \qquad F = \frac{1}{2} \mu v^2 \tag{6}$$

On the other hand, all damping and possible loss of energy must be included in Newton's 2nd law.

$$F = \frac{d}{dt} (mv) = \frac{d}{dt} (\mu \ell v) = \frac{d}{dt} (\mu v^2 t) = \mu v^2 \tag{7}$$

The real value of F is therefore twice as big as in (6), showing that half of the work done by F is lost by damping (deformation, oscillations, impact) when a rope element changes its velocity from zero to v abruptly. The other half of the work is converted to kinetic translation energy.

This is a surprising result; a model neglecting damping is misleading. Similar results are obtained for a rope with a moving sharp bend (figure 5). Neglecting damping as in Hamel (1967) leads, in the limit, to an infinite velocity of the rope end, and to a singularity in the rope tension, so that for Hamel even the Indian rope trick (figure 6) seems possible!

Examples of ropes and chains are, however, not the only ones demanding damping. In the theory of linear non-conservative systems, modelling oscillations of lumped-parameter systems, one can also get paradoxical results by neglecting damping (Thompson, 1987). Self-excited oscillations can even arise due to *internal* damping in viscoelastic materials (Kliem, 1987). The concept of structural stability requires damping too (Thompson, 1982).

Conclusion: Damping is a crucial feature in the modelling of many mechanical systems – in fact, more than one might imagine.

4. MODELLING WITH OR WITHOUT CONSIDERATION OF PHYSICAL FACTS

My third topic could be formulated as a question: Can a weak or even a wrong physical model lead to a good mathematical model?

Let me show you can example. To understand it fully, one must have some insight into plane kinematics and also into simple statics, such as the moment equation and the principle of virtual work.

We want to model the oscillations of an elastic structure attached to the interior of a rotating ring (figure 7)

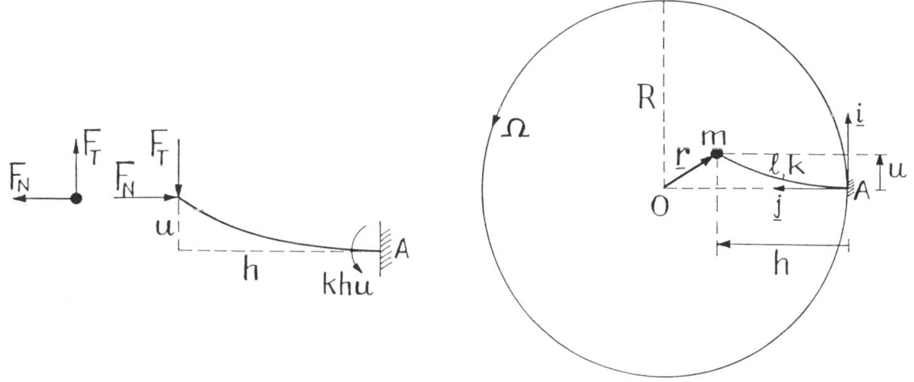

Figure 7 **Figure 8**

Let the structure be a mass m at the end of a massless cantilever beam with length ℓ and stiffness k. A rotating frame of reference is introduced by **i** and **j**, so that $\dot{\mathbf{i}} = \Omega\mathbf{j}$, $\dot{\mathbf{j}} = -\Omega\mathbf{i}$. Since displacements u(t) are assumed small, $h \approx \ell$ = constant seems reasonable, and we get the following results for position and acceleration of the mass in respect to an inertial system.

$$\left.\begin{aligned} \mathbf{r} &= u\mathbf{i} - (R-h)\mathbf{j} \\ \ddot{\mathbf{r}} &= (\ddot{u}-u\Omega^2)\mathbf{i} + ((R-h)\Omega^2+2\dot{u}\Omega)\mathbf{j} \end{aligned}\right\} \qquad (8)$$

Notice especially the Coriolis term $2\dot{u}\Omega\mathbf{j}$. If $\mathbf{F} = F_T\mathbf{i} + F_N\mathbf{j}$ denotes the force from the beam to the mass, Newton's 2nd law implies

$$F_T = m(\ddot{u}-u\Omega^2) \quad , \quad F_N = m[(R-h)\Omega^2+2\dot{u}\Omega] \qquad (9)$$

So far there is nothing wrong.

Now we have to look at the massless beam in equilibrium (figure 8). F_T and F_N are the reactions from the removed mass, and khu is the bending moment in A. Then the moment equation about A yields

$$F_Th - F_Nu + khu = 0 \qquad (10)$$

Putting (9) into (10) results immediately in the nonlinear differential equation

$$\ddot{u} + (k/m-R\Omega^2/h)u - (2\Omega/h)u\dot{u} = 0 \qquad (11)$$

This is exactly the modelling procedure used by two authors of a paper in the Journal of Applied Mechanics (Hernried and Gustafson, 1988). They also emphasise that the Coriolis term $(2\Omega/h)u\dot{u}$, the only nonlinear term, is of vital importance in a stability investigation.

Although (11) is a good equation for starting an interesting bifurcation analysis, it is unfortunately *not* an appropriate model for the present structure, either physically or mathematically, because the Coriolis term appears due to an error in the modelling procedure.

This was pointed out in another paper in JAM (Peters and Hodges, 1988) but not in a convincing pedagogical way since, in a reply, the two authors of the original work still believed they had constructed a good model.

Let me try to convince you, in the first instance by a physical argument. In deriving (8) and (9), h = constant was assumed, the mass moves on a straight line in direction **i** and the Coriolis term has therefore the direction **j**. In (10) we changed the path of the mass to a part of a circle, but we forgot to change the direction of the Coriolis term which, for a circular path, points toward the centre A and therefore can never appear in a moment equation. The whole procedure does not conform with Newtonian mechanics, and is simply bad modelling.

We can also prove this by a correct mathematical derivation. Let h(u) be any smooth function. Then (9) has to be changed to

$$F_T = m(\ddot{u}-u\Omega^2-2\dot{h}\Omega) \quad , \quad F_N = m[(R-h)\Omega^2+2\dot{u}\Omega+\ddot{h}] \tag{12}$$

From the principle of virtual work for the elastic beam we have

$$- F_T\delta u - F_N\delta h - K(u)\delta u = 0 \tag{13}$$

where $- K(u)\delta u$ is the internal work due to the elasticity of the beam. With

$$\delta h = \frac{dh}{du}\,\delta u$$

we get

$$F_T = - K(u) - F_N \frac{dh}{du} \tag{14}$$

Inserting (12) into (14) results in

$$\ddot{u} - u\Omega^2 - \underline{2\dot{h}\Omega} = - \frac{K(u)}{m} - (R-h)\Omega^2 \frac{dh}{du} - \underline{2\dot{u}\Omega \frac{dh}{du}} - \ddot{h} \frac{dh}{du} \tag{15}$$

Since

$$2\dot{u}\Omega \frac{dh}{du} = 2 \frac{dh}{du}\frac{du}{dt}\,\Omega = 2\dot{h}\Omega$$

the underlined Coriolis term in (15) is cancelled! Also, since

$$h = \frac{dh}{du}\,\ddot{u} \ , \ \ddot{h} = \frac{d^2h}{du}\,\dot{u}^2 + \frac{dh}{du}\,\ddot{u}$$

the remaining modelling equation is therefore

$$\ddot{u}\left[1 + \left[\frac{dh}{du}\right]^2\right] + \frac{K(u)}{m} -\Omega^2 u+\Omega^2(R-h(u)) \frac{dh}{du} + \frac{d^2h}{du}\frac{dh}{du}\,\dot{u}^2 = 0 \tag{16}$$

For the choice $K(u) = ku$ and $h(u) =$ constant, (16) implies the usual linear model.

A more advanced model is $K(u) = k_1u + k_2u^3$ and a circular path $h(u) = (\ell^2-u^2)^{1/2}$. As far as terms of third degree, we then get

$$\ddot{u}(1+u^2/\ell^2) + (k_1/m-R\Omega^2/\ell)u + (k_3/m-R\Omega^2/2\ell^3)u^3 + \dot{u}^2u/\ell^2 = 0 \quad (17)$$

A stability analysis of (17) results in the usual pitchfork bifurcation.

One can obtain nonlinearities of second degree in the modelling equation (16), but only by choosing h(u) as a physically unconvincing function like $h(u) = \ell - a|u|^3$.

Conclusion: Since mathematical modelling has to do with a description of the real world, equations like (11), with a term which is simply wrong, can never be good models, although perhaps they are good equations for some mathematical investigations.

5. FINAL REMARKS

Once mechanics was the most important field in which to show how mathematics worked and was effective. Today there are many other interesting fields for mathematical modelling. However, mechanics will always remain popular, because it deals with phenomena from our daily life. I hope I have succeeded in explaining some of the pitfalls every one has to be aware of when modelling in this, not too easy, area.

REFERENCES

Hamel G. (1967). *Theoretische Mechanik*. Springer.

Hernried AG and Gustafson GB. (1988). On the Dynamic Response of a Single-Degree-of-Freedom Structure Attached to the Interior of a Rotating Rigid Ring. *J App Mech*, **55**, 201–205.

Jorgensen AE and Kliem W. (1975). About the Torque of Fictitious Forces. *ZAMM 55*, 534–535.

Kliem W. (1987). The dynamics of viscoelastic rotors. *Dynamics and Stability of Systems*, **2**, No 2, 113–123.

Meriam JL and Kraige LG. (1987). *Engineering Mechanics*, **2**, Dynamics, 224. Wiley.

Peters DA and Hodges DH. (1988). *J Appl Mech*, **55**, 747–748.

Thompson JMT. (1982). *Instabilities and Catastrophes in Science and Engineering*, 6. Wiley.

CHAPTER 41

Modelling Wind-induced Vibrations of Slender Structures

A Battye
Sheffield City Polytechnic, UK

SUMMARY

This paper discusses the vibrations of high mast lighting columns as used in the centre of Sheffield. It is used as a mathematical case study for final year civil engineering students.

1. INTRODUCTION

The aim of this investigation is to determine the fundamental vibration frequency of a tall, non-uniform column, the lighting mast described in section 3. Vibrations in this instance are driven by oscillating lateral forces, caused by the mechanism of vortex shedding occuring when a steady wind blows past an object. The frequency of vortex shedding is a function of the wind speed. So the results can be related to the particular environment containing the structure of interest.

The modelling work described in this paper was particularly associated with the applied numerical methods course taught to final year civil engineering students. Over the years, the professional bodies in engineering have pushed for more evidence of engineering applications in all the separate subjects of degree programmes. To make the final year course directly relevant we had already built in case studies of a very applied nature, and so we found it easy to demonstrate the applications necessary.

An additional spin-off of this study was that it provided part of a motivating bridge between the real world and the mathematics world, in a seminar given to final year education students entitled Industrial Applications of Mathematics. The problem gives a very concrete meaning to a matrix, a vector, an eigenvalue and an eigenvector, thus helping to overcome the basic lack of understanding of and feeling for some of these terms/concepts by many students.

This modelling exercise includes

- collection of data
- elementary finite elements
- eigenvalue computation
- relating to the real world

and can be understood by those having knowledge of calculus and matrix algebra. It can be completed with access to a computer and suitable software.

2. THE REAL WORLD PROBLEM

The efficient illumination of large areas, such as at roundabouts, bus stations, sports pitches, railway sidings, is often achieved today using high mast lighting. Typically, those in use in the centre of Sheffield (see figure 1) are 30m high, and are highly flexible hollow tapered columns.

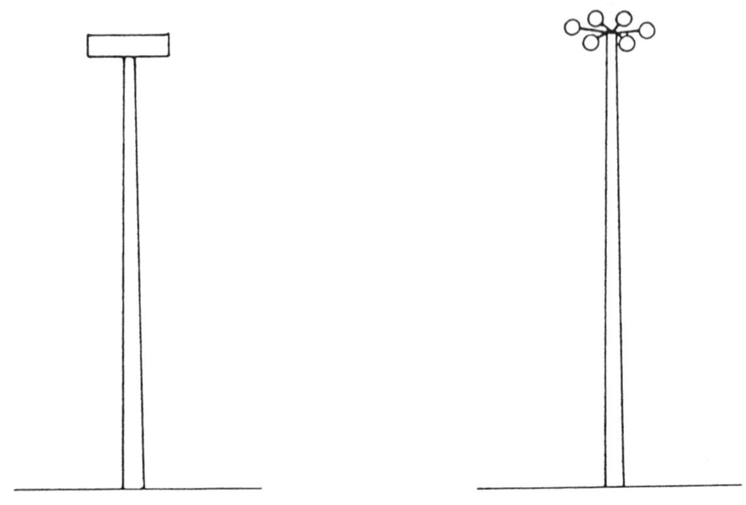

Figure 1

In strong winds, these can be observed vibrating, and immediate questions arising are as follows.

- At what frequency do they vibrate and what implications, if any, has the consequent motion of the structure?
- What mathematics must we use to solve the problem?

3. DATA ACQUISITION OF THIS PROBLEM

I think that modelling ought to be part of problem solving in general, and this inevitably includes collecting the data. How do we start and where do we expect a student to start?

I would recommend including the act of data acquisition as an important part of real problem solving with mathematics, consequently I would expect students to go out and find the information. (I found this fun).

The technical data are not required fully here, but one can discover that the parameters of a typical high mast (the technical and rather unimaginative name) are

30m high
tapered cross-section (made of 20 flats)
galvanised steel (known density and Young's modulus)
440mm dia at base
150mm dia at top
6mm thick walls for 0 → 15m high
5mm thick walls for 15 → 20m high
3mm thick walls for 20 → 30m high
lighting unit weight 180kg

so we see that the classical results for a uniform cantilever beam would not be of much use here!

Finally, Sheffield being fairly windy, it is often possible to observe these masts oscillating and estimate the vibration frequency.

4. THE MATHEMATICAL MODEL

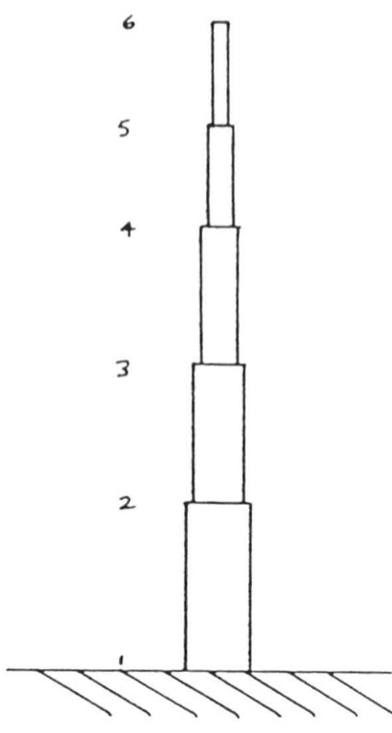

Figure 2

Using the finite element method of analysis of skeletal structures (see for example Dawe 1984) the column is approximated (figure 2) by a number, n, of *uniform* beam elements (Bernoulli–Euler beam elements), not necessarily of equal lengths. Average parameters for each are easily found.

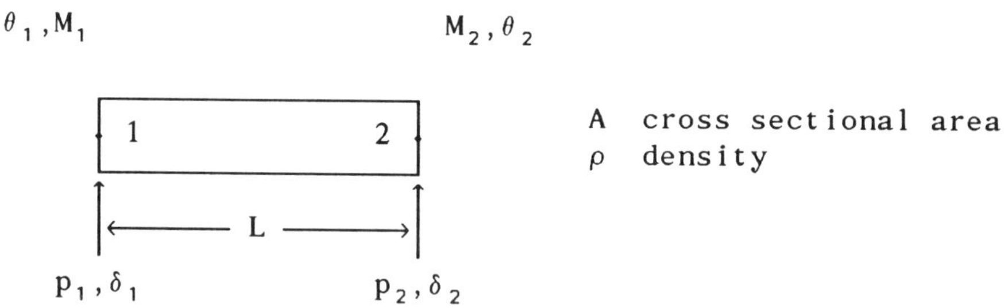

Figure 3

A typical beam element (figure 3) subject to lateral forces P_i and moments M_i causing lateral deflections δ_i and rotations θ_i (at nodes i) has a stiffness matrix **k** and consistent mass matrix **m** (Dawe p45) related by

$$-\mathbf{m}\ddot{\mathbf{y}} = \mathbf{k}\mathbf{y}$$

where the deflection vector y is

$$\mathbf{y} = (\delta_1, \theta_1, \delta_2, \theta_2)^T$$

We thus have a collection of mass and stiffness matrices $\mathbf{m}_j$ and $\mathbf{k}_j$ for each element, j, of the discretised structure. Under the assumption of superposition (Dawe p12, p472) the relationship for the whole structure is

$$-\mathbf{M}\ddot{\mathbf{D}} = \mathbf{KD}$$

where

$$\mathbf{D} = (\delta_1, \theta_1, \delta_2, \theta_2, \ldots, \delta_{n+1}, \theta_{n+1})^T$$

and M and K are the assembled total mass and assembled total stiffness matrices respectively (Dawe p49 onwards).

5. SOLUTION

The structure is approximated by

$$-\mathbf{M}\ddot{\mathbf{D}} = \mathbf{KD}$$

and looking for normal modes of vibration in the form

$$\mathbf{D} = \hat{\mathbf{D}} \sin \omega t$$

we get the familiar eigenvalue problem

$$\mathbf{MD} = \omega^2\mathbf{KD}$$

or $\{\mathbf{M}^{-1}\mathbf{K}\}\ \mathbf{D} = \mathbf{D}/\omega^2$

Using just a few elements (n=4 say) yields a surprisingly accurate answer for the fundamental vibration frequency. It matches closely the observed frequency, and leads very naturally to a wider discussion of vibrations in structures.

Thus, using an appropriate computer-based approach (depending in detail on the students, their aptitude and the emphasis of the training and/or assessment intended) the eigenvalues and eigenvectors can be obtained.

6. INTERPRETATION AND VALIDATION

Vibrations of such a structure can only occur if some force is applied to the structure. This could be catastrophic (earthquake, explosion, crash) or natural (strong steady winds). Structural engineers design for both dynamic wind effects and for resistance to earthquakes, depending on location.

Wind blowing steadily past a structure of this form will form vortices in the wake downstream of the structure. These will be shed periodically from alternating sides of the structure, creating pressure differentials and therefore oscillating lateral forces. The frequency of shedding of vortices is proportional to wind speed (see Cook 1985) and mean wind speeds are known for most areas (see British Standards).

So the results obtained can be evaluated. If typical wind speeds for the locality correspond at all closely to the fundamental vibration frequency, then resonance may occur and the structure could be liable to collapse – remember the Tacoma Narrows bridge disaster of 1940.

In fact, manufacturers are very aware of the potential disasters and design their structures to avoid large energy transfers from the wind.

7. DISCUSSIONS ON THE UNDERLYING MODELLING ASSUMPTIONS

In mathematical modelling it is both usual and edifying to identify the assumptions that are made in obtaining the model. This problem is not one that can be solved with little or no specialised background information, so is not a candidate for the classical modelling approach spelt out in so many modelling and systems books. However, it embodies many assumptions and gives grounds for discussion of them.

Essentially the finite element displacement method used here has assumed

- linear elasticity (deformed element has a cubic profile ≡ cubic spline)
- compatibility (the element displacements match the structure displacements at the nodes)
- equilibrium (at any structure node, the external applied force is equal to the algebraic sum of forces acting on all members meeting at that node)
- structural damping is small and is neglected (current research at the Buildings Research Institute is investigating this).

8. CONCLUDING REMARKS

- Modelling activities are particularly good if topical and accessible. Here in Sheffield the high mast is now much in evidence in city centre construction. These masts are visible from many of our teaching rooms and provide an immediately available real world example that can be used to motivate students in many disciplines.

- Data acquisition is an integral part of modelling, can be intriguing and interesting, and develops communication skills amongst students (and staff!).
- The completing of the modelling loop is important. Here it concerns the fundamental vibration frequency and the occurrence of winds likely to cause resonance of the structure.
- At final year level, technical report writing should be reaching its high point. A problem such as this is eminently suitable as it is very real.
- Such examples as this are taught jointly with civil engineering staff, and provide excellent bridging material that helps in integrating the mathematics and the structural engineering specialist courses.

REFERENCES

Dawe DJ. (1984). *Matrix and finite element displacement analysis of structures.* Clarendon.

Cook NJ. (1985). *The designers guide to wind loading of building structures.* Butterworths.

British Standards Code of Practise 3, chapter 5, part 2 (wind loads).

CHAPTER 42

Configurations of a Roll-on Roll-off Ferry Bridge

JAR Stone
Sheffield City Polytechnic, UK

SUMMARY

This paper is a description of a problem arising out of consultancy between Sheffield City Polytechnic and a local engineering company. The problem was formulated as a mathematical model and the solution of the model was achieved by a mixture of physical insight and numerical technique.

It has formed the basis of a final year project for undergraduate students.

1. INTRODUCTION

Mathematical modelling is now established as part of virtually all tertiary level courses in mathematics in colleges, polytechnics and universities in the UK. It is taught primarily by case study and project work. It is important that students start their modelling education by tackling situations with which they are familiar from the everyday world. Principles of modelling should be emphasised, and complex mathematics avoided at this stage. It should be recognised, though, that modelling is an important tool in scientific, industrial and commercial organisations and students should be exposed to 'real' problems at some stage of their modelling career. Unfortunately, 'real' problems are hard! They require detailed study, sometimes complex mathematics, and demand concepts from outside the students' experience. Projects from industry or commerce which do not require a *substantial* prior knowledge of the area from which the project derives are ideal but scarce. In this paper, such a project is presented. It employs relatively simple mathematics to solve a problem encountered by a small engineering company.

2. THE PROBLEM

A small engineering company is required to design a bridge to allow trains to be loaded on and off a ferryboat of length 150 metres. Trains come down a ramp at an angle of 1.65° to the horizontal onto a bridge which is in three sections, the transition section, the main section and the linkspan section (see figure 1), each of which can rotate about the z–direction relative to its neighbour. On each side of the centre line of the bridge is a rail track, which can carry a train onto or off the ferry. The dimensions of the bridge are given in table 1 and figure 2.

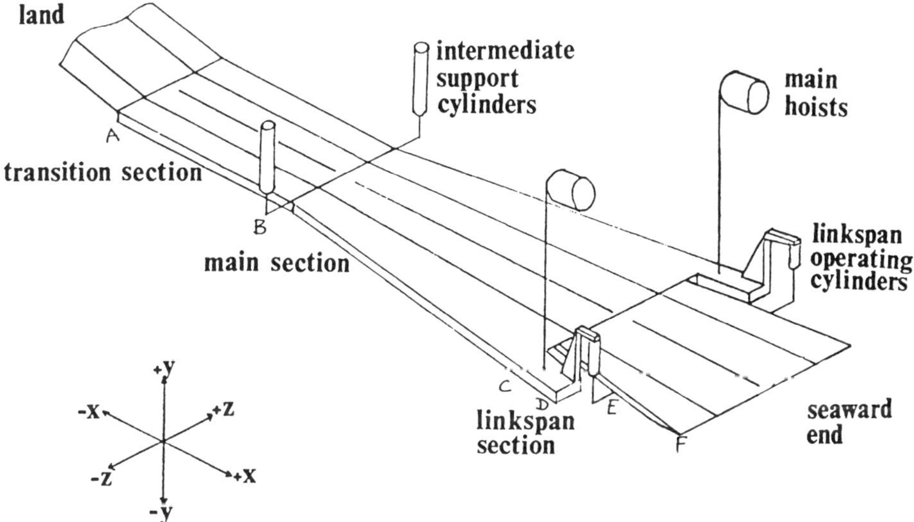

Figure 1

Position	*Description*	*Distance from A*	*Distance from centre line*
A	Rail Track	0	2.47m
B	Rail Track	30m	2.47m
	Support Cylinder	30m	4.50m
C	Rail Track	80m	4.36m
D	Main Hoist	87.6m	7.88m
E	Linkspan Cylinder	95.1m	7.50m
F	Rail Track	105m	5.67m

Table 1

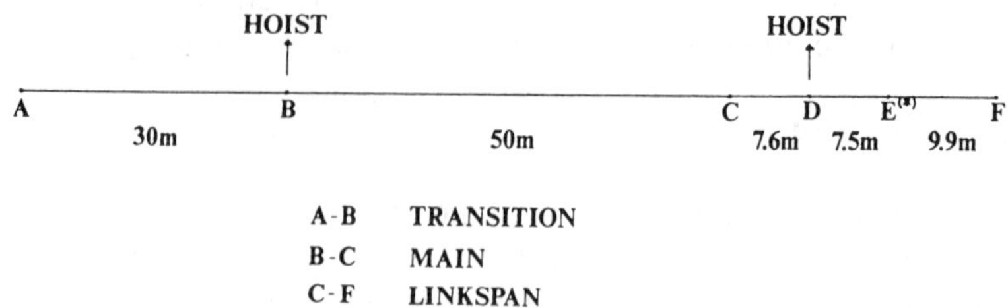

Figure 2

The transition section is supported by a roller bearing at A and joined to the linkspan by a hinge at B. This ensures rotation at both A and B. Intermediate support cylinders at B enable vertical adjustment to be made at this point. The main section is connected to the linkspan at C, although it extends forward of C as shown. Winch units at D allow the main section/linkspan joint to be raised and lowered. The linkspan section is supported by two linkspan operating cylinders mounted by a bracket arrangement (of height 4.71 metres) on the main section at E. The linkspan can be raised or lowered by these cylinders, but when the ferry is berthed the seaward end of the linkspan is supported by the stern of the ferry and the cylinders act as displacement transducers. The linkspan section is articulated and can rotate about its centre line to accommodate the rolling motion of the ship.

Due to tides, ferries approach the dock with loading decks at different heights, ranging from 3.7 metres below the dock level to 4 metres above it. As a ferry approaches, the bridge operator sets the inclination of the bridge sections using the intermediate support cylinders, the main hoist, and the linkspan cylinders, to ensure that the linkspan rests on the ferry when docked (see figure 3).

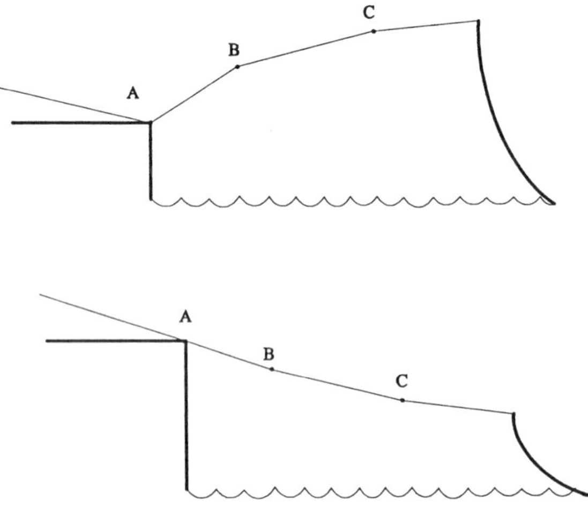

Figure 3

Any configuration is subject to constraint. For safety reasons no bridge section should be inclined at an angle greater than 4° to the horizontal and the maximum angle change between sections, section and dock, or section and ferry should be 2.5°. As the ferry is loaded or unloaded it will rise or fall. The position of the linkspan should be such that it can rise or fall equal amounts before infringing the constraints, and that such a displacement should be as large as possible. The total vertical displacement of the seaward end of the linkspan is called the (linkspan) window. The problem is, then, given the height of the loading deck of the ferry, what should be the settings of the two hoists at B and C and the linkspan cylinders to maximise the window size subject to the inclination constraints?

3. A MATHEMATICAL MODEL

We start by assuming that the ferry is horizontal. The configuration of the bridge is then that of its centre line. We can now carry out a simple two-dimensional analysis. Important variables are clearly the height h of the loading deck above the dock level, and the angles θ_1, θ_2 and θ_3 which the transition section, main section and linkspan respectively make with the horizontal. The operator needs to set, using the hoists, the displacements v_1 and v_2 of B and D respectively. These can easily be calculated once θ_1 and θ_2 are known. The linkspan setting is determined by the distance PE (see figure 4) which is easily calculated from θ_2, θ_3 and the geometry of the bracket supporting the linkspan cylinders.

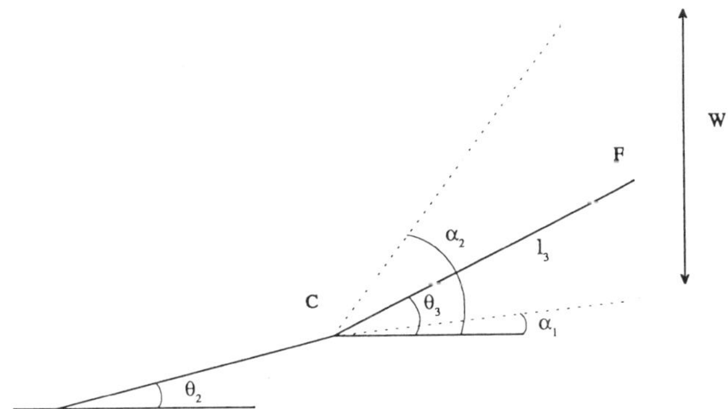

Figure 4

The linkspan window

With reference to figure 4 the window size, W, can be written as

$$W = l_3(\sin\alpha_2 - \sin\alpha_1) \tag{1}$$

where

$$\alpha_1 = \min \theta_3 \tag{2}$$

$$\alpha_2 = \max \theta_3 \tag{3}$$

Since the linkspan is placed in the centre of the allowed angle we also have

$$\theta_3 = (\alpha_1+\alpha_2)/2 \qquad (4)$$

The general model

With reference to figure 5 relationships linking important variables are

$$h = l_1\sin\theta_1 + (l_2-l_4)\sin\theta_2 + l_3\sin\theta_3 \qquad (5)$$

$$-4^\circ \leqslant \theta_i \leqslant 4^\circ \qquad (i=1,2) \qquad (6)$$

$$-2.5^\circ \leqslant \theta_i-\theta_j \leqslant 2.5^\circ \qquad (i,j=1,2,3) \qquad (7)$$

$$-2.5 \leqslant \theta_3 \leqslant 2.5^\circ \qquad (8)$$

$$v_1 = l_1\sin\theta_1 \qquad (9)$$

$$v_2 = l_1\sin\theta_1 + l_2\sin\theta_2 \qquad (10)$$

$$v_3 = [(a^2 + b^2) - 2bc\cos(\theta_2-\theta_3+\epsilon)]^{\frac{1}{2}} \qquad (11)$$

where l_1, l_2, l_3 and l_4 are the lengths of AB, BD, CF and CD respectively and a, b and ϵ are calculated from the geometry of the linkspan cylinder support.

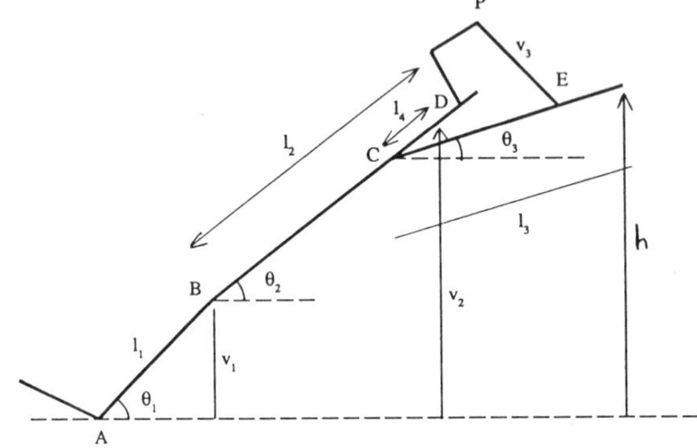

Figure 5

The mathematical problem can now be stated as follows.

For h given, find values of θ_1, θ_2 and θ_3 to maximise

$$W = l_3(\sin\alpha_2 - \sin\alpha_1)$$

subject to (2)-(8).

4. SOLUTION OF THE MATHEMATICAL MODEL

Suppose $\theta_2 = \theta_3 = 0$. Then $\alpha_1 = -2.5^\circ$ and $\alpha_2 = 2.5^\circ$. Consequently $W = 2l_3\sin 2.5^\circ$. This is the largest value that W can take. If θ_2 (and θ_3) $\neq 0$ then the window size is reduced since either α_1 or α_2 have magnitude less than 2.5°. As the magnitude of the inclination of the main section increases the window narrows. Thus in order to maximise W we need to keep the magnitude of θ_2 as small as possible. This will be the strategy for solving the mathematical problem. We identify four phases of configurations as follows.

Phase 1
This phase is defined by

$$0 \leqslant \theta_1 \leqslant 0.85^\circ$$
$$\theta_2 = \theta_3 = 0$$

Note that 0.85° is the largest value that θ_1 can take since the ramp slopes down at an angle of 1.65°. In this phase $W = 2l_3\sin 2.5^\circ$. From (5) this configuration is only possible if the height of the loading deck is between 0 and $l_1\sin 0.85^\circ$ metres above the dock.

Phase 2
This is defined by

$$\theta_1 = 0.85^\circ$$
$$0 \leqslant \theta_2 \leqslant 3.35^\circ$$

Consequently $\alpha_2 = 2.5^\circ$ and $\alpha_1 = \theta_2 - 2.5^\circ$. Relationship (4) gives $\theta_3 = \theta_2/2$ which means that (5) becomes

$$h = l_1\sin 0.85^\circ + (l_2 - l_4)\sin\theta_2 + l_3\sin(\theta_2/2)$$

For a given h this is a nonlinear equation which can be solved for θ_2 using a numerical technique such as the Newton-Raphson method.

Phase 3
This phase is defined by

$$-2.5^\circ \leqslant \theta_1 \leqslant 0$$
$$\theta_2 = \theta_3 = 0$$

The constraints are satisfied and the main section is horizontal, thus maximising W.

As in phase 1, this configuration is only allowable if the loading deck is between 0 and $l_1\sin 2.5^\circ$ metres below the dock level.

Phase 4
This phase is defined by

$$-4^\circ \leqslant \theta_1 \leqslant -2.5^\circ$$
$$\theta_1 = \theta_2 + 2.5^\circ$$

In this configuration B and D move together to satisfy the constraints and to keep the main section as level as possible, thus maximising W.

We now have $\alpha_1 = -2.5^\circ$, $\alpha_2 = \theta_2 + 2.5^\circ$. Relationship (4) gives $\theta_3 = \theta_2/2$ and so (5) gives

$$h = l_1\sin\theta_1 + (l_2-l_4)\sin(\theta_1+2.5^\circ) + l_3\sin((\theta_1+2.5^\circ)/2)$$

For a given value of h this nonlinear equation for θ_1 can be solved as in phase 2.

The phases are summarised in figure 6. Ferrys come into the dock with loading deck levels varying between 3.7 metres below the dock and 4 metres above the dock. A configuration in one of the four phases described above will fit a given loading deck height and the angles θ_1, θ_2 and θ_3 can be found. This leads easily to the determination of the displacements v_1, v_2 and v_3.

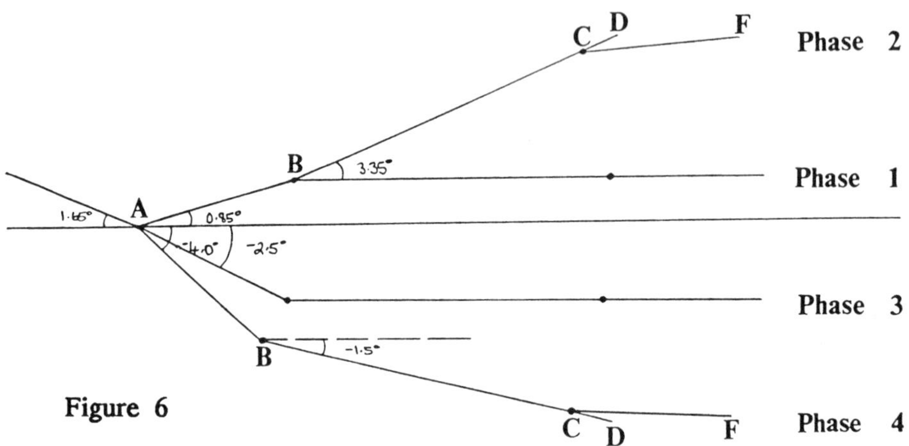

Figure 6

5. RESULTS

The bridge designers required configurations which maximised the window size over a range of loading deck levels. Accordingly a program was written to calculate θ_1, θ_2, θ_3, v_1, v_2, v_3 and W for $-3.7 \leqslant h \leqslant 4$.

Graph 1 shows the variation of window size with loading deck height. The maximum window size is 2.18 metres declining to 1.53 metres at the lowest docking level and to 0.76 metres at the highest docking level.

Graph 2 shows the variation of the displacements of the intermediate hoist (solid line) and main hoist (crossed line). Graph 3 shows the variation of the linkspan cable length (v_3).

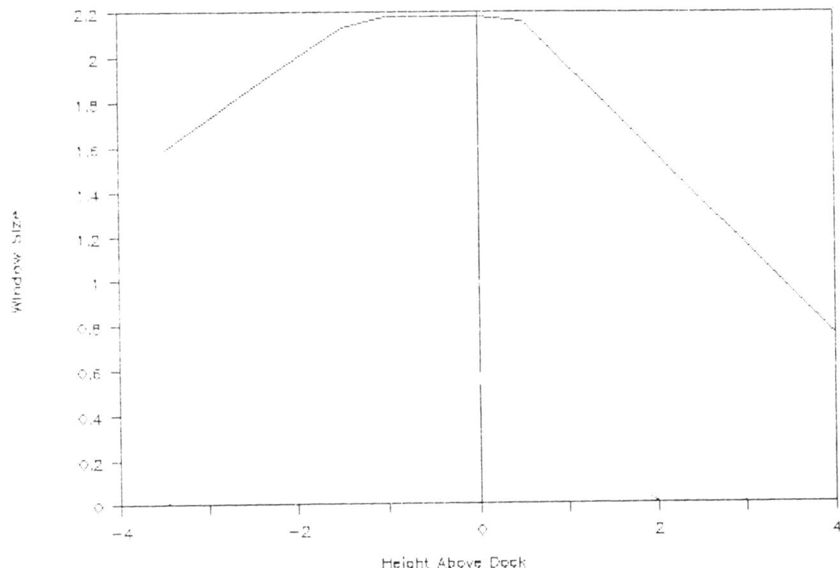

Graph 1

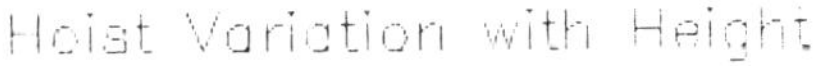

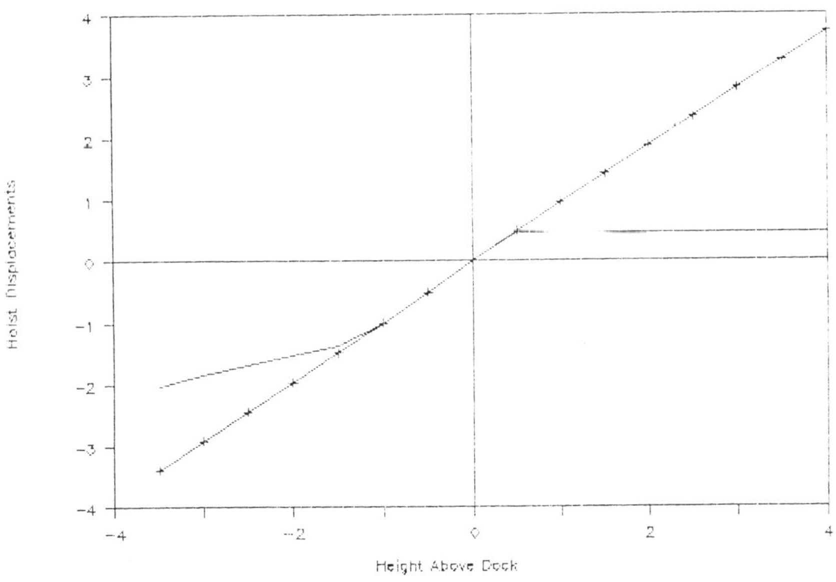

Graph 2

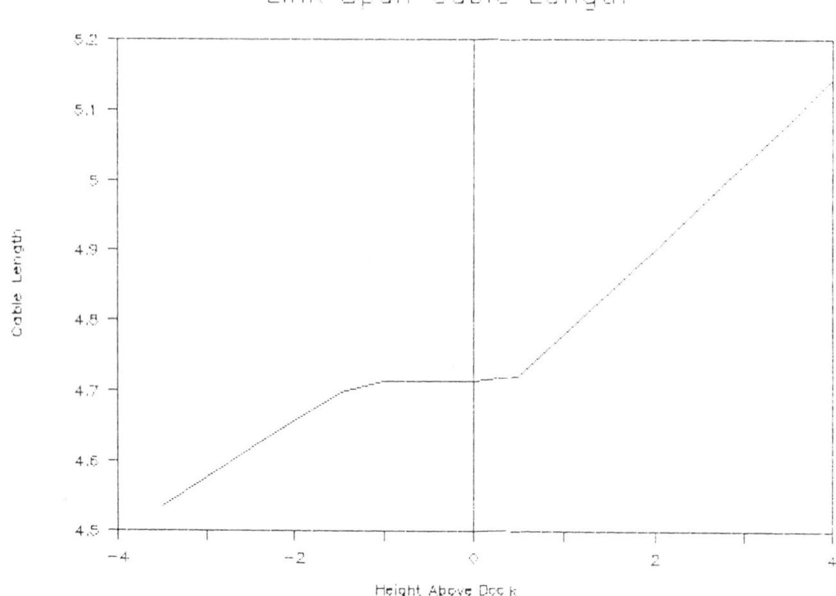

Graph 3

6. FURTHER WORK

Throughout the analysis we have assumed that the ferry is horizontal. It may, however, be subject to a rotation about an axis in the x-direction through the midpoint of the ferry. This is called trim, and the angle of trim could be of the order of 0.5° clockwise or anticlockwise. It may also be subject to heel, that is a rolling motion about its centre line of up to 3° clockwise or anticlockwise. It is possible to include these effects into the analysis. The result of doing this is that the window size reduces to zero for loading decks more than 2 metres above the dock or 3 metres below the dock.

7. CONCLUSION

We have presented in this paper a problem originating from a design problem tackled by a small engineering company. It does not need a great deal of prior experience of engineering to understand the problem and it can be solved by relatively straightforward methods. It is a good example of the sort of modelling problems suitable for inclusion in a course on mathematical modelling.

CHAPTER 43

Why Should Engineers Want to Solve Differential Equations – What the Laplace Transform is *not* for!

NC Steele
Coventry Polytechnic, UK

SUMMARY

Traditionally, engineers have been taught transform methods for the solution of differential equations as a part of their studies in engineering mathematics. This paper looks at a motivation for introducing transform methods, and later discusses the level of competence in solution techniques for differential equations which is actually required by engineering students. The use of the Laplace transform and the Z-transform as design tools will emerge, with particular reference to areas of electrical engineering, with suggestions as to other areas of application. Design, as with modelling, involves circuits of a loop, this time a design loop, with frequent reference to the initial design objective, and to performance relative to this objective.

1. INTRODUCTION

Many mathematics departments have, as a significant proportion of their work, the provision of service courses for engineering students of various specialities. Indeed, some departments, or groups of mathematicians, are dedicated entirely to this activity. In recent years, possibly reflecting an attempt to define a key characteristic of an engineer, education in design has appeared to an increasing extent in the curriculum for the formation of engineers. There are various aspects of engineering design, some more obvious than others; and there are various tools in the form of areas of knowledge or pieces of computer hardware/software available to aid the engineering designer in his task. One purpose of this paper is to illustrate the role of an area of mathematics, specifically the use of the Laplace and related transforms, as one such design tool.

In the experience of the author, there are different degrees of collaboration which take place between engineers and mathematicians in the design of the syllabus for a service course in mathematics. However, it seems to be almost inevitable that the resulting syllabus, from whatever country, will contain a sentence equivalent to 'use of the Laplace transform for the solution of differential equations with constant coefficients'! The reaction of many lecturers on their first encounter with this topic is one of surprise, seeing little advantage in complicating an essentially-straightforward procedure by the introduction of an apparently unnecessary transform process. Later, with an understanding of the concepts of transfer function and frequency response, a slightly more sympathetic view of the use of the transform in *representing* the equation (or system) may be taken. In recent times, discrete-time or digital systems have assumed a great deal of importance in many areas of engineering. This, in turn, has lead to the (re-)appearance of difference equations and the Z-transform method of solution in some engineering mathematics syllabuses. Again, the mathematician may well be surprised that transform methods are specified for the execution of another basically-straightforward procedure.

This paper will argue that compartmentalising the *introduction* and *use* of transform techniques between mathematics and engineering syllabuses, in the traditional way, is not in the best interests of engineering students; in addition, it deprives mathematicians of exhibiting a major application of their material, namely its use in the design or synthesis process.

2. FILTER DESIGN, AN EXAMPLE OF MATHEMATICS AS A DESIGN TOOL

The communication of data or information in one form or another is an essential part of modern life, and much of this communication is in the form of electrical signals, transmitted from a source to a receiver. For efficient utilisation of hardware, in the form of communication links and so on, several signals are combined together using carrier waves of different frequencies to form a single composite signal for transmission purposes. Also, the transmission process itself will introduce noise into the signal, causing the reception of distorted or even incorrect information. One approach to the recovery of a single signal from the composite transmitted signal is by use of devices known as analogue filters, which are frequency selective. This means that such filters will pass signals within certain frequency bands, and reject all others. These devices can also be used to remove unwanted noise from received signals, provided that the noise is restricted to frequency bands which are not being used for the transmission of information.

In a previous paper (Steele et al, 1989), an outline was given of the process of design of such analogue filters, based on a desired frequency-domain performance, with application to communication systems clearly in mind. It is in the field of electrical engineering that most progress seems to have been made in mathematical design, although it is clear that there is scope for the extension of the approach to other fields, for example to the design of suspension or vibration supression systems. This paper concentrates on the implementation of filter designs as electrical devices, and an analysis of the behaviour of LCR circuits (so-called because they contain only inductors L, capacitors C and resistors R), which leads to the expectation that such circuits can be designed to function as analogue filters with specific frequency domain properties.

The design tool to be used will be the Laplace transform; the first stage in the problem of low-pass filter design reduces to determining a stable system Laplace transfer function, such that the amplitude of the corresponding frequency response is a sufficiently accurate approximation to the ideal response of figure 1. In fact this procedure has more general application than is first apparent, since by use of suitable filter transformations, other designs such as high-pass, band-pass and band-reject filters can be produced, based on low-pass designs obtained in this way.

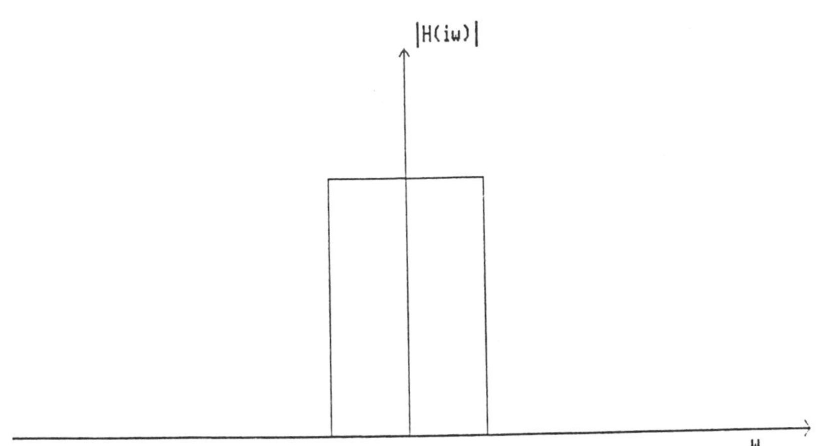

Figure 1 : Frequency response of an ideal low-pass filter

One approach to the problem of modelling the ideal response is to use the Butterworth approximation to this shape, that is, to choose

$$|H(i\omega)|^2 = \frac{1}{1 + (\omega/\omega_c)^{2n}} \tag{1}$$

where n is the filter order. It is clear that, as n increases, the approximation to the ideal shape will improve. From this starting point, we must now attempt to find a system with transfer function H(s) such that the frequency response amplitude is given by (1). On the assumption that our design will be stable, the frequency response will be obtained by setting $s = i\omega$ in H(s), that is by evaluating the transfer function on the imaginary axis. Now,

$$|H(i\omega)|^2 = H(i\omega)H^*(i\omega)$$

and if H(s) is to have real coefficients,

$$H^*(i\omega) = H(-i\omega)$$

so that we must have

$$H(s)H(-s) = \frac{1}{1 + (s/i\omega_c)^{2n}}.$$

For example, in the case when $n = 4$, by using the MATLAB package, the system transfer function H(s) is then easily calculated as

$$\frac{\omega_c{}^4}{s^4 + 2.6131\omega_c s^3 + 3.4142\omega_c{}^2 s^2 + 2.6131\omega_c{}^3 s + \omega_c{}^4}.$$

The second stage in the design process is to deduce the differential equation which will model this system in the time domain. Using the basic input–output relationship of the Laplace transform domain, $Y(s) = H(s)U(s)$, and inverting under the assumption of zero initial conditions, we obtain

$$\frac{d^4y}{dt^4}(t) + 2.6131\omega_c \frac{d^3y}{dt^3}(t) + 3.4142\omega_c{}^2 \frac{d^2y}{dt^2}(t)$$

$$+ 2.6131\omega_c{}^3 \frac{dy}{dt}(t) + \omega_c{}^4 = \omega_c{}^4 u(t).$$

Again, the MATLAB package can be used to exhibit the frequency response of the design; the result is shown in figure 2.

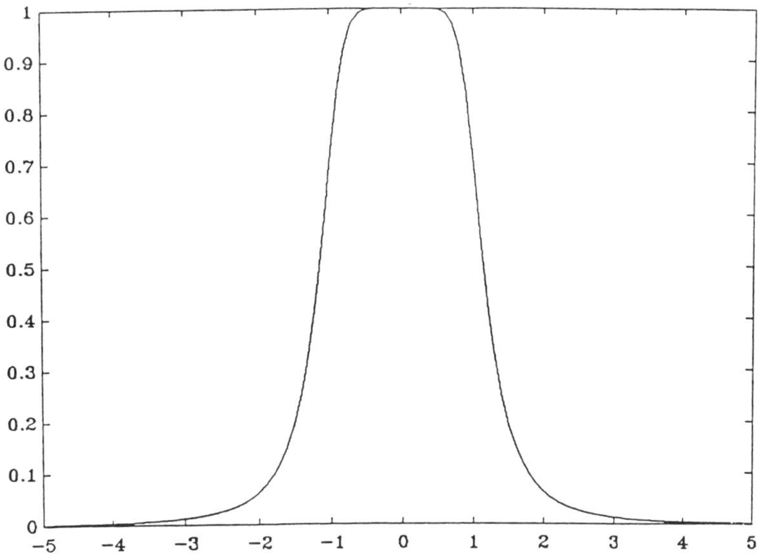

Figure 2 : Frequency response of the low-pass filter design

In carrying out the design process, a value has to be assigned to the filter order n. In the illustration, above, n was set to 4, but in reality this would have to be determined in relation to some response criteria specified as the design objective. It is clear, from an elementary knowledge of the analysis of electrical circuits, that the filter order n dictates the number of components necessary to realise the design, and hence the cost of the circuit, and perhaps also its reliability. Thus, this short and simple design exercise has highlighted the essential compromise which must be made by all designers in whatever field - that between performance, reliability and cost.

In this section the Laplace transform has been demonstrated in a role other than that of a technique for solving differential equations. Instead, the differential equation has emerged as the end product of a process which commenced with a frequency-domain specification for a system, with the design being achieved by use of the transform as a key tool. In the next section, an attempt is made to answer a question which will occur to all mathematicians - can every differential equation designed by this process be realised as an electrical circuit?

3. REALISATION OF DESIGNS

In order to discuss possible realisation methods, it is necessary to draw upon experience gained through the analysis of electrical networks.

Figures 3 and 4 illustrate two different network constructions, called Type 1 and Type 2 networks respectively. It can be shown that a Butterworth filter of *odd* order can be realised using a Type 1 network, whilst a filter of *even* order can be realised using a Type 2 network. In fact it is also possible to give an algorithm for determining the component values in the realisation of each filter type, based on the process of continued fraction expansion. A full discussion may be found in, for example, Jones and Steele (1989); however, for demonstration purposes, an example is included here.

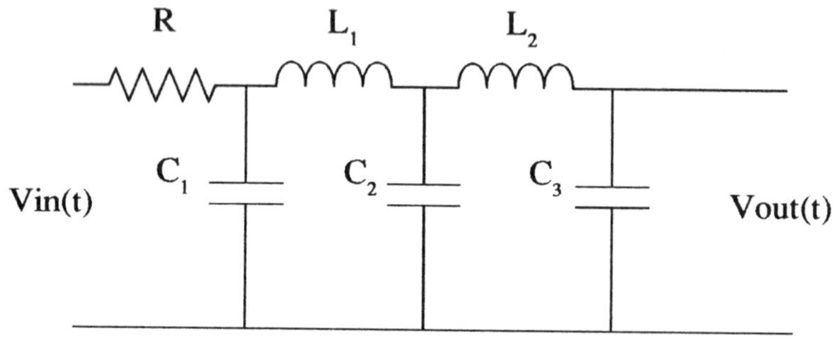

Figure 3 : A Type 1 network

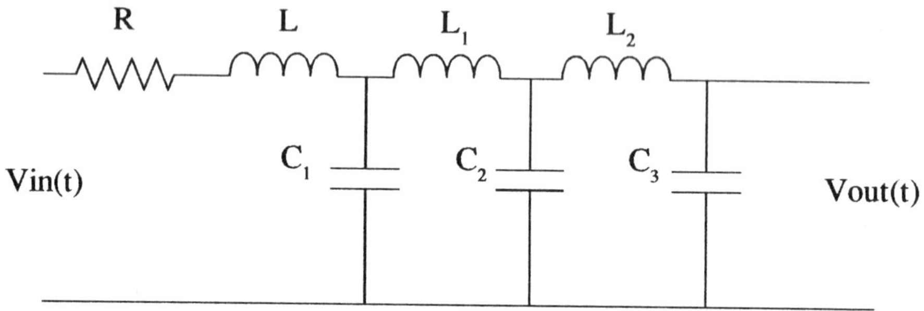

Figure 4 : A Type 2 network

The transfer function of the third-order Butterworth filter is

$$\frac{\omega_c^{\,3}}{s^3 + 2\omega_c s^2 + 2\omega_c^{\,2} s + \omega_c^{\,3}}.$$

The characteristic polynomial of the system, that is the denominator of the transfer function, is written as

$$g(s) = M(s) + N(s),$$

where

$M(s) = 2\omega_c s^2 + \omega_c^3$ is the even part, and

$N(s) = s^3 + 2\omega_c^2 s$ is the odd part.

The procedure now consists of finding the continued fraction development of RM(s)/N(s), where R is the value of the leading resistance in the network.

From this expansion we deduce that the Type 1 circuit for the realisation of this filter will have component values

$$C_1 = 1/2R\omega_c$$

$$L_1 = 4R/3\omega_c$$

$$C_2 = 3/2R\omega_c \; .$$

Thus our first circuit of the modelling/design loop is complete. The process commenced with a specification of a desired performance, and we have produced a filter design, which approximates to the design specification. If the approximation is good enough for the purpose intended, then our task is complete; if this is not the case, then a new design has to be produced, with as a first strategy an increase in the size of n, the filter order. As technology advances, new criteria for the acceptability of system designs follow, and we now embark on a second circuit of the loop in recognition of the development of reliable digital technology.

4. DESIGNING DIGITAL REPLACEMENT SYSTEMS

In recent years, the digital computer has made an enormous impact on the field of electronics and electrical engineering generally. In the area of communications, particularly, digital devices have replaced analogue designs in a variety of applications, with improved performance and reliability. The second circuit of the modelling/design loop involves recognising these developments and designing accordingly. As a further demonstration of the use of mathematics as a design aid, it is possible to show how the analogue design of section 3 above may be replaced by a discrete-time system, and hence by a segment of computer code. This replacement may take place in a number of ways, see for example Poularikas and Seely (1988), or Jackson (1986), but attention here is restricted to a single illustrative method. The approach selected is the so-called bilinear transform method, and is based upon the replacement of the integration process, denoted by 1/s in the Laplace transform domain, with a representation in the Z-transform domain of the trapezoidal method. Although other approaches are possible, this strategy has the significant advantage that, provided the prototype analogue design is itself stable, then the resulting digital device will also be stable. It is easy to show that this design strategy implies that the input-output difference equation representing this digital design is

$$\alpha y(k+3) + \beta y(k+2) + \gamma y(k+1) + \delta y(k) =$$

$$T^3 \omega_c{}^3 [u(k+3) + 3u(k+2) + 3u(k+1) + u(k)]$$

where T is the inter-sample time, and α, β, γ, δ are constants depending on T and u(k).

Clearly, the computer coding for the implementation of the filter is trivial, and our second trip around the modelling loop is apparently complete! However, as in all modelling and design, the results must be checked. Obviously an examination of the frequency response of the system is the first task, and it is here that we encounter a problem which is inherent in digital system design. Although the frequency response of our system is a reasonable approximation to the desired response at low frequencies, we find that there are periodic repeats of the pass-band as frequency increases. In a paper of this nature it is not possible to investigate the causes of this effect, nor deduce measures which can be taken to prevent the errors introduced, known as aliasing errors, destroying the usefulness of the digital design. However, it is interesting to note that practical digital devices involve the use of an analogue pre-filter in their overall specification, meaning that the first trip round the modelling loop was by no means wasted!

5. WHERE IS THE BOUNDARY BETWEEN MATHEMATICS AND ENGINEERING?

The last section contained an example of the use of the Z-transform as a design tool, this time for the production of a digital design based on an analogue prototype. Other methods exist which do not rely on the use of a prototype, but which do involve the Z-transform representation of the digital system in the same way. We feel that the material presented here can be regarded as an application of mathematics which has lead to a point at which engineering judgements must be made. In the experience of the author, the development of the material discussed in sections 2, 3 and 4 often receives a fairly cursory treatment in the engineering 'compartment' of a programme in preparation for the genuine engineering (practical) considerations. Using such an approach, the student may well be left to bridge the gap between the above ideas and the use of Laplace transforms as a technique for the solution of differential equations, as described in the mathematics 'compartment'.

The emergence of a sound bridge may be problematical, and if it is not established, then the power of mathematical description may never be fully appreciated and exploited in a later career.

The author has surreptitiously introduced most of this material into engineering mathematics courses, to his own enjoyment and with appreciable success in terms of the motivation of many students. A tangible result of this, and a similar design-driven approach to communication networks, has been the supervision of engineering project work by mathematics staff, and a request for additional mathematics lectures in the design element of the programme.

These developments have lead to a certain clouding of the traditional boundaries between aspects of mathematics and of engineering, a situation which seems to have considerable advantages for academic staff and students alike.

6. CONCLUSION

This paper has demonstrated a motivation for the introduction of transform techniques in the consideration of linear, time-invariant systems, both in continuous and discrete time. This motivation has depended on the use of the transforms as an aid to the *design* of such systems, rather than in the analysis of given systems. However, it is important to note, that as the discussion has developed, there has been frequent (sometimes implicit) reference to experience gained through the *analysis* of such systems. In seeking an answer to the question 'Why do engineers want to solve differential equations?', it is clear that a major purpose must be to gain such experience. This seems to imply that the ability to identify a stable system, and to describe the nature of its solution as the system output, is a major required skill. Thus, whilst it is important to inculcate manipulative ability and confidence through skill-building 'solution' exercises, it would be unfortunate if solution was perceived by the student as the sole purpose of the mathematical treatment of linear systems. This sentiment becomes particularly important with the discussion of the Laplace and Z-transforms, where the value as a solution aid is not particularly apparent, especially to the more able student. However, the demonstration of the use of these transforms, as an essential tool in the design of usable systems, illustrates a fascinating application of mathematics in the real world.

In this paper, attention has been focussed on an application of mathematics in communication theory. Many of the ideas are capable of extension to other fields of engineering, with some applications more obvious than others. The design of vibration-suppression systems would seem a productive area for attention.

REFERENCES

Jackson LB. (1986). *Digital Filters and Signal Processing*. Kluwer, Boston.

Jones RH and Steele NC. (1989). *Mathematics in Communication Theory*. Ellis Horwood, Chichester.

Poularikas AD and Seely S. (1988). *Elements of Signals and Systems*. PWS–Kent, Boston.

Steele, NC, Jones RH and Round A. (1989). Communication Theory – A Rich Field for Applications of Mathematics. In Blum W *et al* (eds), *Applications and Modelling in Learning and Teaching Mathematics*. Ellis Horwood, Chichester.

CHAPTER 44

ADCT -
The Future Standard for Still-image Compression

B Niss and JV Andersen
Copenhagen Telephone Company, Denmark

SUMMARY

The paper describes a new algorithm (mainly developed at the Copenhagen Telephone Company, KTAS) for compressing digital images. The algorithm, which has been chosen to form the basis of the future ISO standard for still-picture compression, allows for subjective picture quality to be maintained although more than 90% of the data that constitute the original picture has been discarded. On closer inspection, the algorithm relies on a few simple mathematical models which are accessible to first or second year college/university students of mathematics-based subjects.

1. INTRODUCTION

High quality photographic pictures of digital form will be an important means of communication in the future. For many applications - within commercial, professional or scientific sectors - the optimum carrier of information is pictures.

Digital pictures may be stored and transmitted like any other data. There is, however, one drawback - the amount of data per picture is enormous. As an example, the digital TV studio standard, rendering pictures of a quality significantly higher than the quality familiar from broadcast television, specifies that each picture consists of 720 picture elements (pixels) in each of 576 lines. The colour of a pixel is represented by 16 bits. In total this amounts to 810 Kbytes – or the equivalent of 200 pages of text. Transmitting such pictures over present day public networks (1200 bits per second) will take an hour and a half. Even with the introduction of the digital telephone network – Integrated Services Digital Network (ISDN) – which operates at 64 Kbits per second, transmission will take 100 seconds. The need for compressing data is thus evident. Two approaches for reducing the data may be considered.

- An error-free algorithm which retains data exactly. However, normal pictures may be compressed by a factor of two, at the most.
- A data-reduction algorithm. When deviations from the original pixel values are allowed, pictures may be compressed significantly without losing subjective quality.

The research and development department of the Copenhagen Telephone Company (KTAS) has, for some years, worked internationally on the development of picture compression algorithms. The algorithm to be described in the following – Adaptive Discrete Cosine Transform (ADCT) – was mainly developed within KTAS and it has been chosen to form the basis for the coming international (ISO) standard for still picture compression. The algorithm actually consists of a few simple mathematical models of image representation, image perception and image coding which are accessible to college or university students with a basic knowledge of linear algebra, calculus and a little statistics. The authors believe that this compression algorithm provides an example which demonstrates that mathematical modelling is sometimes a powerful tool for handling real-world problems.

2. THE ALGORITHM

One characteristic of a digital photographic picture is that neighbouring pixels look alike, that is their values are strongly correlated. This correlation makes it possible, with only a few pieces of information, to tell something about a lot of pixels. Think of the extreme (but rather dull case) where all pixel values are identical – then you only need to state the value and the number of pixels.

All picture compression algorithms are based on removal of this inherent correlation within a picture. The way ADCT removes the correlation starts with dividing the picture into square blocks of eight-by-eight pixels. These blocks are then treated as independent sub-pictures.

2.1 Basis functions

The eight-by-eight blocks may be regarded as points in a 64 dimensional vector space. One possible orthogonal basis for this space consists of 64 two-dimensional discrete cosine functions of the form

$$B_{ij}(x,y) = F(i)\ F(j)\ \cos(i(x+0.5)\pi/8)\ \cos(j(y+0.5)\pi/8)$$

where the horizontal and vertical frequencies i and j, as well as the spatial coordinates x and y are in the set {0...7}.

F is the normalisation factor given by:

$$F(m) = \begin{cases} 1/\sqrt{8}, & m = 0 \\ 1/2, & m > 0 \end{cases}$$

Any eight-by-eight block of pixel values may be formed as a linear combination of these basis functions. The amplitudes C(i,j) of the 64 cosine functions are the coefficients in the linear combination. The task of calculating the values C(i,j) for a given block of pixels is called the Discrete Cosine Transformation (DCT).

2.2 Discrete Cosine Transformation

One formulation of the DCT is

$$[C] = [T][P][T]^T$$

where [T] is the transformation matrix, $[T]^T$ is its transpose, and [P] is the pixel block, regarded as a matrix. The elements of [T] are given by

$$T_{ij} = F(j)\ \cos(j(i+0.5)\pi/8)$$

Given the coefficient matrix [C], the original pixel block may be reconstructed by using the inverse DCT, which may be calculated by

$$[P] = [T]^T[C][T].$$

2.3 Quantisation

To substitute 64 pixel values with 64 coefficients does not in itself constitute any reduction in data. What then is the advantage in moving from the pixel domain to the transform domain? Firstly, unlike the pixel values, the coefficients are virtually uncorrelated. Secondly, the sensitivity of the human eye to the basis functions drops with the frequency of the function. While an observer placed in a standardised viewing distance from the screen (which is between four and six times the screen height) is extremely sensitive to the low-frequency basis function, which must then be represented quite accurately, the high-frequency basis functions may often be omitted altogether or at the most be represented quite inaccurately. Via psycho-visual experiments, it is possible to establish the scale by which the amplitudes of each basis function should be measured. Following the argument above, coefficients of the low-frequency functions can be measured on a fine scale, whereas the high-frequency coefficients can be measured on a crude scale. This is done, in practice, by representing a coefficient as the nearest integer multiple of a number characteristic of each basis function. This number is known as the quantisation steps, and the process as quantisation. Once again – the quantisation steps are small for the low-frequency functions and large for the high-frequency functions.

Note now that a coefficient with an absolute value below half the quantisation step will be allotted the value of zero after quantisation. Looking at real life pixel blocks, it turns out that a very large fraction – on average more than 90% – of the coefficients will be zero after quantisation. Not surprisingly, only the low-frequency coefficients will normally come out non-zero. For a high-frequency amplitude to survive quantisation, it must have a very large absolute value.

It is very important to realise that quantisation is a data corrupting process, in the sense that it is impossible to reconstruct the original pixel values from quantised coefficients. On the other hand, the quantisation steps have been chosen so that the observer, in the standard viewing distance, is unable to tell the difference between the original and the corrupted picture. If, however, the observer moves closer to the screen, or zooms in on the picture, the differences will emerge.

2.4 Coding

What has been achieved so far? Firstly, we have converted 64 pixel values to 64 quantised coefficients of which a large number are zero. Where then is the data reduction? So far there is none. Reduction is introduced by the manner in which the quantised coefficients are stored, or transmitted to the receiver. The reduction obtained stems from two types of coding – run-length coding and entropy coding.

We know that practically all non-zero coefficients belong to low-frequency basis functions, while virtually all the rest are zero. We therefore choose to transmit the low frequency coefficients first. Suppose coefficient number m has been sent. Next we send the number k of zero coefficients before the next non-zero coefficient. Along with the value k we send the number of bits ℓ necessary to represent the subsequent non-zero coefficient. Small coefficients need only few bits, whereas large coefficients need many bits. Below is a table showing the bit allocation for the smaller coefficients

bits	coefficient
0	1
1	-1
00	2
01	3
10	-2
11	-3
000	4
001	5

and so on.

Knowing *a priori* that the subsequent coefficient is non-zero, no code is necessary for the value of zero, thus expanding the dynamic range of values per transmitted bit. The numbers k (run-length code for coefficients to be skipped) and ℓ (number of bits in next coefficient) form a two-dimensional piece of information (k,ℓ). A special end of block pair is used to tell the receiver that the rest of the coefficients in the block are zero. The statistical distribution of pairs in the $k-\ell$ plane is highly concentrated towards (0,1). Having introduced the end-of-block code, by far the most values of k are small and, having introduced quantisation, most quantised coefficients may be represented by only a few bits. Data exhibiting such a skewed distribution possess a low entropy. In this connection, entropy means the average number of bits necessary to represent one item from the sample. Suppose that N different values are represented in the sample, and that the probability of the ith value is p(i). Then the entropy of the sample is given by

$$E = -\sum_{i=1}^{N} p(i)\log_2 p(i)$$

Example: N = 256, p(i) = 1/256 for all i. Result: E = 8 bits!

If no correlation exists between the items in the sample, it is impossible to represent the values with fewer bits on average than is given by the entropy. On the other hand, it is possible to construct codes for the item values that bring the average number of bits quite close to the entropy. Again, assuming no inter-data correlation, the optimum coding strategy is Huffman coding. Without entering into details about this coding scheme, the basic principle is that values with a high probability are represented by short codes whereas infrequent values are given by long codes. Furthermore Huffman codes are constructed such that no start or stop bits are necessary. The receiver examines the first bit - Is this a valid code? If not append the next bit. Is it now a valid code? - and so on until a valid code has been assembled. Next, it is translated to the desired value via the code table, and the procedure starts all over with the subsequent bits. A more detailed presentation of the principles of Huffman coding may be found, for example, in Gonzales and Wintz (1977).

Huffman coding the (k,ℓ) pairs is significantly cheaper, in terms of bits spent, than using fixed length codes. So the net result is that, instead of transmitting 64 pixel values per block with 16 bits per value, we only send, on average, around four quantised coefficients per block with, again on average, 12 bits per value. This constitutes an overall data reduction by a factor of about 20 - and the observer is virtually unable to see the difference.

To illustrate this, figure 1 shows an example of a digital picture. Figure 2 shows the same picture, but compressed by a factor of 21 and then reconstructed. The subjective impressions of the two pictures are virtually identical. Figure 3 shows a detail in the original picture, enlarged eight times linearly. The individual pixels are seen as squares. Finally, figure 4 shows the same area in the reconstructed picture. At this magnification, the corruption of data is evident.

Figure 1

Figure 2

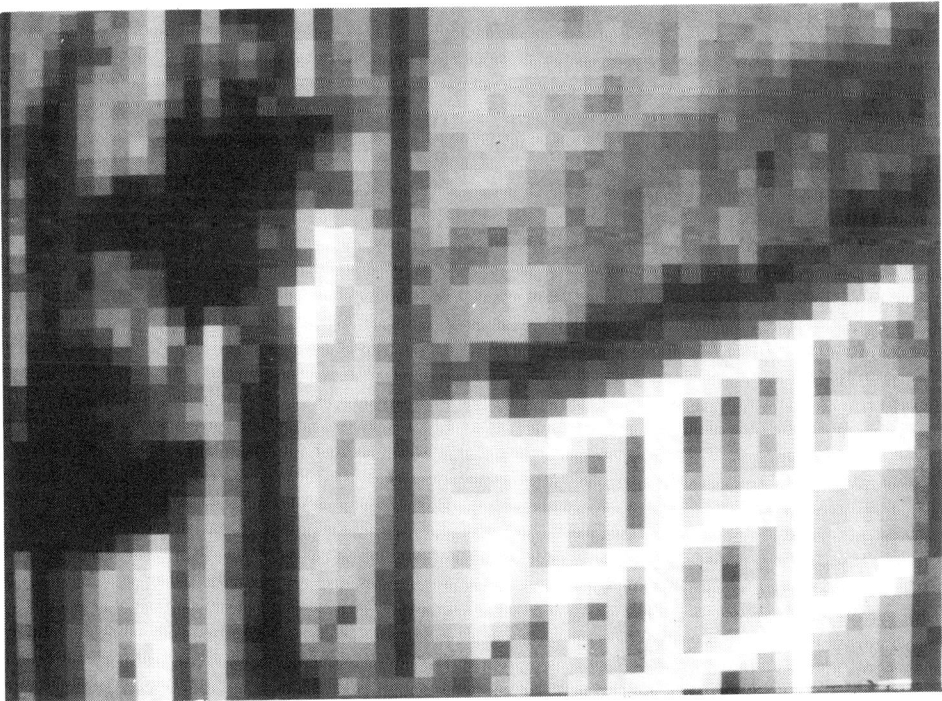

Figure 3

Figure 4

The numbers for possible compression factors cited above are average numbers. The actual numbers depend on the picture content. Pictures characterised mostly by extended areas of smooth variations are cheaper to code, while pictures with a substantial content of fine, high-frequency details require more bits to obtain the same subjective quality. This makes the algorithm adaptive.

The encoding/decoding process is summarised in the block diagram below.

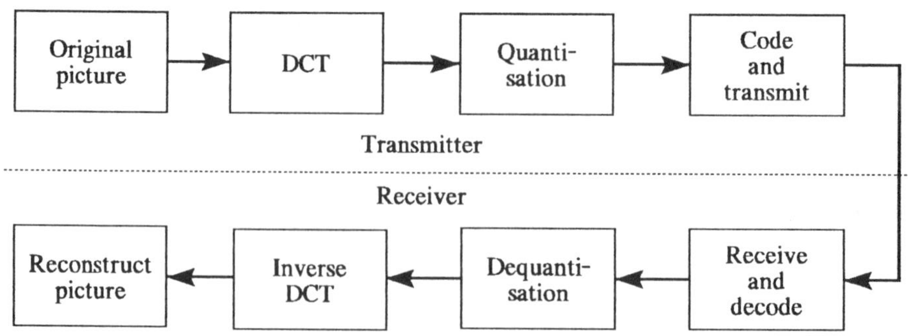

2.5 Progressive and sequential build-up

Transmitting coefficients as described above are used for what is called a sequential picture build-up. Data for each eight-by-eight block are grouped such that the receiver will reconstruct the picture block by block, starting in the upper left hand corner and finishing in the lower right hand corner five seconds later (given the digital TV studio standard plus ISDN, as mentioned in the introduction). It is possible, however, to shuffle data in a variety of ways, thus allowing for a so-called progressive picture build-up. This shuffle could be only one coefficient per block at a time, it could be only the most significant bits of each coefficient at a time, or it could be a combination of these. The result of such a re-ordering of data would be the arrival of a very crude, but certainly intelligible, picture extremely fast (0.5 seconds), a picture which grows better and better with time unless the user stops transmission. Such a build-up is especially well suited for browsing through picture databases.

2.6 Inter-block correlation

It was mentioned in the beginning that the blocks in which the picture is divided are treated independently. Actually inter-block correlation exists. The mean pixel values of two neighbouring blocks - which is the amplitude of the constant basis function $B_{00}(x,y)$ - are highly correlated. Special consideration, taking this correlation into account, is given to these coefficients, which in the algorithm are transmitted before any other coefficients. Knowledge of the mean values for each block makes it possible, to a certain extent, to predict some of the coefficients of the low-frequency basis functions. This prediction has two advantages. Firstly, it improves the quality of the early crude versions of the pictures in the progressive build-up. Secondly, it improves the overall compression factor, because only differences between the correct and the predicted coefficients need be sent.

3. CONCLUSION

An adaptive algorithm for compressing digital still pictures based on the discrete cosine transform has been presented. Compressing data by a factor of 20 is possible, dependent on picture content, without reducing the subjective impression of the picture. Thereby, the cost of storing pictures in picture databases, as well as the time needed for transmission, are reduced by the same factor.

REFERENCES

Gonzales RC and Wintz P. (1977). *Digital Image Processing.* Addison-Wesley Publishing Company.